ASTROPHYSICAL PLASMAS AND FLUIDS

ASTROPHYSICS AND SPACE SCIENCE LIBRARY

VOLUME 235

ASTROPHYSICAL PLASMAS AND FLUIDS

by

VINOD KRISHAN

*Indian Institute of Astrophysics,
Bangalore, India*

KLUWER ACADEMIC PUBLISHERS
DORDRECHT / BOSTON / LONDON

Library of Congress Cataloging-in-Publication Data

```
Krishan, V. (Vinod)
   Astrophysical plasmas and fluids / Vinod Krishan.
      p.   cm. -- (Astrophysics and space science library ; 235)
   Includes bibliographical references and index.
   ISBN 0-7923-5312-9 (alk. paper)
   1. Plasma astrophysics. 2. Fluid dynamics.   I. Title.
  II. Series: Astrophysics and space science library ; v. 235.
  QB462.7.K75  1998
  523.01--dc21                                                98-31126
```

ISBN 0-7923-5312-9 (HB)
ISBN 0-7923-5490-7

Published by Kluwer Academic Publishers,
P.O. Box 17, 3300 AA Dordrecht, The Netherlands.

Sold and distributed in North, Central and South America
by Kluwer Academic Publishers,
101 Philip Drive, Norwell, MA 02061, U.S.A.

In all other countries, sold and distributed
by Kluwer Academic Publishers,
P.O. Box 322, 3300 AH Dordrecht, The Netherlands.

Printed on acid-free paper

Printed in the Netherlands.

Dedication

To everyone, I have known.

CONTENTS

Preface

Life was simple when the dynamic, the spectral and the resolving powers of our instruments were small. One observed whole objects – planets, stars, sunspots, galaxies, often in rainbow colours. Then the revolution occurred: we acquired the centimetric eyes, the millimetric eyes, the infrared eyes, the ultraviolet eyes, the X-ray eyes and the γ-ray eyes. With these we see mottles on the surface of stars, streams in sunspots, and spirals in nuclei of galaxies. We see regions of multiple mass densities and temperatures in a precarious balance, losing it occasionally, exhaling flares. The universe is timed, cosmic phenomena are clocked; eternity is lost and variability is bought. Microarcsecond resolutions revealed stirring and sizzling interiors underneath serene surfaces. Short durations and small scales demanded employing a discipline with similar attributes – the discipline of Plasmas and Fluids – known more for its complexity than for its felicity. Some would like to wish it away.

We shall learn about plasmas for it is too little familiarity that breeds fear. Complexity can be systemized, to a large extent, by looking for a common denominator among apparently disparate phenomena.

It is not immediately obvious what the contents and the style of a graduate level course on plasmas and fluids aimed at understanding astrophysical phenomena should be. Plasmas and fluids are huge subjects by themselves. The cosmic phenomena where plasmas and fluids play a definite role are equally diverse and numerous. It is not possible to achieve proficiency of comparable levels both in the physics of plasmas and fluids as well as in astrophysical phenomena in a one-semester course. Since a graduate student of astronomy has opportunities to learn about astrophysical phenomena in other astronomy courses, we have chosen to emphasize the physics of plasmas and fluids in this text, illustrating it with examples from the observed astrophysical phenomena. There is no dearth of good text books on plasma physics and fluid dynamics; however, to the best of my knowledge and assessment, no treatment designed to teach a graduate student the essentials of plasmas and fluids in the context of astrophysical phenomena in a span of one semester, exists under one cover. The choice of topics and the pedagogical treatment, I hope, will prove to be one of the attractive features of the· present book.

I have strongly felt that the physics of plasmas and fluids could be presented in a more logical sequence than has usually been done. In looking at some books, it appears to me that some authors, not being sure of the competence of the readers, tend to go back and forth in dealing with presumably complex topics. For example, the fluid and the magnetohydrodynamic descriptions of plasmas are presented in earlier parts of a book whereas their equations are derived in later parts, after describing the kinetic treatment. One wonders, if it is not more desirable to systematically establish the mathematical frame work in a fairly complete manner and then study the various topics as special cases.

With this view, I begin with the Liouville equation from which originates all our knowledge of a large system. In this approach, there is absolutely no confusion whether the single particle picture is a special case of a fluid picture or vice-versa. All descriptions emerge from the Liouville equation, depending upon the assumptions and simplifications thrown in from the physics of a system. This style also helps us to remain aware of the inter-relationships of the single-fluid, the two-fluid, and kinetic treatments. The rest of the contents of the book, essentially, highlight the two major astrophysical issues — the configurational concerns and the radiative requirements — which often influence each other. Another way in which this book differs from a standard text book on plasma physics is by containing a lengthy chapter on nonconducting fluids. Since a zeroth order comprehension of many astrophysical phenomena is attempted by studying fluids in gravitational fields, a familiarity with the basics of fluid dynamics, including the all pervading turbulence, is a must. I have also aspired to make chapter one as informative as possible. Some of the topics, such as strongly coupled plasmas and dusty plasmas have been dealt with in chapter one, since these topics are still in an exploratory stage, and perhaps it is premature to include them with the core course. The reader may find that some concepts appear more than once at different places in the book; this, I believe, cannot hurt. Having said what this book is, I must say what it isn't. Throughout the writing of this book, I had to often remind myself that it is neither a book exclusively on astrophysics, nor on plasma physics and fluids, so the temptation to specialize was curbed. After taking a course based on this text, I expect that those who are interested will be able to easily venture into the world of rigour, while those not so inclined to do so will get much more out of seminars on plasma instabilities and magnetohydrodynamics.

It is now time to express my heartfelt gratitude to my friends and critics, who over the last twenty years that I have been in the field, have always demanded more from me than I delivered. This constant phase lag has kept me on my toes without toppling. I am most grateful to Professor Paul J. Wiita, an astrophysicist at the Georgia State University, USA, for reading every word in this book and suggesting changes in its substance and style. I also thank Professor Som Krishan, a plasma physicist at the Indian Institute of Science, Bangalore, who has also read every word in this book, for exacting standards of conceptual clarity in my presentation. I have tried. All lapses, however, are mine.

I am extremely grateful to Dr. Baba Anthony Varghese of the Indian Institute of Astrophysics, Bangalore, without whose help, my thoughts could not have become printed words. I also thank Ms Pramila N.K. for her help in typing and correcting the manuscript. And lastly, as I have enjoyed writing this book, I hope you will enjoy reading it.

Vinod Krishan

CHAPTER 1

PLASMA – THE UNIVERSAL STATE OF MATTER

1.1. How Should We Describe a Plasma?

On being asked how should one describe one's field of activity, the Nobel Laureate Subramanyan Chandrasekhar replied: The description should answer the following questions (1) What is the nature of the system? (ii) towards what purpose is its study and (iii) what are the techniques for its investigation? Taking inspiration from the 'man who knew the best', we shall in this chapter, begin with the nature of a plasma and continue to explore it in the rest of the book, pausing in Chapter 2 to develop the techniques. The purpose unfolds itself as we learn more and more about the nature of a plasma. However, if you like one-liners, we can say: the nature of a plasma is **Hyperactive,** the purpose is to **Comprehend the Universe** and the techniques employed are **Many-Body**.

We are fairly well familiar with the three states of matter - the solid, the liquid and the gas. Some of us have some familiarity with a fourth state known as the plasma. This plasma state of matter is so prevalent throughout the cosmos that it is the **Plasma Universe**, a term coined by the Nobel Laureate Hannes Alfven, that we observe and are in awe of. We may find it disconcerting that our expertise and experience in the three common states of matter are barely adequate to comprehend only about one percent of the universe, the remaining 99% being in the plasma state. Of course, if 90% of the universe is actually made up of the so called **Dark Matter** (gravitating but nonradiating matter), as some people believe, then the remaining 10% is the radiating and reflecting and, therefore, observable matter; plasmas then form the silver lining of this dark universe. This, in no way exempts us from the study of plasmas since it is the visible radiating matter that telleth of the unseen! Let us forget about fractions and start exploring the nature of a plasma. So, what is a plasma?

A solid can be converted into a liquid and a liquid can be converted into a gas by heating them. With an increase in the temperature of a solid or a liquid, the freedom of movement of the constituent atoms or molecules increases. What happens if we continue to heat a gas? The gas is ionized. The atoms or molecules lose some of their electrons and a sizzling sea of positively charged ions and negatively charged electrons is created. The net electric charge of the system is zero. Some neutral atoms and molecules may

1

still be present. The electric charges in motion produce electric currents. These are the makings of a plasma. A plasma is an electrically conducting gas, though the converse need not be true.

Astrophysical plasmas consist mainly of protons, electrons, He ions and traces of heavy ions and atoms. In equilibrium, these particles obey the **Boltzmann Distribution**, so that the number density n_e of particles in a given energy state E_l, at a temperature T is determined from:

$$n_l = g_l \exp\left[-E_l/K_B T\right], \tag{1.1}$$

where g_l is the degeneracy factor describing the number of states with the energy E_l and K_B is the Boltzmann constant. The ratio p of the number density of particles in states l and m is, therefore, given by:

$$p = \frac{n_l}{n_m} = \frac{g_l}{g_m} \exp\left[-\left(E_l - E_m\right)/K_B T\right]. \tag{1.2}$$

If we identify the state l with the electron – ion pair and state m with a neutral atom, we can find the fractional ionization from equation (1.2) provided the energy difference $(E_l - E_m)$ is equal to the ionization energy I of the atom. Thus the number density of ions n_i is found to be:

$$n_i = n_m \left(\frac{g_i}{g_m}\right) \exp\left[-I/K_B T\right]. \tag{1.3}$$

The values of g_i and g_m are found from quantum - mechanical calculations and here, we give an approximate formula for the ratio

$$\begin{aligned}
\frac{g_i}{g_m} &\simeq \left(\frac{2\pi m_e K_B T}{h^2}\right)^{3/2} \frac{1}{n_i}, \\
&\simeq 2.4 \times 10^{21} T^{3/2} n_i^{-1},
\end{aligned} \tag{1.4}$$

where h is Planck's constant, m_e the electron mass, and the temperature T is in degrees Kelvin. Substituting Equation (1.4) in Equation (1.3), we obtain **Saha's Ionization Equation**:

$$\frac{n_i}{n_m} = 2.4 \times 10^{21} T^{3/2} n_i^{-1} \exp\left[-I/K_B T\right]. \tag{1.5}$$

Problem 1.1: Plot the fractional ionization $[n_i/n_T]$ as a function of temperature T for hydrogen for various values of the total density $n_T = n_i + n_m$. Determine the ratio p at $T = 10^4 K$ and $n_T = 10^{21}$ cm^{-3}.

We find that a significantly large degree of ionization is achieved for hydrogen even at temperatures much below that corresponding to the ionization energy of 13.6 eV. This is **Thermal Ionization**. Stellar plasmas are mostly produced through thermal ionization.

Another way of producing ionization is by applying electric fields for example through an **Electric Discharge**, in terrestrial laboratories or during lightening in planetary atmospheres. (Figure 1.1)

Figure 1.1. Ionization of Air During Lightening.

Matter can also be ionized by the action of electromagnetic radiation. The ultraviolet radiation from hot stars ionizes most of the interstellar medium. The solar ultraviolet radiation creates our ionosphere.

In high density regions, collisions among electrons and atoms can also cause ionization. At even higher densities, the phenomenon of **Pressure Ionization** occurs. During this process, the matter is so tightly packed that electrons are squeezed out of their energy levels if their energy, the **Fermi Energy** exceeds the ionization energy. Interiors of large gaseous planets such as Jupiter, and the ultradense stars called white Dwarfs are some of the probable sites of pressure-ionized plasmas.

Ionization is a necessary but not a sufficient condition for a plasma. In order to qualify as a plasma, an ionized gas must admit **Quasineutrality** and exhibit **Collective** behavior (Frank–Kamenetskii 1972).

1.2. Collective And Quasi-Neutral

Two isolated charges separated by a distance experience the Coulomb force. In an ionized gas, there are many other charges between these two charges, influencing and being influenced by them. Thus, the motion of a given charge is determined collectively by the entire system of charges. There are no free charges in a plasma. As these charges move around, local concentrations of positive and negative charges can develop, producing electric fields. This charge separation, though, can occur only on microscopic scales, for otherwise forbiddingly large electric fields would be generated. The existence of a very small amount of charge separation over a very short spatial scale for a very short time interval is what is meant by the **Quasineutrality** of a plasma. It is for this ability of a plasma to sustain a tiny difference in positive and negative charge densities that its study has acquired the status of an independent discipline. The charge separation arises due to the thermal motion of electrons and ions. Thus, electric fields of strengths such that the associated electrostatic energy per particle does not exceed the thermal energy per particle are produced spontaneously. This condition restricts the amount, the extent and the duration of charge separation in a plasma. The electric field E due to a net charge density Q existing over a region of linear extent l is $E = 4\pi Q l$ and the electrostatic energy of an electron or proton is $4\pi Q e l^2/2$. The average thermal energy per particle with one degree of freedom is $(K_B T/2)$. Therefore,

$$4\pi Q e \frac{l^2}{2} \leq \frac{K_B T}{2}, \tag{1.6}$$

and l is the spatial scale of charge separation. For complete charge separation in a plasma of electron density n, $Q = en$, so that

$$l \leq \left(\frac{K_B T}{4\pi n e^2} \right)^{1/2} \equiv \lambda_{De}, \tag{1.7}$$

where λ_{De} is called the **Debye Length** of electrons. It is also known as the screening distance, as it screens the charge separation from the rest of the system. The time duration, τ_e, for which this charge separation can exist is the time taken by an electron to travel the distance l with the mean thermal speed V_e, so that

$$\tau_e = \frac{l}{V_e} = \left(\frac{m_e}{4\pi n e^2} \right)^{1/2} \equiv \omega_{pe}^{-1}, \tag{1.8}$$

where

$$V_e^2 = \frac{K_B T}{m_e},$$

and ω_{pe} is called the **Electron-Plasma Frequency**. Thus, the higher the density n, the shorter is the time scale τ_e and the spatial scale λ_{De}. We realize that we can have a spatial scale λ_{Di} and a time scale τ_i for ions and that $\lambda_{Di} \leq \lambda_{De}$ and $\tau_i >> \tau_e$. Therefore τ_e is the shortest time scale for which charge separation can exist and strict neutrality can be violated. It is in this sense that a plasma is **Quasi-Neutral**; on a spatial scale λ_{De} and a time scale τ_e, departures from charge neutrality exist.

In this sea of electrons and ions, there must be Coulomb collisions among these particles. The collisions produce mixing of charges, and therefore reduce the spatial scale of charge separation. In order that there still exists a finite spatial scale of charge separation, the mean free path λ_e of the electrons must be much larger than the scale of charge separation λ_{De}. This condition gives:

$$\lambda_e = \frac{V_e}{\nu_{ei}} >> \left(\frac{K_B T}{4\pi n e^2}\right)^{1/2}, \qquad (1.9)$$

or

$$\omega_{pe} >> \nu_{ei},$$

i.e. the electron plasma frequency must be much larger than the electron – ion collision frequency ν_{ei}. Another important consequence of a low collision frequency ν_{ei} is that the electrons and ions take a long time to thermalize with each other and reach a common temperature. For durations smaller than the collision time $(\nu_{ei})^{-1}$, electrons and ions can remain at different temperatures T_e and T_i respectively. An ionized gas with vanishingly small λ_{De} does not qualify to be a plasma as such a system is strictly neutral and not quasineutral. Further, in order to treat the part of the system enclosed in a region of linear dimension λ_{De} as a statistical system with a temperature T and number density n, it must contain a large number of particles. The number of particles N_d in a sphere of radius λ_{De} – called the **Debye Sphere**, is

$$N_D = \frac{4\pi}{3} n \lambda_{De}^3, \qquad (1.10)$$

and

$$N_D >> 1.$$

We have now found all the conditions that an ionized gas must satisfy before it can be called a plasma. These are given by equations (1.9) and (1.10) along with the obvious requirement that the size of the system must be larger than all the characteristic spatial scales such as λ_e and λ_{De}. We shall learn more about the consequences of quasi-neutrality in Chapter 5.

Problem 1.2: Estimate N_D for (1) earth's ionosphere, (2) the solar corona, (3) Solar wind, (4) cometary tails, (5) atmosphere of a pulsar, (6) interstellar medium, (7) galactic HII regions (Figure 1.2), (8) Nuclei of active galaxies (9) extragalactic jets (Figure 1.3) and (10) intracluster medium of a supercluster of galaxies.

1.3. Electrostatic Potential in a Plasma – Debye Screening

Suppose we insert a positive test charge q in a plasma. We expect that it will immediately attract a cloud of negative charges around it. The effect of the test charge is not felt outside this cloud. The rest of the plasma is screened from the test charge by the cloud. What is the size of this cloud and what

Figure 1.2. HII Regions are "Strung Out Like Pearls Along the Spiral Arms of Galaxies". The Galaxy M51 Together with its Companion Galaxy NGC5195.

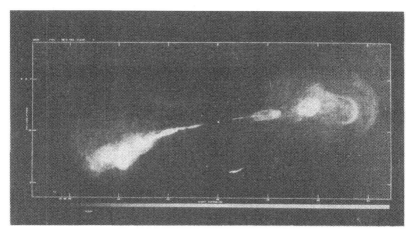

Figure 1.3. The Extragalactic Jets in the Object Hercules A Imaged at Radio Wavelength of 6-cm Using VLA Radio Telescope.

is the potential due to q in a plasma? According to Poisson's equation, the electric potential φ can be determined from

$$\nabla^2 \varphi = 4\pi e \left(n_e - n_i\right), \qquad (1.11)$$

where n_e and n_i are respectively the electron and singly charged ion densities in the plasma. The two densities become unequal due to the presence of the potential φ caused by the test charge q. The particles, in addition to their kinetic energy, now possess potential energy $|e\varphi|$. If we assume that at a temperature T_e the electrons follow the Boltzmann distribution given by

$$n_e = N \exp\left[-W_e/K_B T\right], \tag{1.12}$$

where the total energy $W_e = \frac{1}{2}m_e V^2 - e\varphi$, we find

$$n_e = n_0 \exp\left[\frac{e\varphi}{K_B T}\right], \tag{1.13}$$

where n_0 is the electron density in the absence of the potential. The ions at their temperature T_i also follow a similar distribution:

$$n_i = n_0 \exp\left[-\frac{e\varphi}{K_B T}\right]. \tag{1.14}$$

These Boltzmann distributions are discussed again in Chapter 5. For $|e\varphi/K_B T_e| << 1$ and $|e\varphi/K_B T_i| << 1$ we can expand the exponentials in Equations (1.13) and (1.14), substitute in Equation (1.11), and find the spherically symmetric potential φ at a distance r from the charge q obeys the linear equation:

$$\frac{1}{r^2}\frac{\partial}{\partial r}\left(r^2\frac{\partial \varphi}{\partial r}\right) = \frac{4\pi n_0 e^2}{K_B}\left(\frac{1}{T_e} + \frac{1}{T_i}\right)\varphi \equiv \frac{\varphi}{\lambda_D^2}. \tag{1.15}$$

The solution to Equation (1.15) can be easily seen to be (Figure 1.4):

$$\varphi = \frac{q}{r}\exp\left[-r/\lambda_D\right], \tag{1.16}$$

where

$$\frac{1}{\lambda_D^2} = \frac{1}{\lambda_{De}^2} + \frac{1}{\lambda_{Di}^2}, \tag{1.17}$$

and λ_{De} and λ_{Di} are the electron and ion Debye lengths respectively. So, the electrostatic potential φ (Equation 1.16) is no longer like the Coulomb potential which varies as (r^{-1}), but has additional exponential decay with distance r and diminishes to $1/e$ of its Coulomb value at $r = \lambda_D$, the **Effective Debye Length**. Thus, for $r >> \lambda_D$, the potential becomes vanishingly small. The potential is felt by the plasma particles within a distance $r \simeq \lambda_D$ from the position of the charge q. We, therefore, find that the size of the screening or **Debye Cloud** is $\simeq \lambda_D$. It increases with an increase of temperature, since electrons with high kinetic energy can withstand the attraction of the positive charge q up to larger distances. On the other hand, λ_D decreases with an increase of density n_0, since a larger number of charges or electrons can now be accommodated in a shorter region to annul the effect of q. As T_e and $T_i \to 0$ or $n_0 \to \infty$, $\lambda_D \to 0$ and all the interesting plasma

phenomena disappear. A finite λ_D is the source of the entire game of electromagnetic phenomena in a plasma. We have already seen the collective
behavior – the potential due to a single change q in a plasma is a function
of the electron or ion density and their temperatures and does not depend
on the individual properties of the plasma particles.

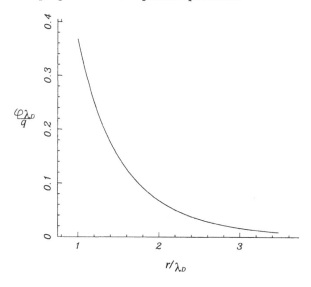

Figure 1.4.   The Screened Electrostatic Potential φ Due to a Charge q in a Plasma.

1.4. Coulomb Collisions Among Plasma Particles

There are several characteristic time scales in a plasma. We have already
encountered three of them, τ_e and τ_i corresponding to the electron-plasma
frequency ω_{pe} and the ion-plasma frequency ω_{pi}, and the third corresponding to the electron -ion collision frequency ν_{ei}. We shall encounter more time
scales when we study a plasma in a magnetic field. Since the collision frequency is one of the defining characteristics of a plasma, it must be known
before we proceed to explore the nature of a plasma any further.

In a plasma, charged particles continuously feel the Coulomb force due
to other particles. Therefore, the actual trajectory of a particle is not a
sum of discrete random paths. However, it is found that some aspects of
thermal motion and transport processes can be well accounted for using
the description of Coulomb collisions and defining an effective Coulomb
cross-section. We can treat collisions resulting in large scattering angles
and those resulting in small scattering angles separately. Here, we present
an approximate analysis of Coulomb collisions, a more formal treatment of
collisions can be seen in Chapter 2.

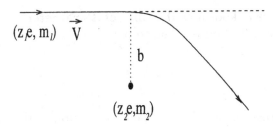

Figure 1.5. Coulomb Collision Between Two Charged Particles.

A particle of charge $(z_1 e)$ and mass m_1 undergoes a deflection in its trajectory due to the effect of the Coulomb force exerted by another particle of charge $z_2 e$ and mass m_2. The magnitude of the Coulomb force F when the two particles are a distance r apart is given by

$$F = \frac{z_1 z_2 e^2}{r^2}.$$ (1.18)

This two-body collision problem can be reduced to the problem of a single particle of reduced mass μ and relative velocity $\vec{V}$ moving in the force field $\vec{F}$. The particle feels this force for the duration t it spends in the neighbourhood of the other particle. This is the time needed to cross the closest distance of approach b. Thus

$$t = \frac{b}{V},$$ (1.19)

where $\vec{V}$ is the relative velocity of the particles. The change ΔP in the momentum of the particle is, therefore

$$\Delta P = \delta(\mu V) = \left(\frac{z_1 z_2 e^2}{b^2}\right)\left(\frac{b}{V}\right),$$ (1.20)

with

$$\mu = \frac{(m_1 m_2)}{(m_1 + m_2)}.$$

Now, we know that for a scattering angle of 180° the particle reverses its direction of motion and the change in the momentum is twice the original momentum. Therefore, we expect that for large deflections the change in the momentum of a particle is of the order of its momentum, i.e.,

$$\frac{b}{V}\frac{z_1 z_2 e^2}{b^2} \simeq \mu V.$$ (1.21)

Thus the distance b, known as the **Impact Parameter**, between the particles for a large deflection is found to be:

$$b = \frac{Z_1 Z_2 e^2}{\mu V^2}.$$

(1.22)

The effective cross section for a binary interaction is the area of a circle of radius b, i.e. πb^2. The effect of all small angle scatterings is collectively included in a parameter known as the **Coulomb Logarithm** $\ln \Lambda$. The Coulomb logarithm is the ratio of the maximum and minimum impact parameters. The maximum impact parameter is of the order of Debye screening distance λ_D because for distances larger than this, the Coulomb potential becomes vanishingly small and the particle does not undergo any deflection in its path. The minimum impact parameter b_{min} corresponds to the maximum deflection and can, therefore, be taken as the value of b given in Equation (1.22). It is inversely proportional to V^2. In the classical regime, b_{min} must be larger than the de Broglie wavelength of the particle. In the quantum-mechanical regime, b_{min} can be taken to be the de Broglie wavelength given by

$$b_{min} = \frac{\hbar}{\mu V}.$$

(1.23)

Thus, the Coulomb logarithm in the classical regime is:

$$\ln \Lambda = \frac{\lambda_D \mu V^2}{Z_1 Z_2 e^2},$$

(1.24)

and the total Coulomb crossection σ_c is given by:

$$\sigma_c = \left(\pi b^2 \right) \ln \Lambda.$$

(1.25)

A particle moving with velocity V, in a medium of density n undergoes $(nV\sigma_c)$ collisions per unit time. The collision frequency ν_c, is therefore, given by

$$\nu_c = nV\sigma_c,$$

(1.26)

and the mean free path by

$$l_c = \frac{1}{n\sigma_c}.$$

(1.27)

If the particle undergoes collisions with several different types of particles, the total mean free path is the sum of all the free paths and the total collision frequency is the sum of all the collision frequencies.

Three types of collisions can occur in a fully ionized plasma - among like particles such as electron – electron and ion-ion and among unlike particles such as electron – ion. Thus, the **Electron – Electron Collision Frequency** is given by

$$\nu_{ee} = \frac{4\pi n e^4}{m_e^{1/2} \left(3K_B T\right)^{3/2}} \ln\Lambda_{ee}, \tag{1.28}$$

in a plasma of temperature T. Here V has been replaced by its root mean square velocity. The **Ion – Ion Collision Frequency** is given by

$$\nu_{ii} = \frac{4\pi n z_1^2 z_2^2 e^4}{m_i^{1/2} \left(3K_B T\right)^{3/2}} \ln\Lambda_{ii}. \tag{1.29}$$

Finally, the **Electron – Ion Collision** frequency ν_{ei} is found to be

$$\nu_{ei} = \frac{\pi n z^2 e^4}{(m_e)^{1/2} \left(3K_B T\right)^{3/2}} \ln\Lambda_{ei}. \tag{1.30}$$

for the charge (ze) on the ion of mass $m_i \gg m_e$. Now, collisions bring about thermal equilibrium among the various particles. During each collision, there is an energy transfer from the more energetic particle to the less energetic particle. The laws of momentum and energy conservation imply that during a large angle collision between particles of identical masses, the particles have nearly equal energies after the collision. Therefore $\tau_{ee} = (\nu_{ee})^{-1}$ and $\tau_{ii} = (\nu_{ii})^{-1}$ represent the time durations during which electrons and ions reach a common temperature T. The case of electron – ion thermalization is different. Again from kinematics, it can be easily checked that during a large angle collision an electron transfers only a fraction (m_e/m_i) of its energy to an ion. Therefore, equilibration among electrons and ions would take that much longer, i.e. the thermalization, also called the relaxation time τ_{ei} for electrons and ions is given by

$$\tau_{ei} \simeq \frac{m_i}{m_e} \left(\nu_{ei}\right)^{-1}. \tag{1.31}$$

Problem 1.3: Estimate the Coulomb logarithms $\Lambda_{ee}, \Lambda_{ii}$ and Λ_{ei} and the Coulomb frequencies ν_{ee}, ν_{ii} and ν_{ei} for an electron - proton plasma of density $n = 10^{10} cm^{-3}$ and temperature $T = 10^6 K$.

We find that for an electron-proton plasma $\tau_{ei} \gg \tau_{ii} \gg \tau_{ee}$ which implies that electrons and ions can remain at different temperatures for much longer than can ions and ions and electrons and electrons. This is why a plasma is often characterized by two temperatures, one for ions and the other for electrons. As we shall see, a magnetized plasma can have even more than two temperatures. How do the three Coulomb logarithms compare? There could be one more type of collision process – Coulomb collisions of ions with electrons with a collision frequency ν_{ie}. It may not be very clear at this stage that ν_{ie} is different from ν_{ei}. We shall learn in Chapter 2 that, in the magnetohydrodynamic description of a plasma, where it is assumed that the rate of momentum transfer between the electron fluid and the ion fluid is proportional to their relative velocity, the two collision frequencies are related as (Equation 2.111):

$$\rho_e \nu_{ei} = \rho_i \nu_{ie},\qquad(1.32)$$

where ρ_e and ρ_i are the mass densities of the electron and the ion fluids respectively. Thus, in a plasma with equal particle densities, we find $\nu_{ie} << \nu_{ei}$ and therefore, electron-ion collisions are the main process of thermalization among electrons and ions. These collision frequencies determine the rates of transport processes, such as diffusion and dissipation processes, such as Ohmic heating.

1.5. Diffusion in a Plasma

Associated with a collisional process is a mean free path $l_m = V\tau$ that a particle with velocity V traverses between two collisions in a collision period τ. The particle diffuses in the system from one position to another through this random walk. The diffusion coefficient D is defined as:

$$D = \frac{l_m^2}{\tau}.\qquad(1.33)$$

If a plasma is magnetized, the particles are not so free to walk randomly. We shall learn, in detail, about the motion of charged particles in a magnetic field in Chapter 3, but for the present, it would suffice to remember that charged particles execute a circular motion in a direction perpendicular to the magnetic field and move freely along the magnetic field. The diffusion rates, consequently, differ in these two directions. Perpendicular to the magnetic field, a particle diffuses from one orbit to another only if it undergoes collisions with other particles. The mean free path, here, is then of the order of the radius of the orbit, which is the cyclotron radius $R_B = V/\Omega_B$ where Ω_B is the cyclotron frequency (Equation 3.6). The diffusion coefficient $D_\perp$ in the perpendicular direction is given by:

$$D_\perp = \frac{R_B^2}{\tau} = \frac{V^2}{\Omega_B^2 \tau}.\qquad(1.34)$$

Along the magnetic field, the parallel diffusion coefficient $D_\parallel$ is same as in an unmagnetized plasma (Equation 1.33). The ratio of the two diffusion coefficients is found to be:

$$\frac{D_\perp}{D_\parallel} = \frac{1}{\Omega_B^2 \tau^2}. \tag{1.35}$$

In the absence of the magnetic field $D_\perp = D_\parallel$. Therefore, we can write a general expression as:

$$\frac{D_\perp}{D_\parallel} = \frac{1}{1 + \Omega_B^2 \tau^2}, \tag{1.36}$$

so that $D_\perp = D_\parallel$ for $\Omega_B = 0$ and Equation (1.35) is recovered for $\Omega_B^2 \tau^2 \gg 1$. This provides us with a definition of a magnetoplasma i.e. one where the period of circular motion is much less than the collision period. The motion of the particles is dominantly circular and not a drunkard's doodle. The perpendicular diffusion coefficient is inversely proportional to the square of the magnetic field B. This dependence on B is also followed by other transport coefficients such as thermal conductivity and electrical conductivity.

It is easy to see that in an electron-proton plasma, the parallel diffusion coefficient $D_{\parallel e}$ for electrons is much larger than $D_{\parallel i}$, the parallel diffusion coefficient for ions. Does this mean that electrons will diffuse away from a region and ions will be left behind there? No. The quasineutrality condition forbids that. The slow diffusing ions will pull back the fast diffusing electrons and yield conditions for a joint diffusion of both electrons and ions with a diffusion coefficient $D_{A\parallel}$ such that $D_{\parallel e} > D_{A\parallel} \gtrsim D_{\parallel i}$. This phenomenon is known as **Ambipolar Diffusion**. In a plasma, therefore, the net diffusion rate is not decided by the fast diffusing particles, but rather by the slow ones.

In a magnetized plasma, the perpendicular diffusion coefficient for electrons, $D_{\perp e}$ is much smaller than $D_{\perp i}$ (Equation 1.34) for identical electron and ion temperatures T. Therefore, the joint diffusion coefficient $D_{A\perp}$ is determined dominantly by the slow diffusing species, the electrons. The physical reason for the slower perpendicular diffusion of electrons is their smaller mean free path, which is nothing but their cyclotron radius. We shall derive an expression for the ambipolar diffusion coefficient in Chapter 5, since it requires the two-fluid description of a plasma.

In a fully ionized plasma, the diffusion is predominantly governed by Coulomb collisions among unlike particles. The collisions among like particles do not affect a change of the centre of mass and hence cause little or no diffusion.

Problem 1.4: Estimate the different types of $D_\perp$ and $D_\parallel$ for electrons and protons in a pulsar atmosphere with the electron density $n_e \simeq 10^{10}$ cm^{-3}, the temperature $T \simeq 10^5$ K and the magnetic field $B \simeq 10^6$ Gauss.

1.6. Electrical Resistivity of a Plasma

For moderate values of the electric field $\vec{E}$, a plasma obeys Ohm's law:

$$\vec{E} = \eta \vec{J}, \tag{1.37}$$

where $\vec{J}$ is the current density and η is the resistivity. Electrons, the major carriers of the current, move under the combined actions of the applied field $\vec{E}$ and the frictional force due to collisions with ions. Under steady state conditions $(\partial/\partial t = 0)$. We get:

$$-en\vec{E} - \vec{\Gamma}^{ei} = 0, \tag{1.38}$$

where $\vec{\Gamma}^{ei}$ is the frictional force density. For a simple collisional model where the frictional force is proportional to the relative velocity between electrons and ions, we can write (see Section 2.17):

$$\vec{\Gamma}^{ei} = m_e n \nu_{ei} \left(\vec{V}_e - \vec{V}_i \right). \tag{1.39}$$

The current density $\vec{J}$ is by definition:

$$\vec{J} = -en \left[\vec{V}_e - \vec{V}_i \right], \tag{1.40}$$

and we find the resistivity

$$\eta = \frac{m_e \nu_{ei}}{ne^2}. \tag{1.41}$$

Now, substituting for ν_{ei} from Equation (1.30), we obtain the surprising result that the resistivity is almost independent of the electron density n. Therefore, the current density $\vec{J}$ driven by the electric field $\vec{E}$ is independent of the concentration n of the charge carriers. If a plasma contains neutral atoms or molecules, the situation, however, changes. Recall that the electron – ion collisions are infrequent at high temperatures. The reason for this is that at high thermal velocities an electron spends a rather short time in the vicinity of an ion and therefore loses only a small quantity of momentum. Thus, at sufficiently high temperatures, electrons do not feel the frictional drag due to ions and the motion of the two species decouples. The plasma resistivity becomes vanishingly small. Under such conditions, electrons can gain energy from the applied electric field, unhindered. There is a critical value of the electric field, known as the **Dreicer Field**, E_D, above which electrons feel only acceleration and no frictional force. The value of E_D can be determined from equation (1.38) and is found to be:

$$E_D \simeq \frac{m_e}{e} \nu_{ei} \left(\frac{K_B T}{m_e} \right)^{1/2}, \tag{1.42}$$

where V_e has been replaced by the root mean velocity and $V_i \ll V_e$. This is one way of generating high energy electron beams. Of course, the acceleration cannot go on indefinitely. The charge separation, resulting from the decoupling of electron and ion motion, builds up to a value that makes the system unstable, since a plasma always tends to remain quasi-neutral. These instabilities give rise to waves with amplitudes growing with time. The electrons now feel a kind of frictional drag due to these waves. It is found that the resistivity under such circumstances becomes anomalously large, sometimes larger by several orders of magnitude. The effective Ohmic heating, correspondingly, takes place at a much higher rate than that due to Coulomb resistivity. The anomalously large resistivity turns out to be extremely useful in explaining phenomenon such as solar flares, where large amounts of energy are released in a very short time interval (Figure 1.6). Combining Faraday's induction law and Ohm's law, we find that the magnetic field $\vec{B}$ obeys an equation which is characteristic of a diffusion process:

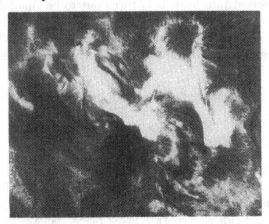

Figure 1.6. A Solar Flare Photographed in H_α Light – Releasing Power $\simeq 10^{30}$ erg sec^{-1}.

$$\frac{\partial \vec{B}}{\partial t} = \frac{\eta c^2}{4\pi} \nabla^2 \vec{B}. \tag{1.43}$$

This equation can be solved by using the 'separation of variables technique'. Assuming L to be the characteristic spatial scale of variation, we find:

$$B = B_0 \exp\left[-t/t_d\right], \tag{1.44}$$

where

$$t_d = \frac{4\pi L^2}{\eta c^2}.$$

Thus, the magnetic field decays or diffuses out in a plasma over a characteristic time t_d. For a perfectly conducting medium, $\eta = 0$, so the magnetic field remains put within the plasma forever. If the plasma moves, so does the magnetic field. This state has been described as the **Frozen-in Magnetic Field**. The diffusion time can also become large if L is large. This is specially true of astrophysical systems. The diffusing magnetic field, according to Ampere's law, produces a current density J given approximately by:

$$J \simeq \frac{cB}{4\pi L}.\tag{1.45}$$

The rate of dissipation of energy due to the Ohmic (or the Joule) heating due to the current density J is

$$\eta J^2 = \frac{\eta c^2 B^2}{(4\pi L)^2} = \frac{1}{t_d}\left(\frac{B^2}{4\pi}\right),\tag{1.46}$$

which shows that t_d is the characteristic time in which magnetic energy density is converted into heat. If t_d turns out to be smaller than the age of an astronomical magnetized object, a planet, a star or a galaxy, we will have to devise ways and means to replenish its magnetic field. This necessity has lead to the invention of dynamos of various types, none very satisfactory at present.

Problem 1.5: Estimate t_d for the Earth and the Sun.

1.7. Plasma as a Dielectric Material

In a plasma, the charged particles not only move in response to the externally applied electric and magnetic fields, but during their motion, they also continuously produce electric and magnetic fields. Thus, a very complex interplay of fields and motion takes place and we face a difficult task of determining these fields in a self-consistent manner.

In a material of macroscopic dimensions, the average charge density is made up of two parts: (i) that due to the average charge of the atomic or molecular ions or the average free charge residing on the macroscopic body; (ii) that due to induced charges produced by polarization. In the absence of external fields, an atom or a molecule may or may not have a permanent electric dipole moment. Even if they have, due to their random thermal motion, the electric dipole moments are directed in a random manner, so that the average dipole moment of the entire system is zero. In the presence of an external field, there is a net dipole moment, which tends to align itself with the external field. This net dipole moment produces a charge density since the external field causes displacement and redistribution of the charges. If the polarization is uniform, there is no net change in the charge density. This is why the polarization charge density is expressed as the divergence of the dipole moment per unit volume $\vec{P}_E$. Poisson's equation including free ρ_f and induced ρ_I charge densities becomes:

$$\vec{\nabla} \cdot \vec{E} = 4\pi\rho_f + 4\pi\rho_I,$$

but

$$\vec{\nabla} \cdot \vec{P}_E = -\rho_I, \tag{1.47}$$

so that

$$\vec{\nabla} \cdot \left[\vec{E} + 4\pi\vec{P}_E\right] \equiv \vec{\nabla} \cdot \vec{D} = 4\pi\rho_f,$$

where $\vec{D}$ is the displacement vector. Since, $\vec{P}_E$ is produced only by the application of $\vec{E}$, we can write $\vec{P}$ as a power-series in $\vec{E}$ as:

$$P_{Ei} = \sum_j \alpha_{ij} E_j + \sum_{j,k} \beta_{ijk} E_j E_k + \cdots . \tag{1.48}$$

Experiments tell us that the linear term in $\vec{E}$ is quite adequate at moderate temperatures and electric fields. For isotropic conditions, Equation (1.48) can be recast as:

$$\vec{P}_E = \chi_e \vec{E}, \tag{1.49}$$

where χ_e is known as the **Electric susceptibility** of the medium. The displacement vector $\vec{D}$ is related to the electric field $\vec{E}$ through the dielectric constant ϵ of the medium as:

$$\vec{D} = \epsilon\vec{E}, \tag{1.50}$$

or

$$\vec{E} + 4\pi\chi_e \vec{E} = \epsilon\vec{E},$$

so that

$$\epsilon = 1 + 4\pi\chi_e. \tag{1.51}$$

For a uniform medium ϵ is independent of space, Poisson's equation takes the form:

$$\vec{\nabla} \cdot \vec{E} = \frac{4\pi}{\epsilon}\rho_f. \tag{1.52}$$

Thus, for a given free charge density ρ_f, the electric field inside a plasma is reduced by the factor ϵ. The reduction results because the direction of the electric field produced by the induced charge density is opposite to that of the applied field. A plasma with a large dielectric constant screens AC electric fields the way a plasma with small Debye length screens DC electric fields.

1.8. Plasma as a Magnetic Material

In a macroscopic medium, the atomic electrons, due to their intrinsic magnetic dipole moment as well as due to their motion, produce fluctuating currents. These currents give rise to a magnetic field. Thus, in analogy with charge density, the total current density consists of two parts: (1) current density due to circulating bound or induced charges and (ii) conduction current density J_c due to actual transport of charges. If $\vec{M}$ is the magnetic dipole moment per unit volume, the corresponding current density $\vec{J}_M$ is given by:

$$\vec{J}_M = c\left(\vec{\nabla} \times \vec{M}\right). \tag{1.53}$$

and Ampere's law modifies to:

$$\vec{\nabla} \times \vec{B} = \frac{4\pi}{c}\vec{J}_c + 4\pi\left(\vec{\nabla} \times \vec{M}\right), \tag{1.54}$$

which can recover its original form

$$\vec{\nabla} \times H = \frac{4\pi}{c}\vec{J}_c, \tag{1.55}$$

if we define the total magnetic field $\vec{H}$ as:

$$\vec{H} = \vec{B} - 4\pi\vec{M}. \tag{1.56}$$

Now, if we assume that the magnetic dipole moment per unit volume $\vec{M}$ is related to $\vec{H}$ as:

$$\vec{M} = \chi_m\vec{H}, \tag{1.57}$$

and write

$$\vec{B} = \mu_m\vec{H}. \tag{1.58}$$

We find

$$\mu_m = 1 + 4\pi\chi_m. \tag{1.59}$$

Here, χ_m is the magnetic susceptibility and μ_m is the magnetic permeability.

In a plasma with magnetic field $\vec{B}$, each particle executes a circular motion and produces a magnetic moment (Equation 3.37)

$$\mu = \frac{mV_\perp^2}{2B}. \tag{1.60}$$

Thus, in a plasma, we do not have a linear relation between magnetic dipole moment and magnetic field as for a magnetic material (Equation 1.57). Therefore, it is not of much utility to treat a plasma as a magnetic material.

1.9. Plasma as a Refracting Medium

The propagation of electromagnetic waves through a plasma can be understood easily, if we could define a refractive index of a plasma. We will show that the dielectric constant ϵ can be related to the refractive index N of the medium. Let us consider the propagation of electromagnetic waves through a medium of dielectric constant ϵ, electrical conductivity σ and magnetic permeability μ_m. Maxwell's equations in such a medium become:

$$\vec{\nabla} \cdot \vec{D} = \vec{\nabla} \cdot \left[\epsilon \vec{E} \right] = 0,$$

$$\vec{\nabla} \cdot \vec{B} = \vec{\nabla} \cdot \left[\mu_m \vec{H} \right] = 0,$$

$$\vec{\nabla} \times \vec{E} = = -\frac{\mu_m}{c} \frac{\partial \vec{H}}{\partial t}, \qquad (1.61)$$

$$\vec{\nabla} \times \vec{H} = \frac{\epsilon}{c} \frac{\partial \vec{E}}{\partial t} + \frac{4\pi\sigma}{c} \vec{E},$$

along with Ohm's law

$$\vec{J} = \sigma \vec{E}.$$

Assuming that the transverse fields $\vec{E}, \vec{B}, \vec{D}, \vec{H}$ vary as $\exp[i(\vec{k} \cdot \vec{r} - wt)]$ representing electromagnetic waves of wave vector $\vec{k}$ and frequency ω, substituting in Equations (1.61) and eliminating either $\vec{H}$ or $\vec{E}$, we find that $\vec{H}$ or $\vec{E}$ satisfies the equation:

$$\left[k^2 - \left(\mu_m \epsilon \frac{\omega^2}{c^2} + 4\pi i \frac{\mu_m \omega \sigma}{c^2} \right) \right] \vec{E} = 0, \qquad (1.62)$$

which implies

$$\frac{k^2 c^2}{\omega^2} = \mu_m \epsilon + 4\pi i \frac{\mu_m}{\omega} \sigma. \qquad (1.63)$$

We see that the left hand side is the square of the ratio of the phase velocity c of light in vacuum and the phase velocity (ω/k) in the medium and is therefore equal to the square of the refractive index N, by definition. Therefore, if we treat a plasma as a dielectric ($\sigma = 0$) and a nonmagnetic material ($\mu_m = 1$) the refractive index N is given by:

$$N^2 = \epsilon. \qquad (1.64)$$

The resistive properties of a plasma are included in the dielectric function ϵ. All aspects of electromagnetic wave propagation through a plasma are studied using Equation (1.64). We can thereby learn about the frequency pass bands, cutoffs, absorption, refraction and reflection properties of a plasma.

When high intensity radiation propagates through a plasma, it modifies the plasma characteristics completely. The reflection region may become absorbing and a transparent plasma may become a scattering medium. All this is accomplished through the dielectric constant ϵ, which in addition to depending upon plasma parameters, becomes a function of the intensity and the frequency of the incident radiation. Calculation of the dielectric constant, which is anything but a constant, is a major occupation of plasma physicists. We shall calculate ϵ for plasmas under different conditions in Chapters 5 and 6.

1.10. Plasma as a Source of Coherent Radiation

A source of radiation is coherent if its size is smaller than the wavelength of the radiation it emits, for then, the differences in the retarded times of the different parts of the source can be neglected. Plasmas, due to their cooperative nature, are found to be very efficient sources of strong and coherent radiation over a huge range of the electromagnetic spectrum. The source of energy lies in the nonthermal distributions of particles such as an electron beam traversing a plasma, or in an anisotropic velocity distribution, such as a loss-cone distribution arising in an inhomogeneous magnetic field. These distributions, in an attempt to relax to an equilibrium, produce electric and magnetic fields which may be in the form of electrostatic or electromagnetic waves. Since these waves are produced by the induced charge densities and electric currents, the bulk of the plasma particles participate in this process. We have already learned that in a plasma, charge densities can exist only over distances of the order or less than the Debye length. The condition for the production of coherent radiation is that its wavelength must be greater than the Debye length. Under these conditions, all the particles contained in the Debye sphere are in phase with each other and participate collectively in the emission process. Thus, the typical size of a coherent plasma source is of the order of the Debye length. Of course it will never be possible to resolve a source of this size in cosmic circumstances. That there is a coherent process in action, is inferred, for example, from observations of time variability (Figure 1.7) of intensity and polarization of radiation among other possible consequences (Krishan 1997 and references therein).

1.11. Strongly Coupled Plasmas

The coupling in plasmas is associated with Coulomb interaction among the plasma particles. A plasma with one species of charged particles, for example electrons, embedded in a uniform background of the oppositely charged species, for example ions, is known as an **One Component Plasma (OCP)**. The coupling constant, Γ, of an OCP is defined as the ratio of

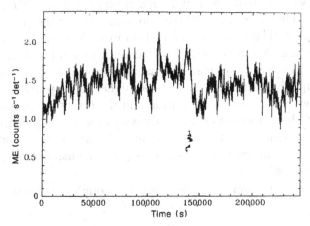

Figure 1.7. Rapid Time Variability of X Rays Observed From the Active Galactic Nucleus NGC5505.

the average Coulomb energy to the average kinetic energy. Plasmas with $\Gamma \geq 1$ are known as strongly coupled plasmas. For a system of charged particles at temperature T obeying Boltzmann statistics, the kinetic energy per particle is approximately $K_B T$. For an electron gas obeying the Fermi-Dirac distribution, the average energy per particle is the **Fermi Energy**, E_F, given by

$$E_F = \frac{\hbar^2}{2m_e} \left(3\pi^2 n\right)^{2/3}, \qquad (1.65)$$

where $\hbar = h/2\pi$ and h is Planck's constant and n is the particle density. The Coulomb energy E_c per particle of an OCP is estimated to be:

$$E_c = \frac{(ze)^2}{a}, \qquad (1.66)$$

where (ze) is the particle charge and

$$a = \left(\frac{3}{4\pi n}\right)^{1/3}, \qquad (1.67)$$

is the radius of a sphere with volume $(1/n)$ and is variously called, the **Interparticle Distance**, the **Ion-Sphere Radius** or the **Wigner-Seitz Radius**. The coupling constant Γ for a classical system is therefore given by

$$\Gamma = \frac{(ze)^2}{a K_B T}, \qquad (1.68)$$

from which, we find that for a hydrogen plasma at $T = 10^6 K$, $n \simeq 10^{26}$ cm^{-3}, $\Gamma \simeq 1$. Therefore, most of the commonly occurring plasmas are not strongly coupled. The Coulomb interaction in such plasmas can be treated as a perturbation. A system with $\Gamma \geq 1$ cannot be studied using perturbation methods. A strongly coupled plasma shows novel traits not exhibited by a weakly coupled plasma. For $\Gamma \geq 1$, the higher order particle correlations play a decisive role in the configuration and the transport properties of the plasma. It has been found that the resistivity of a strongly coupled plasma is much larger than that of a weakly coupled plasma. A strongly coupled plasma is found to undergo a spontaneous phase transition, releasing latent heat in the process.

For a system of degenerate electrons, the coupling constant Γ is found to be:

$$\Gamma = \frac{e^2/a}{E_F} = 0.543 r_s,$$
(1.69)

where $r_s = \left(\dfrac{3}{4\pi n}\right)^{1/3} (m_e c^2/\hbar^2)$ is the Wigner-Seitz Radius of the electrons divided by the Bohr Radius. For conduction electrons in metals, $r_s = 2 - 6$, so that $\Gamma = 1 - 3$. Thus, the electron gas in metals is a strongly coupled OCP.

Strongly coupled plasmas are a more common occurrence in celestial compact objects than in terrestrial systems. An example close to home is provided by the planet Jupiter. The interior of the planet is made up of hydrogen with a few percent helium with average mass density $\rho = 1 - 10 gm cm^{-3}$ at a temperature $T \sim 10^4 K$. It is found to be a strongly coupled plasma with $r_s = 0.6 - 1$ and $\Gamma = 20 - 50$. The observation that Jupiter and Saturn emit 2-3 times more infrared radiation than they absorb from the sun, has been interpreted to be the result of the energy release due to a phase transition of the strongly coupled plasma in the interiors of these planets.

The interiors of sun–like stars have $\Gamma \simeq 0.05$ and therefore do not qualify as strongly coupled plasmas, but strong coupling effects need to be included in the study of the state of heavy elements and their mixing.

A neutron star contains about the mass of the Sun in a sphere of radius of about 10 km. It is, perhaps, the most condensed state of matter. Theoretical models show that a crust of 1-2 km radius and a mass density of $10^4 - 10^7$ gm cm^{-3} consisting mostly of iron exists just below the surface of a neutron star. With an expected coupling constant $\Gamma \simeq 10 - 10^3$, the tendencies of this OCP of iron nuclei towards Wigner - Crystallization and a glassy transition are being investigated since the knowledge of the state of crystal matter is crucial for understanding the cooling rate of a neutron star.

The interior of a white dwarf consists of dense material with mass density and temperature similar to what exists in the crust of a neutron star. White dwarfs with interiors made up of carbon-oxygen mixture are believed to be the progenitors of some type 1 supernovae. The possibilities of phase separation and formation of alloys in this carbon-oxygen mixture have important bearing on the cooling rates, mechanism of supernovae explosion transport processes and neutrino emission processes. A lot of work, theoretical and experimental, is being carried out in the field of strongly coupled plasmas and it will not be an exaggeration to say that the motivation comes from the 'Heavens' and from our eternal search for new materials (Ichimaru 1990).

1.12. Dusty Plasmas

Though, so far, we have been mainly referring to a two component plasma with electrons and ions, in astrophysical situations, a third component, called the dust, is often present. This three component system is quite different from the one with three species of charged particles. For, one thing, the dust particles acquire electric charge, like a capacitor, when they are inserted in a plasma. Their charge is not a constant. It is a function of plasma parameters, varies with time, and it takes finite time for a dust particle to acquire charge. Dust particles are of macroscopic dimensions, of the order of a micron and smaller. The composition of the astrophysical dust varies from one environ to another. It could be carbonaceous, silicates, ferrites or any alloy of them. Heavy molecules and frozen ices are also a part of the family of dust grains. The composition of dust is determined from its response to the radiation that falls on it. The absorbed, the scattered and the reemitted radiation carries diagnostics of the dust grains. So, what happens when a plasma of electrons and protons is impregnated with dust grains?

The grains suffer collisions with electrons and protons and acquire electric charge in the process. The electrostatic potential due to the charged grains affects the electron and proton density distributions, which in turn modify the electron and proton fluxes impinging upon the grains. Thus, the charging of the grain and plasma particle distributions must be determined self-consistently. The system is complex, therefore, we need to make some simplifying assumptions. The first assumption is that the number density n_d of dust grains is much less than the electron or proton densities. Two cases are identified in this connection: a plasma is called a **Dusty Plasma** if the number of grains N_d in a Debye Sphere is much larger than unity. The opposite case with $N_d \leq 1$ is referred to as **Dust In A Plasma**. In addition to the charging of a grain by electron and ion currents directed on to it, its charged state may also change due to photoemission if subjected to radiation. The secondary emission of electrons when a grain is bombarded by electrons as well as the field emitted electrons, further, deprive a grain of its negative charge. The rate of change of the charge Q of a grain due to all possible causes can be expressed as:

$$\frac{dQ}{dt} = \sum_s I_s, \tag{1.70}$$

where I_s is the current due to a process s. Let us first consider the charging of a grain only due to electron and proton fluxes. Thus, if $\vec{J_e}$ and $\vec{J_i}$ are the current densities due to electrons and protons respectively, then the total currents I_e and I_i are given by:

$$I_e = -4\pi a^2 n_e e V_e \alpha_e, \tag{1.71}$$

and

$$I_i = 4\pi a^2 n_i e V_i \alpha_i,$$

where a is the radius of a spherical grain, V_e and V_i are the velocities of the electrons and protons relative to that of the grain velocity V_d, and α_e and α_i are known as the sticking coefficients. In a Maxwellian plasma, V_e and V_i can be replaced by the corresponding thermal velocities. If V_d is larger than the thermal speed of protons, than V_i is replaced by V_d. For constant n_e, n_i, and temperatures T_e, T_i, we see from Equations (1.70) and (1.71), that the charge Q increases linearly with time. Of course, this cannot continue for long. As the charge Q builds up, the negative charge on the grain begins to repel electrons and attract protons, so that in the neighborhood of the grain, n_e decreases and n_i increases. A steady state is reached when $I_e = I_i$ and $Q = Q_0$. Realizing that at a common temperature T, the electron thermal velocity $V_{Te} >> V_{Ti}$, the ion thermal velocity, we find that the time taken by the grain to accumulate a charge Q_0 is given by:

$$t_0 = \frac{Q_0}{4\pi a^2 e n_0 \left(K_B T/m_e\right)^{1/2}}, \tag{1.72}$$

where we have taken $n_e = n_i = n_0$ and $\alpha_e \simeq 1$. We know that electrostatic potential $\varphi_0 \simeq -K_B T/e$ can exist in a plasma. Therefore, the grain charge Q_0 produces the potential φ_0 due to which the electron current to the grain ceases. If the grain is treated as a capacitor of capacitance C_0 then the charge Q_0 is related to C_0 and φ_0 as

$$Q_0 = C_0 \varphi_0, \tag{1.73}$$

and $C_0 = a$ for a spherical capacitor. Thus, we find:

$$t_0 = \frac{\lambda_D}{a} \omega_{pe}^{-1}, \tag{1.74}$$

and

$$Q_0 = -\frac{K_B T a}{e}.$$

Thus the charge relaxation time in a plasma is $\simeq \omega_{pe}^{-1}$, and the smaller the size of the grain, the smaller the crossection for an electron encounter, and the longer charging time t_0, the larger temperature, the larger electron flux to the grain and, finally the larger capacitance a, the larger charge Q_0. For a more accurate derivation of these results, we need to know the particle distribution in the presence of the grain potential and determine I_i and I_e. It is found that for equal electron and ion temperatures, the grain potential is given by: $\varphi \simeq -2.5 K_B T/e$ which is not too different from the value obtained from the approximate treatment given above.

In astrophysical situations such as comets whizzing through the solar wind or interstellar grains braving the ultraviolet radiation of stars, photoemission of electrons from the grains must be included in the list of charging processes. The electron emission takes place via the well known photoelectric effect described by the Einstein relation:

$$\hbar\omega = W + E,$$

where ω is the frequency of the radiation falling on a material of work function W, and E is the kinetic energy of the emitted electron. For astrophysical grains, the work function is of the order of a few electron-volts, consequently the radiation frequency ω corresponds to the ultraviolet part of the electromagnetic spectrum. Thus a knowledge of ω and W provides us with an estimate of E and therefore of the current I_p. The electron emission endows the grain with a positive charge which may try to pull back the emitted electrons. Again, in principle, we could determine the steady state by requiring the total current

$$I_e + I_i + I_p = 0. \tag{1.75}$$

The steady state potential due to the charged grain could be positive or negative. In reality, however, a lot of work and a lot more guess work goes into the determination of the work-functions for which a good idea of the composition is a prerequisite. The sizes and shapes and their distributions are the other important parameters, which are intimately connected with the formation mechanisms of dust grains.

Far from a charged grain, the electrostatic potential in a plasma tends to vanish due to screening effects. What if there are many grains within the Debye sphere? Each grain has a potential $\sim (K_B T/e)$ with an e-falling distance of the order of the Debye length. But now, due to the presence of other charged grains, the potential far from the grain is not vanishingly small, but has a finite value, say, φ_F. So, the net potential $(\varphi - \varphi_F)$ will now be supported by a lesser value of Q given by:

$$Q \simeq a(\varphi - \varphi_F), \tag{1.76}$$

where the capacitance of the spherical grain still remains close to a for $a \ll \lambda_D$. As we realize by now that charging of grains constitutes a host of complex processes and therefore, can be addressed with any thoroughness only in a specific circumstance.

There is a limit to the quantity of charge that a grain can hold. The electrostatic repulsion among different parts of a grain can tear it apart, leading to **Electrostatic Disruption** of the grain. Let a small surface element ΔS of the grain moves a distance Δr due to electrostatic repulsion. This will reduce the potential energy by an amount

$$(-\Delta\varphi) = \frac{E_g^2}{8\pi} \Delta S \Delta r, \tag{1.77}$$

where $(E_g^2/8\pi)$ is the electrostatic energy density of the grain. The force acting on the element ΔS is $(-\Delta\varphi/\Delta r)$ and the force per unit area f_s is

$$f_s = \left(-\frac{\Delta\varphi}{\Delta r}\right)\frac{1}{\Delta S} = \frac{E_g^2}{8\pi} = \frac{Q^2}{8\pi a^4}, \tag{1.78}$$

where Q is the charge on the grain. If f_t is the tensile strength of the grain material, then the grain will not disrupt or split apart due to electrostatic repulsion if

$$f_t > f_s,$$

or

$$a^4 > \frac{Q^2}{8\pi f_t} \equiv a_c^4. \tag{1.79}$$

Thus, grains with charge Q and radius $a < a_c$, the critical radius, undergo disruption. The detached fragment moves away from the parent grain due to electrostatic repulsion. But it turns out that it is not always so. Let the fragmented charge q be at a distance x from the centre of the remaining part of say radius a (Figure 1.8). If this sphere of radius a is at zero potential then we know that an image charge q' inside the sphere at a distance x' along with the charge q at x satisfies this boundary condition:

$$\frac{q}{x - a} + \frac{q'}{a - x'} = 0, \tag{1.80}$$

or

$$\frac{q}{a\left(\frac{x}{a} - 1\right)} + \frac{q'}{x'\left(\frac{a}{x'} - 1\right)} = 0, \tag{1.81}$$

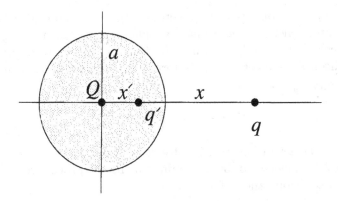

Figure 1.8. The Parent Grain with Charge Q and the Fragmented Point Charge q.

which gives

$$q' = \frac{-qa}{x} \text{ and } xx' = a^2.$$

If, now, we wish the sphere of radius a to have a surface potential φ, we can place a charge $Q = a\varphi$ at the centre of the sphere. The potential energy $(q\varphi_g)$ of the fragment of charge q is, therefore, given by:

$$q\varphi_g = \frac{qQ}{x} + \frac{qq'}{x - x'} = \frac{aq\varphi}{x} - \frac{aq^2}{x^2 - a^2}, \tag{1.82}$$

which shows that the potential energy has two contributions: the repulsive component due to q and Q being of the same sign and the attractive component due to the image charge q'. If x_0 is the position at which $\varphi_g = 0$, then it is easy to check that $(q\varphi_g) < 0$ for $a < x \le x_0$ and $(q\varphi_g) > 0$ for $x > x_0$.

Problem 1.6: What happens at $x = a$?

If both the parent and the fragmented grains are point particles, then there is no image charge and the potential φ_g is purely repulsive. However, if we include gravitational forces, the potential energy could again become attractive. This behavior of the parent and the fragmented grains plays a decisive role in creating a dusty atmosphere around large bodies. In realistic situations, additional effects due to thermal and radiation pressures must be included.

The role of electrically charged dust is perhaps best studied in planetary atmospheres. Erosion of comets, asteroids, volcanic activity, for example, on Jupiter's moon Io and electrostatic blow off from other solar system bodies produce large amounts of dust in planetary magnetospheres. The interplanetary dust collides with planetary rings and moons at velocities $\sim$ 30 km/sec and generates more dust and plasma. A grain moving through a plasma feels Coulomb drag akin to frictional drag due to electron – ion collisions discussed earlier. The drag force $\vec{F}_g$ on a grain of mass m_g, charge Q and velocity $\vec{V}_g$ can be written as:

$$\vec{F}_g = -m_g \nu_{gi} \left(\vec{V}_g - \vec{V}_i \right),$$

(1.83)

where ν_{gi} is the grain-ion collision frequency and $\vec{V}_i$ is the ion velocity. The collision frequency can be determined using the procedure outlined for electron-ion collisions and is found to be:

$$\nu_{gi} = \left\langle n_i \left(\vec{V}_i - \vec{V}_g \right) \sigma_{gi} \right\rangle,$$

(1.84)

where the Coulomb crossection σ_{gi} is determined from the balance between relative kinetic energy of the ion and its electrostatic potential energy. The angular brackets denote the averaging over the ion velocities. It is not difficult to arrive at the following expression for $\vec{F}_g$:

$$\vec{F}_g \simeq \frac{4\pi Q^2 e^2 m_g}{m_i^2} \int f(\vec{V}_i) \frac{\left(\vec{V}_i - \vec{V}_g \right)}{\left| \vec{V}_i - \vec{V}_{g,} \right|^3} d\vec{V}_i$$

(1.85)

where $f(\vec{V}_i)$ is the ion distribution function and m_i is the ion mass. The radial motion of dust grains from satellites of planets or from the interplanetary medium towards planets for formation of planetary rings is possible due to drag forces; otherwise, the dust particles continue to execute Keplerian circulation (Figure 1.9). Collective effects may change the expression (1.85) substantially.

It may happen that a dust grain acquires a large number of charges. These charges, can, then constitute, what is known as a **Non-neutral Plasma**. In such a plasma, either there are charges only of one sign or there is an excess of charges of one sign. Such a plasma can be in equilibrium under the action of an electric and magnetic field. The radial force balance in steady state gives (Equation 5.2):

$$-\frac{m_e V_\theta^2}{r} = -eE_r - \frac{e}{c} V_\theta B_z,$$

(1.86)

where V_θ is the azimuthal velocity of the electron fluid, E_r, the radial electric field, and B_z the axial magnetic field in a cylindrical plasma made up of only electrons. The radial electric field is determined from Poisson's equation as:

$$\frac{1}{r} \frac{\partial}{\partial r} (rE_r) = -4\pi e n_e,$$

(1.87)

Figure 1.9. The Dusty Rings of Saturn, Photographed by the Hubble Space Telescope.

where n_e is the electron number density. For constant n_e, we find

$$E_r = -2\pi e n_e r. \tag{1.88}$$

Substituting for E_r in Equation (1.86), and writing $V_\theta = \omega r$, we get a quadratic in ω^2, whose two solutions are:

$$\omega_\pm = \frac{\omega_{ce}}{2}\left[1 \pm \left(1 - \frac{2\omega_{pe}^2}{\omega_{ce}^2}\right)^{1/2}\right], \tag{1.89}$$

where $\omega_{ce} = e(B_z/m_e c)$ is the electron cyclotron frequency. Thus, all the electrons rotate with angular velocities given by Equation (1.89). For low densities, such that $(2\omega_{pe}^2/\omega_{ce}^2) << 1$, the fast rotation described by $\omega_+ \simeq \omega_{ce}$ is just the cyclotron motion of all electrons around the magnetic field; the slow rotation described by $\omega_- \simeq (\omega_{pe}^2/2\omega_{ce}) = c(\vec{E} \times \vec{B})/B^2$ is known as the $E \times B$ drift (see Section 3.5). The highest density permissible for a real value of ω, and therefore for the equilibrium of the electron plasma is given by

$$\omega_{pe}^2 = \frac{\omega_{ce}^2}{2}, \tag{1.90}$$

or

$$n_e = \frac{B_z^2}{8\pi m_e c^2},$$

For densities greater than that given by Equation (1.90), there is no equilibrium and the full equation of motion of electrons shows that the plasma acquires a time dependent radial velocity and begins to expand. Micronsize interstellar grains have been attributed electronic charges $\simeq 100$. This corresponds to an electron density $n_e \simeq 2 \times 10^{13}$ cm^{-3}. Equation (1.90) tells us that the grain must have a magnetic field $\simeq 10^4$ Gauss for confining these 100 electrons. This is too high a value of B_z. Actually Equation (1.90) describes the equilibrium of electrons in vacuum. When electrons reside on grains, there are other attachment forces. Non-neutral plasmas share many of the properties of neutral plasmas, such as Debye shielding and collective effects. They also differ in substantial ways – a non-neutral plasma can easily attain the regime of strongly coupled plasma whereas in a neutral plasma, oppositely charged particles begin to recombine under high pressure.

It is very important to analyze the motion of dust grains, especially consequences of their electric charge in the vicinity of a planet in order to understand the existence of planetary rings. The equation of motion of a grain of charge $Q(t)$, mass m_g and velocity V_g in a frame rotating with a planet's angular velocity Ω can be written as (see Equations 4.14 and 7.1):

$$\frac{d\vec{V}_g}{dt} = \frac{Q(t)}{m_g c}\vec{V}_g \times \vec{B}(\vec{r}) + 2\vec{V}_g \times \vec{\Omega} + \frac{Q(t)}{m_g}\left(\vec{E}(\vec{r},t) + \frac{\left(\vec{\Omega} \times \vec{r}\right) \times \vec{B}(\vec{r})}{c}\right)$$
$$-\vec{\Omega} \times \left(\vec{\Omega} \times \vec{r}\right) + \vec{g}, \tag{1.91}$$

where $\vec{g}$ is the acceleration due to the planet's gravity, and the inductive electric field $\vec{E}$, in the frozen-in approximation ($\sigma \to \infty$), is given by:

$$\vec{E} = -\frac{\left(\vec{\Omega} \times \vec{r}\right) \times \vec{B}}{c}. \tag{1.92}$$

In the rotating frame, $\vec{B}$ is no longer a function of time. Substituting for $\vec{E}$ in Equation (1.91) and using cylindrical coordinates, we get

$$\frac{d\vec{V}_g}{dt} = \frac{Q\vec{V}_g}{m_g c} \times \left[\vec{B} + \frac{2m_g c}{Q}\vec{\Omega}\right] + \nabla\left(\frac{r^2\Omega^2}{2} - \psi_g\right), \tag{1.93}$$

where we have used $\vec{\Omega} \times (\vec{\Omega} \times \vec{r}) = -\Omega^2\vec{r} = -\vec{\nabla}\left(\frac{\Omega^2 r^2}{2}\right)$ and $\vec{g} = -\vec{\nabla}\psi_g$.

Equation (1.93), is the equation of motion of a grain of mass m_g and charge Q in an effective magnetic field

$$\vec{B}_e = \vec{B} + \frac{2m_g c}{Q}\vec{\Omega}, \tag{1.94}$$

and an effective electric field

$$\vec{E}_e = \frac{m_g}{Q}\vec{\nabla}\left[\frac{1}{2}r^2\Omega^2 - \psi_g\right]. \tag{1.95}$$

This type of motion is discussed in detail in Chapter 3. The grain executes circular motion about $\vec{B}_e$ and the centre of the circular motion drifts along $\vec{B}_e$, if Q is independent of time. Q is a function of time, the cyclotron motion is accompanied by an additional rocking motion in a direction perpendicular to the magnetic field. For a value of the time dependent part of Q smaller than its time independent part we can use perturbation methods illustrated in Chapter 3 to analyze the motion of the grain in terms of circular motion and the drift velocities of its guiding centre. These motions tell us if the grains would settle in a circular orbit about the planet or have an additional radial drift towards or away from the planet. This has consequences for the ages of the planetary rings.

Problem 1.7: Find the cyclotron frequency of an iron grain of one micron radius and 100 electronic charges at the surface of Earth in the effective magnetic field given by Equation (1.94).

The excitation of waves in a dusty plasma can be studied using standard methods described in Chapters 4, 5 and 6. Charged grains along with electrons and ions form a three component plasma, though not in the strictest sense. In addition to mass, momentum and energy conservation laws for each species of particles, we must include the charge density conservation laws:

$$-en_e + (ze)n_i + Qn_g = 0, \tag{1.96}$$

where $(-e)$, (ze) and Q are the charges on an electron, an ion and a grain, respectively of densities n_e, n_i and n_g; Q may be positive or negative. There is a new time scale associated with the grain plasma frequency $\omega_{pg} = (4\pi n_g Q^2/m_g)^{1/2}$ which could be much larger than those associated with electron- and ion-plasma frequencies. The role of these very low frequency oscillations in astrophysical situations has not been fully explored. One consequence could be that electromagnetic radiation traversing a dusty plasma shows large period modulations. The scattering of electromagnetic radiation in a dusty plasma, as for example in the earth's ionosphere, has been investigated in order to account for the observed scattering, which is enhanced over the usual Thomson scattering by electrons. Inclusion of additional charged density fluctuations due to grains of variable charges appear to increase the scattering crossections.

A particularly interesting repercussion of electrically charged dust can be seen in the situation of gravitational collapse where an initial mass density enhancement attracts more and more material due to its gravitational force and may grow to be a macroscopic structure such as a star or a galaxy. If the collapsing material has charged grains, the collapse may be inhibited first due to electrostatic repulsion. Then because of the electrostatic disruption of grains, the gravitational forces may become weak, further retarding the rate of collapse. These investigations form a part of the study of Jean's Instabilities with which the issues of structure formation in the universe are addressed (Bliokh, Sinitsin and Yaroshenko 1995).

1.13. Study of Plasmas: Towards what Purpose?

Since most of the visible universe is in the plasma state, knowing about plasmas would help us to understand some of the workings of the universe. Of course, it is the controlled thermonuclear fusion – that pollution free, nearly free, eternal source of energy – that is the ultimate goal of most of the laboratory plasma physicists. In the meantime, plasmas have been serving mankind through a host of technological applications – from communication to dyeing, to deposition of ions on metals, and fabrication of new materials. Plasma physics has also drawn the attention of people seeking the ultimate accelerators, the keyholes to the structure of matter and the origin of the universe.

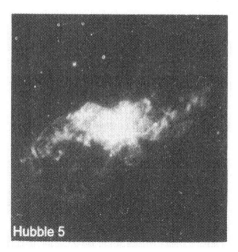

Figure 1.10.   Planetary Nebula Exhibiting Bipolar Filamentary Structures.

Filamentary structures of all sizes and shapes are observed on all scales in the universe – be it on planetary and stellar atmospheres, supernovae ejecta, planetary nebulae, galactic environs or extragalactic realms (Figure 1.10). The macroscopic stability of these structures is studied using single and two-fluid descriptions of a plasma. These descriptions relate the size, the pressure, the fields and the flows in a plasma structure. In addition, we would like to know how do the characteristics of radiation that propagates or originates in these structures depend upon their defining parameters, such as, density, temperature magnetic and velocity fields. For example, quasiperiodic time variations in the radiation flux may indicate that the emitting region is in a state of oscillation.

In the same manner, we can learn about the medium through which the radiation propagates. For example, the observed delay in the arrival times of pulses of different frequencies from a pulsar, is attributed to the dispersion properties of the interstellar plasma. This time the delay can be related to the electron density, the magnetic field and the size of the intervening interstellar medium. The observations and modeling of the nonthermal radio emission from the sun provides (Figure 1.11) us with estimates of density, temperature, magnetic field and geometric configuration of the solar corona. Through the absorption and scattering of electromagnetic radiation in the emission line regions of a quasar we hope to learn about the invisible central object, suspected to be a black hole.

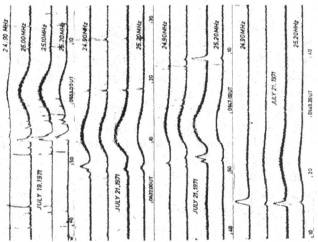

Figure 1.11. Nonthermal Radio Emission From the Sun Known as Type III Radio Bursts Krishan et.al. (1978).

It is in the realm of coherent sources of electromagnetic radiation that plasmas exhibit their versatility the most. Plasmas are good at fast and large releases of energy. This is possible as they can store free energy in several forms, as gradients in configuration and/or in velocity space. Thus, large departures from equilibrium are first allowed to grow; this is the state of instability. After attaining a critical stage the plasma undergoes relaxation, either in an explosive manner, (Fukai, Krishan &Harris 1969, 1970), or in a more gentle way. Solar flares are one such phenomenon where a complex configuration of magnetic and velocity fields becomes unstable and relaxation takes place with the release of electromagnetic and mechanical energy. Most of the strong extragalactic radio sources are associated with nonthermal (non-Maxwellian) distributions of energetic particles which thermalize through single particle and collective plasma process, the latter being always more efficient and faster, if and when they happen. Often, the radiation observed from astrophysical sources has several components in it. There may be a steady emission over which is superimposed a rapidly varying or a quasi-periodic component; or the contribution of thermal to nonthermal processes may vary in different parts of a single source; or the emission may appear as absorption at some parts of the spectral region. All these situations can be a result of wave-particle and wave-wave interaction processes which can enhance or eliminate certain spectral regions. The generation and propagation characteristics of cosmic radiation bring us the diagnostics of the physical conditions in distant objects.

1.14. Techniques of Studying Plasmas

As for any many body system, the statistical methods are the most suitable for studying plasmas. The use of statistical methods have provided us with three levels of description of plasma particles and their attendant and externally imposed electromagnetic and other fields. For moderately dense plasmas, the particle-particle correlations can be ignored and the N-body system can be described though an N – particle distribution function, which is a function of positions and velocities of N-particles at a given instant of time and describes the number density of particles with given velocities at given space-time points. The only condition that this function has to satisfy is contained in the **Liouville Equation** which expresses its constancy in the phase space of positions and velocities. By integrating over all positions and velocities except one we get a single particle distribution function, and the condition of its constancy in the phase space is nothing but the **Boltzmann Equation**. Combined with Maxwell's equations, the collisionless Boltzmann or the **Vlasov Equation** describes the entire range of plasma phenomena including stability, heating and radiative processes – essentially the microscopic aspects of the plasma. This constitutes the **Kinetic Description** of plasmas.

Further simplification is achieved by taking the velocity moments of the Vlasov equation. Averaged macroscopic quantities, such as number density, velocity and pressure are obtained for each species of plasma particles. Each species is now treated as a fluid. A plasma consisting of electrons and protons has two interpenetrating fluids – the electron fluid and the proton fluid. Each fluid moves with a single velocity, has one single temperature and behaves like a conducting fluid in the presence of electromagnetic fields. This is known as the **Two-Fluid Description** and is very handy for describing phenomena in which electrons and ions play differential roles.

A third level of description is obtained by combining the equations of electron fluid and ion fluid. Here, the electrons and ions lose their identities. Instead, a single fluid with a specific mass density, velocity, a current density and pressure is the outcome. This description of a plasma covers a wide variety of phenomena and has earned itself an independent title - **Magnetohydrodynamics** (MHD). At the root of MHD lies the mutual interaction of the fluid flow and the magnetic field. The magnetic field and its associated current produce a Lorentz force which accelerates the fluid across the magnetic field, which in turn creates an electromotive force resulting in currents that modify the field. Macroscopic configurational stability, generation of magnetic fields – in fact all phenomena not dependent upon charge separation are studied using MHD. Each of the three descriptions has its region of applicability and can be deployed for linear and nonlinear problems.

1.15. Waves in Plasmas

After ascertaining the equilibrium of a plasma, its response to a small disturbance must be investigated. A plasma is said to support linear waves if the space-time variations of its defining parameters, such as density, velocity, magnetic field, etc take sinusoidal forms. Linear – because under small disturbances, the equations describing these oscillations are linear. Their nontrivial solutions provide us with a relation, known as the **Dispersion Relation**, between the frequency and the propagation wavevector of a wave. The dispersion relation may have more than one root. Each root represents a wave with definite phase and group velocities. By substituting the dispersion relation back into the plasma equations, we can determine the relative magnitudes of the various parameters and fields, albeit, not the absolute strengths, as well as the polarization of the waves. This much knowledge of waves is enough to classify them as **MHD Waves**, **Drift Waves** or **Electromagnetic** and **Electrostatic Waves**. These waves have finite lifetimes, they suffer damping due to Coulombic collisions and other non-ideal effects. But a plasma can be remarkably collisionless. Under the circumstances, wave-particle and wave-wave interactions can drain out the energy from a given wave. These processes, together, are clubbed as collisionless damping mechanisms, of which the **Landau Damping** in unmagnetized and **Cyclotron Absorption** in a magnetized plasma are the most effective.

1.16. Instabilities In Plasmas

In a plasma, the waves can go unstable, i.e., their amplitudes grow with time, usually in an exponential way. There are three broad classes of plasma instabilities. The first class consists of instabilities which are studied using single or two-fluid descriptions. The driving force and energy for the excitation of this class of instabilities is contained in the nonequilibrium or nonthermal arrangement of the plasma fluid and the magnetic field in configuration space. For example, the Rayleigh-Taylor instability, which comes into play when a heavy fluid lies over a light fluid. The velocity shear between fluids of different mass densities could also excite instabilities as in the solar wind – comet tail interaction. A current carrying conducting fluid may undergo bending or twisting due to the excitation of these instabilities.

The second major class of instabilities involves a collisionless transfer of energy and affects the plasma at a microscopic level. These instabilities are studied using the kinetic description. The driving force and energy are contained in the nonthermal or non-Maxwellian velocity distribution functions of the electrons and the ions. For example, energetic electron and ion streams traveling through an ambient plasma medium could excite electrostatic waves, which then get converted into electromagnetic waves through several possible nonlinear processes. The other sources of energy are the density gradient, temperature anisotropy, and the current flow. Given the source of energy, the excitation of an instability requires an intermediary – one or the other of the many possible waves that a plasma supports, in order to tap the free energy. For example, it may be necessary to have an otherwise stable wave of phase velocity lying in the nonthermal part of the velocity distribution function. There is generally a threshold condition which must be satisfied before the waves go unstable. Then, once they start growing, their saturation levels and mechanisms need to be determined. These instabilities affect the transport phenomena in a very substantial manner.

The third major class of instabilities constitutes the **Parametric Instabilities**. Here, the driving energy is contained in a finite amplitude wave, electrostatic or electromagnetic, that impinges on a plasma. It then couples with other waves in the plasma and drives them unstable. These unstable waves may eventually undergo dissipation and heat the bulk plasma or accelerate some plasma particles. This class of instabilities also plays an important role in the generation of high frequency radiation from low frequency radiation, e.g., through a process called Stimulated Raman Scattering, where low frequency radiation, by scattering on the electron plasma wave of a high energy electron beam is converted into a high frequency radiation. This process is akin to inverse Compton scattering but with the important difference that a single electron in the Compton scattering is substituted by an electron plasma wave in the Raman scattering. Both these processes have been applied to explain the properties of Quasar nonthermal radiation. Parametric instabilities have been widely studied in the earth's ionospheric plasmas, the solar corona and extragalactic plasmas (Liu & Kaw 1976; Krishan 1997).

1.17. Plasmas in Curved Space – Time

Astrophysical plasmas are accelerated to relativistic speeds in the vicinity of compact objects such as pulsars and black holes. This necessitates the inclusion of effects due to **Special Theory of Relativity**. In addition, the strong gravitational fields of the compact objects may produce curvature in the space immediately around them. Under such circumstances, it becomes essential to study fluids and plasmas in the curved space – time using the **General Theory of Relativity** (GTR), formulated by Albert Einstein in 1916. The special theory of relativity operates in inertial frames of reference which are related to one another by the Lorentz Transformations of space-time coordinates, velocities and electromagnetic fields. The general theory of relativity is the generalization to include noninertial frames of reference which are not related to one another by any fixed transformation laws. **The Principle of Equivalence** tells us that the properties of the motion in a noninertial frame are the same as those in an inertial frame in the presence of a gravitational field. This implies that a system in an accelerated frame of reference is equivalent to its being in a gravitational field. We know that additional forces such as – the centrifugal and the coriolis forces arise when we go into a rotating frame of reference, for example, in the frame in which the earth is rotating. We call these forces as fictitious forces, since they disappear as soon as we go back to a nonrotating frame or inertial frame of reference. So, although, the motion in an accelerated frame of reference can be simulated by an equivalent gravitational field, this gravitational field has very different properties from a real gravitational field. The equivalent gravitational field may increase indefinitely at large distances. Whereas, the real gravitational field, as we know, must vanish at infinity. Thus, we can eliminate the gravitational field only locally, in a limited region of space. There is no transformation to a noninertial frame by which the field can be eliminated over all space.

Recall that the interval ds between two space – time points, in an inertial frame of reference in the Cartesian coordinate system is given by

$$ds^2 = c^2 dt^2 - dx^2 - dy^2 - dz^2, \qquad (1.97)$$

and it retains the same form in any inertial frame. In a noninertial frame, the form of ds changes, it is no longer a sum of the squares of the four coordinate differentials. There are cross terms involving products of the different coordinate differentials.

Problem 1.7: Show that in a coordinate system, uniformly rotating with angular velocity Ω such that $x = x' \cos \Omega t - y' \sin \Omega t$, $y = x' \sin \Omega t + y' \cos \Omega t$

and $z' = z$, the interval ds is given by

$$ds^2 = \left[c^2 - \Omega^2 \left(x'^2 + y'^2\right)\right] dt^2 - dx'^2 - dy'^2 - dz'^2. \qquad (1.98)$$

By no transformation of the coordinate t, can Equation (1.98) take the form of Equation (1.97). Thus, in a noninertial system, we write a general expression for ds as:

$$ds^2 = -g_{ik}\, dx^i dx^k, \quad i = 0, 1, 2, 3, \quad k = 0, 1, 2, 3.$$

where g_{ik} are functions of space coordinates x^1, x^2, x^3 and the time coordinate x^0, which form a curvilinear coordinate system. The matrix g_{ik} contains all the properties of the curvilinear coordinate system and is known as **the Space-Time Metric**. We can easily see that for a Cartesian coordinate system $g_{00} = -1$ and $g_{11} = g_{22} = g_{33} = 1$ and all off-diagonal terms vanish. These are known as the **Galilean Values**. If a given g_{ik} defining a certain space-time cannot be made to have Galilean values by any transformation, then this g_{ik} describes what is known as a **Curved Space-Time**. For a flat space-time, g_{ik} can always be reduced to its Galilean values. Thus, real gravitational fields produce curved space-time in which electromagnetic fields as well as fluid flows must be defined and described. This will be done in Chapter 4.

1.18. Nonconducting Fluids

Strictly speaking, there is no region in the universe which is completely devoid of ionized matter. The omnipresent high energy cosmic rays produce ionization even in the coldest molecular clouds. Nevertheless, the techniques and the physics of nonconducting fluids come very handy while addressing certain aspects of structure formation and stability which are predominantly a consequence of the gravitational forces. Further, the developments in hydrodynamic turbulence, a rather matured field, are of a great instructive value for studying magnetohydrodynamic (MHD) turbulence, a field in the making. At present, however, MHD turbulence has attained seniority at least as far as the issues of creating ordered structures in otherwise disordered media are concerned for example through dynamo like mechanisms. And it is the three dimensional fluid turbulence in which such a possibility of creating coherent structures is still in an exploratory stage. The two communities pursuing fluid turbulence and MHD turbulence have a lot to learn from each other and an astrophysicist has a lot more to learn from both of them, as we shall see in Chapter 7.

References

Bliokh, P., Sinitsin, V. and Yaroshenko, V., 1995, Dusty and Self-Gravitational Plasmas in Space, Kluwer Academic Publishers.

Frank-Kamenetskii, D.A., 1972, Plasma The Fourth State of Matter, Plenum Press, New York.

Fukai, J. Krishan, S. and Harris, E.G., 1969, Phys. Rev. Lett., **23**, 910.

Fukai, J. Krishan, S. and Harris, E.G., 1970, Phys. Fluids., **13**, 3031.

Ichimaru, S. (Editor): 1990, Strongly Coupled Plasma Physics, North-Holland, Elsevier Science Publishers B.V.

Krishan, V., el al. 1980, Solar Physics, **66**, 347.

Krishan, V., 1997, Space Science Reviews, **80**, 445.

Liu, C.S. and Kaw, P.K., 1976, Advances in Plasma Physics, **6**, 83.

CHAPTER 2

STATISTICAL DESCRIPTION OF A MANY-BODY SYSTEM

2.1. Tracking Them Down!

Molecules in a gas or a liquid, men in a metropolis and stars in a galaxy are in a state of incessant motion resulting from the action of various forces : internal, due to other molecules or (wo)men or stars, and external, due to the rest of the universe. It would be a frustrating task if we had to know the position and velocity of every molecule or star at every instant of time, in order to deal with the gas or the galaxy. The saving grace is that we need not know every move of every molecule or star and will still be able to manipulate the system to our advantage. For most purposes, it suffices to know the average properties like density, momentum, pressure, and energy, transport properties like thermal and electrical conductivities, and mechanical and electromagnetic stresses of a large system. In this chapter, we set up a mathematical framework to describe the behaviour of a macroscopic system. After establishing a transport equation for a discrete system in the phase space, continuum limits are taken to facilitate the study of fluids. The mass, momentum and energy conservation laws are derived for each species of particles in a fluid. This multifluid description is further simplified to a single fluid description with a two component fluid of electrons and ions taken as an example. This chapter contains the entire set of mathematical tools needed to investigate 'a single particle', 'a multifluid' and 'a single fluid' characteristics of an electrically conducting as well as an electrically nonconducting system. In the subsequent chapters, these descriptions will be further explored under various simplifying and tractable circumstances.

2.2. The Phase Space

A many-body system consists of a large number of particles, say N. The time evolution of such a system is contained in Hamilton's equations of motion for the specified initial conditions. The Hamiltonian H is a function of the canonically conjugate variables : the coordinates $\vec{q}_1, \vec{q}_2, ..\vec{q}_N$ and the corresponding momenta $\vec{p}_1, \vec{p}_2, ..\vec{p}_N$ of the N particles.

Hamilton's equations of motion are :

$$\frac{d\vec{q_i}}{dt} = \frac{\partial H}{\partial \vec{p_i}}, \qquad \text{and} \qquad \frac{d\vec{p_i}}{dt} = -\frac{\partial H}{\partial \vec{q_i}}, \qquad (2.1)$$

where $i = 1, 2, ..N$,
and $H(\vec{q_1}, \vec{q_2}, \vec{q_3}..\vec{q_N}; \vec{p_1}, \vec{p_2}, \vec{p_3}, ..\vec{p_N})$ is the Hamiltonian. Thus, once H and the initial conditions are known, the system is completely determined for all times. At any given time, the system is completely defined if N coordinates $\vec{q_1}..\vec{q_N}$ and N momenta $\vec{p_1}..\vec{p_N}$ are known, or in other words, the mechanical state of such a system can be represented by a single point in a 2N dimensional space. The movement of the point in the 2N dimensional space is governed by the 2N vector equations contained in Equation (2.1). The 2N dimensional space made up of N coordinates $\vec{q_1}..\vec{q_N}$ and N momenta $\vec{p_1}..\vec{p_N}$ is called the **Phase Space**. The time derivative of any function $f(\vec{q_1}..\vec{q_N}, \vec{p_1}..\vec{p_N}, t)$ of these 2N variables can be written as :

$$\frac{df}{dt} = \sum_{i=1}^{N} \frac{\partial f}{\partial \vec{q_i}} \cdot \frac{d\vec{q_i}}{dt} + \frac{\partial f}{\partial \vec{p_i}} \cdot \frac{d\vec{p_i}}{dt} + \frac{\partial f}{\partial t}, \qquad (2.2)$$

which, on using Equation (2.1), becomes :

$$\frac{df}{dt} = \sum_{i=1}^{N} \frac{\partial f}{\partial \vec{q_i}} \cdot \frac{\partial H}{\partial \vec{p_i}} - \frac{\partial f}{\partial \vec{p_i}} \cdot \frac{\partial H}{\partial \vec{q_i}} + \frac{\partial f}{\partial t} \qquad (2.3)$$

$$= [f, H] + \frac{\partial f}{\partial t}.$$

Here $[f, H]$ is called the Poisson bracket of f with H. If f is a constant of motion of the system then $[f, H] = 0$ (Goldstein 1950).

Problem 2.1: Prove that the total energy is a constant of motion of a system.

2.3. The Gibb's Ensemble and the Liouville Equation

Although positions and momenta of all the particles of a large system can be completely determined by making use of the set of Equations (2.1), often this information is not required for describing the characteristic properties of this system. Besides, it is impossible to know accurately the initial conditions of every particle in a large system. We usually describe a system by its energy or temperature for which we need not know the trajectory of every atom or molecule in it. Thus, an exact solution of a large system is neither tractable nor desirable. Instead, we can use the methods of **Statistical Mechanics** to predict certain average properties of a system. This can be done by studying a large number of identical systems. They are as identical as the uncertainty in their initial conditions permits them to be. They may differ from each other in their location in the phase space, but have the same averaged properties. Thus, many different microscopic configurations may correspond to the same macroscopic state of the system. A collection of such systems is known as **an Ensemble, the Gibb's Ensemble**. Since, each system is represented by a single point in the phase space, an ensemble is represented by a group of points in the phase space. Each member of an ensemble follows Hamilton's equations (2.1). For example, the total energy of a simple harmonic oscillator is given by $(p^2 + q^2)$ and remains invariant throughout the motion, even though p and q keep changing. Thus, every such pair (p, q) which preserves the invariance of energy is a member and these members collectively form an ensemble. We can then associate a density ρ with these members or points in the phase space. The phase space density ρ is a function of N coordinates $\vec{q}_1 \ldots \vec{q}_N$ and N momenta $\vec{p}_1 \ldots \vec{p}_N$. The number of members in a volume $d\vec{q}_1 \ldots d\vec{q}_N \, d\vec{p}_1 \ldots d\vec{p}_N$ of the phase space at an instant t are:

$$\rho(\vec{q}_1 \ldots \vec{q}_N, \vec{p}_1 \ldots \vec{p}_N, t) d\vec{q}_1 \ldots d\vec{q}_N \, d\vec{p}_1 \ldots d\vec{p}_N.$$

This is the statistical weight of the system. The total number of members is found by integrating the density ρ over the entire volume of the phase space. Thus the probability of finding a system in the volume $d\vec{q}_1 \ldots d\vec{q}_N \, d\vec{p}_1 \ldots d\vec{p}_N$ of the phase space is given by:

$$\frac{\rho(\vec{q}_1..\vec{q}_N, \vec{p}_1..\vec{p}_N, t) d\vec{q}_1..d\vec{q}_N \, d\vec{p}_1..d\vec{p}_N}{\int \rho(\vec{q}_1..\vec{q}_N, \vec{p}_1..\vec{p}_N, t) d\vec{q}_1..d\vec{q}_N \, d\vec{p}_1..d\vec{p}_N}. \tag{2.4}$$

We can always normalize ρ such that

$$\int \rho(\vec{q}_1..\vec{q}_N, \vec{p}_1..\vec{p}_N, t) d\vec{q}_1..d\vec{q}_N \, d\vec{p}_1..d\vec{p}_N = 1. \tag{2.5}$$

The phase space density ρ varies with time due to changes in the coordinates $\vec{q}$ and momenta $\vec{p}$ with time as the system evolves. In addition, ρ may have an explicit dependence on time even at a fixed point in the phase space. Thus, the total time derivative of ρ can be written as

$$\frac{d\rho}{dt} = \frac{\partial\rho}{\partial t} + \sum_{i=1}^{N} \frac{\partial\rho}{\partial \vec{q}_i} \cdot \frac{d\vec{q}_i}{dt} + \frac{\partial\rho}{\partial \vec{p}_i} \cdot \frac{d\vec{p}_i}{dt},$$

or

$$\frac{d\rho}{dt} = \frac{\partial\rho}{\partial t} + [\rho, H]. \tag{2.6}$$

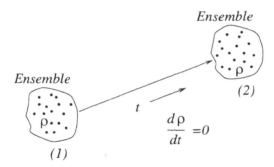

Figure 2.1.   Time Evolution of a Many-Body System in the Phase Space.

We may recall that the motion of a system from point (1) in the phase space to another point (2) represents the time evolution of the canonical transformation which connects these two points (Figure 2.1). A volume element in the phase space remains invariant under a canonical transformation. All the members in a given volume around point (1) will end up in the same volume around point (2) following Newton's laws. Thus both the number of members as well as the volume element they lie in, remain constant as the system evolves. Hence, the density ρ also remains an invariant so that,

$$\frac{d\rho}{dt} = 0,$$

or

$$\frac{\partial \rho}{\partial t} = -[\rho, H]. \tag{2.7}$$

This is the **Liouville Equation**.

It has been shown that finding the solution of the Liouville equation is the same as the integration of the canonical equations of motion. Therefore, for large systems, it is much more advantageous to deal with a single Liouville equation, which describes the entire system. In statistical equilibrium the density ρ does not depend on time explicitly i.e., $(\partial \rho / \partial t) = 0$. The equilibrium can, therefore, be described as $[\rho, H] = 0$. Thus a system is in equilibrium if its phase space density ρ is a function only of the constants of the motion which do not explicitly depend upon time. For a conservative system in equilibrium, ρ is only a function of energy. Specific choices of the functions will determine the properties of a system. An example is a microcanonical ensemble for which ρ is a constant for a system having a given energy and zero, otherwise. The linearity of the Liouville equation enables the use of the **Superposition Principle**; i.e., if ρ_1 and ρ_2 are solutions then their linear combination will also be a solution of the Liouville equation.

2.4. Distribution Functions

With the normalization condition (Equation 2.5) the function $\rho(\vec{q}_1..\vec{q}_N, \vec{p}_1.. \vec{p}_N, t)\, d\vec{q}_1..d\vec{q}_N..d\vec{p}_1..d\vec{p}_N$ becomes the probability of finding the system in a volume element $d\vec{q}_1..d\vec{q}_N\, d\vec{p}_1..d\vec{p}_N$ of the phase space at time t. We defined ρ as the density of points representing a system in the phase space. How are these points distributed in the phase space? This information is contained in the specific way in which ρ depends upon the invariants of the system at a given time t. Thus, these points may be distributed in a Gaussian manner or a power law form or a microcanonical form as mentioned in the previous section. When ρ is given a specific form through its dependence on the constants of motion, it is known as the **Distribution Function**. Thus $\rho_N(\vec{q}_1..\vec{q}_N, \vec{p}_1..\vec{p}_N, t)d\vec{q}_1..d\vec{q}_N\, d\vec{p}_1..d\vec{p}_N$ is the probability of finding the system of N particles in the volume element $d\vec{q}_1..d\vec{q}_N\, d\vec{p}_1..d\vec{p}_N$ in the phase space, at a time t and ρ_N is known as the **N-Particle Distribution Function**.

We can also define the reduced distribution functions for a part of a system e.g. for s particles in an N particle system. Thus $\rho_s(\vec{q}_1..\vec{q}_s, \vec{p}_1..\vec{p}_s, t)$, the probability of finding s specific particles 1,2 .. s at a given time t with coordinates $\vec{q}_1..\vec{q}_s$ and momenta $\vec{p}_1..\vec{p}_s$ can be found by integrating the N-particle distribution function ρ_N over all the coordinates of (N-s) particles i.e. (Prigogine 1962):

$$\rho_s(\vec{q}_1..\vec{q}_s, \vec{p}_1..\vec{p}_s, t) = \int \rho_N(\vec{q}_1..\vec{q}_N, \vec{p}_1..\vec{p}_N, t)d\vec{q}_{s+1}..d\vec{q}_N\, d\vec{p}_{s+1}..d\vec{p}_N. \quad (2.8)$$

Problem 2.2: Show that

$$\rho_s(\vec{q}_1..\vec{q}_s, \vec{p}_1..\vec{p}_s, t) = \int \rho_{s+1}(\vec{q}_1..\vec{q}_{s+1}, \vec{p}_1..p_{s+1}, t)d\vec{q}_{s+1}d\vec{p}_{s+1}.$$

Functions that describe the distribution of specified particles are called **Specific Distribution Functions**. We generally work with a **Reduced Distribution Function** f for an arbitrary set of particles. Thus, the probability f_s of finding s arbitrary particles at positions $\vec{q}_1..\vec{q}_s$ and momenta $\vec{p}_1..\vec{p}_s$ is given by:

$$f_s(\vec{q}_1..\vec{q}_s, \vec{p}_1..\vec{p}_s, t) = \frac{N!}{(N-s)!}\rho_s(\vec{q}_1..\vec{q}_s, \vec{p}_1..\vec{p}_s, t), \quad (2.9)$$

since a sequence of s particles out of N identical particles can be chosen in $N!/(N-s)!$ ways.

Problem 2.3 : Show that

$$f_s = \frac{1}{(N-s)} \int f_{s+1}d\vec{q}_{s+1}d\vec{p}_{s+1},$$

$$= \frac{1}{(N-s)!} \int f_N \, d\vec{q}_{s+1}..d\vec{q}_N \, d\vec{p}_{s+1}..d\vec{p}_N,$$

$$= \frac{N!}{(N-s)!} \int \rho_N \, d\vec{q}_{s+1}..d\vec{q}_N \, d\vec{p}_{s+1}..d\vec{p}_N.$$

The functions f_s are also called **Generic Distribution Functions** and are used in the statistical description of a many body system.

Problem 2.4 : Show that the normalization condition for an s particle distribution function is given by

$$\int f_s(\vec{q}_1..\vec{q}_s, \vec{p}_1..\vec{p}_s, t) d\vec{q}_1..d\vec{q}_s d\vec{p}_1..d\vec{p}_s = \frac{N!}{(N-s)!}.$$

From the distribution function f_N in the phase space, one can also define distribution functions φ_s in the momentum space and distribution functions n_s in the coordinate space as well as the mixed distribution functions $f_{s,m}$ of s positions and m momenta as :

$$\varphi_s(\vec{p}_1..\vec{p}_s, t) = \frac{(N-s)!}{N!} \int f_s(\vec{q}_1..\vec{q}_s, \vec{p}_1..\vec{p}_s, t) d\vec{q}_1..d\vec{q}_s, \quad (2.10)$$

$$n_s(\vec{q}_1..\vec{q}_s, t) = \int f_s(\vec{q}_1..\vec{q}_s, \vec{p}_1..\vec{p}_s, t) d\vec{p}_1..d\vec{p}_s, \quad (2.11)$$

and

$$f_{s,m}(\vec{q}_1..\vec{q}_s, \vec{p}_1..\vec{p}_m, t) = \frac{1}{(N-s)!} \int f_N(\vec{q}_1..\vec{q}_N, \vec{p}_1..\vec{p}_N, t) d\vec{q}_{s+1}..d\vec{q}_N$$

$$d\vec{p}_{m+1}..d\vec{p}_N, \quad (2.12)$$

with normalizations given as :

$$\int \varphi_s(\vec{p}_1..\vec{p}_s, t) d\vec{p}_1..d\vec{p}_s = 1, \quad (2.13)$$

$$\int n_s(\vec{q}_1..\vec{q}_s, t) d\vec{q}_1..d\vec{q}_s = \frac{N!}{(N-s)!}, \quad (2.14)$$

and

$$\int f_{s,m} d\vec{q}_1..d\vec{q}_s d\vec{p}_1..d\vec{p}_m = \frac{N!}{(N-s))!}. \quad (2.15)$$

Using the reduced distribution functions the averages of all the physical quantities like pressure, energy etc. can be determined, provided the reduced distribution functions are well behaved. That is, under the limits $N \to \infty$

and the volume $L^3 \to \infty$ but the concentration $N/L^3 \to$ a finite constant; i.e., the reduced distribution function must approach a well defined limit. This assumption, made initially, is preserved throughout the time evolution of the distribution function.

2.5. One Particle Distribution Function

It is impractical to work with ρ_N. Instead, one or two or three particle distribution functions are used. The time evolution of reduced distribution functions can be studied beginning with the Liouville equation for the N particle distribution function f_N:

$$\frac{d}{dt} f_N(\vec{q}_1..\vec{q}_N, \vec{V}_1..\vec{V}_N, t) = 0, \tag{2.16}$$

where we have used velocities $\vec{V}$ instead of momenta $\vec{p}$. Integration of f_N over all coordinates $(\vec{q}_2..\vec{q}_N, \vec{V}_2..\vec{V}_N)$ gives the reduced **One Particle Distribution Function** $f_1(\vec{q}_1, \vec{V}_1, t)$ such that $f_1(\vec{q}_1, \vec{V}_1, t)d\vec{q}_1, d\vec{V}_1$ is the probability of finding one particle in the volume element $d\vec{q}_1 d\vec{V}_1$ of the phase space; integration of f_N over all coordinates $(\vec{q}_3..\vec{q}_N, \vec{V}_3..\vec{V}_N)$ gives the reduced **Two Particle Distribution Function** $f_2(\vec{q}_1, \vec{q}_2, \vec{V}_1, \vec{V}_2, t)$ such that $f_2(\vec{q}_1, \vec{q}_2, \vec{V}_1, \vec{V}_2, t) d\vec{q}_1 d\vec{q}_2 d\vec{V}_1 d\vec{V}_2$ is the joint probability of finding one particle in the volume element $d\vec{q}_1 d\vec{V}_1$ and another particle in the volume element $d\vec{q}_2 d\vec{V}_2$ of the phase space, and so on. The corresponding Liouville equation for each of the distribution functions $f_1, f_2..$ can be derived by writing Equation (2.16) in its expanded form and performing the appropriate integrations over the coordinates (Nicholson 1983).

The expanded form of Equation (2.16) can be written as:

$$\frac{\partial f_N}{\partial t} + \sum_{i=1}^{N} \frac{\partial f_N}{\partial \vec{q}_i} \cdot \frac{d\vec{q}_i}{dt} + \frac{\partial f_N}{\partial \vec{V}_i} \cdot \frac{d\vec{V}_i}{dt} = 0. \tag{2.17}$$

Integrating Equation (2.17) over all the coordinates except $(\vec{q}_1, \vec{V}_1)$ gives:

$$\frac{\partial f_1}{\partial t} + \vec{V}_1 \cdot \frac{\partial f_1}{\partial \vec{q}_1} + \int \frac{d\vec{V}_1}{dt} \cdot \frac{\partial f_N}{\partial \vec{V}_1} d\vec{q}_2..d\vec{q}_N \, d\vec{V}_2..d\vec{V}_N = 0, \tag{2.18}$$

since,

$$\sum_{i=2}^{N} \int \vec{V}_i \cdot \frac{\partial f_N}{\partial \vec{q}_i} d\vec{q}_2..d\vec{q}_N \, d\vec{V}_2..d\vec{V}_N = 0.$$

Now the acceleration $\dfrac{d\vec{V}_1}{dt}$ of a single particle results from the externally applied force $\vec{F}_{ext}$ as well as from the internal force $\vec{F}_{int}$ due to the mutual interactions among all the particles of the system, so that

$$\frac{d\vec{V_1}}{dt} = \frac{\vec{F}_{ext}}{m} + \frac{\vec{F}_{int}}{m},$$

(2.19)

where m is the mass of that particle. The internal force $\vec{F}_{int}$ on particle one due to all the other particles can be written as:

$$\begin{aligned}
\vec{F}_{int} &= \sum_{l=2}^{N} \vec{F}(\vec{q_1}, \vec{V_1}; \vec{q_l}, \vec{V_l}), \\
&\equiv \sum_{l=2}^{N} \vec{F}_{1l}.
\end{aligned}$$

(2.20)

Substituting for acceleration from Equations (2.19) and (2.20), Equations (2.18) becomes:

$$\frac{\partial f_1}{\partial t} + \vec{V_1} \cdot \frac{\partial f_1}{\partial \vec{q_1}} + \frac{\vec{F}_{ext}}{m} \frac{\partial f_1(\vec{q_1}, \vec{V_1}, t)}{\partial \vec{V_1}} = -\int \sum_{l=2}^{N} \frac{\vec{F}_{1l}}{m} \cdot \frac{\partial f_2(\vec{q_1}, \vec{V_1}; \vec{q_l}, \vec{V_l}, t)}{\partial \vec{V_1}} d\vec{q_l} d\vec{V_l},$$

$$\equiv \left. \frac{\delta f_1}{\delta t} \right|_c.$$

(2.21)

It is clear from Equation (2.21) that the time evolution of a single particle distribution function f_1 depends on the two-particle distribution function f_2 and that of f_2 depends on the three-particle distribution function f_3 and so on and so forth. This hierarchy of equations, though first obtained by Yvon (1935, 1937), is popularly known as **BBGKY Hierarchy** after Bogolioubov (1946), Born and Green (1946-1947) and Kirkwood (1946-1947) who studied it extensively.

Equation (2.21) describes the time evolution of the single particle distribution function due to diffusion effects (the term $\frac{\partial f_1}{\partial \vec{q_1}}$), external forces which change the velocities of particles and therefore the distribution function (the term $\frac{\partial f_1}{\partial \vec{V_1}}$) and particle interactions (the term $\left. \frac{\delta f_1}{\delta t} \right|_c$). The various representations chosen for the interaction term lead to different evolution equations.

The simplest thing to do is to neglect all interactions. This gives us the **Collisionless Boltzmann Equation**:

$$\frac{\partial f_1}{\partial t} + \vec{V_1} \cdot \frac{\partial f_1}{\partial \vec{q_1}} + \frac{\vec{F}_{ext}}{m} \cdot \frac{\partial f_1}{\partial \vec{V_1}} = 0.$$

(2.22)

This equation, for example, describes the distribution of stars in a galaxy since collisions between stars are rare; the mutual gravitational fields of stars are represented as an external potential. Equation (2.22) also describes a system of charged particles, like a low density plasma in the presence of external electromagnetic fields. The low density ensures that the internal force is not significant and the system does not modify the external force. The advantage of studying the system with Equation (2.22) over studying the motion of each particle is that we can investigate statistically the trajectories of a large number of particles under a large number of different initial conditions residing in the initial value of the distribution function.

At moderately high particle densities, the interactions among particles must be taken into account. Recall that $f_2(\vec{q}_1, \vec{V}_1, \vec{q}_l, \vec{V}_l, t)$ is the joint probability density of finding one particle at $(\vec{q}_1, \vec{V}_1)$ and another at $(\vec{q}_l, \vec{V}_l)$. If the particles move independently of each other, the joint probability is equal to the product of the individual probabilities, i.e.

$$f_2(\vec{q}_1, \vec{V}_1, \vec{q}_l, \vec{V}_l, t) = f_1(\vec{q}_1, \vec{V}_1, t) f_1(\vec{q}_l, \vec{V}_l, t). \tag{2.23}$$

Substituting Equation (2.23) in the right hand side of Equation (2.21), we get :

$$\frac{\partial f_1(\vec{q}_1, \vec{V}_1, t)}{\partial t} + \vec{V}_1 \cdot \frac{\partial f_1(\vec{q}_1, \vec{V}_1, t)}{\partial \vec{q}_1} + \left(\frac{\vec{F}_{ext} + \vec{F}_s}{m} \right) \cdot \frac{\partial f_1(\vec{q}_1, \vec{V}_1, t)}{\partial \vec{V}_1} = 0, \tag{2.24}$$

where

$$F_s = \int \sum_{l=2}^{N} \vec{F}_{1l} f_1(\vec{q}_l, \vec{V}_l, t) d\vec{q}_l d\vec{V}_l, \tag{2.25}$$

represents the mean force on particle one due to all the other particles and is called the **Self - Consistent Force**. Equation (2.24) is known as **Vlasov Equation** and is used to study waves and instabilities in moderately coupled plasmas in which velocities and positions of the particles still remain uncorrelated. The Vlasov Equation represents the collective behaviour of a plasma in which the charged particles experience electromagnetic forces due to their motions in addition to the external force.

In a high particle density system, mutual interactions among the particles may give rise to a correlated behaviour. Under such circumstances, the probability of finding particles in the phase space are no longer independent. The probability of finding particle one in a certain volume of the phase space depends upon the location of other particles in the phase space. The joint probability f_2 can then be written as:

$$f_2(\vec{q}_1, \vec{V}_1, \vec{q}_l, \vec{V}_l, t) = f_1(\vec{q}_1, \vec{V}_1, t) f_1(\vec{q}_l, \vec{V}_l, t) + f_c, \tag{2.26}$$

where f_c now represents the pair correlations. The term $\left.\dfrac{\delta f_1}{\delta t}\right|_c$ of Equation (2.21) includes both correlated and uncorrelated particle collisions. We now discuss a few models of particle collisions.

2.6. The Krook Collision Model

Collisions thermalize a system and drive it towards some equilibrium. A simple representation of the collision term can be seen in the Krook's model as:

$$\left.\frac{\delta f_1}{\delta t}\right|_c = -\frac{1}{\tau}\left(f_1 - f_{10}\right), \tag{2.27}$$

where f_{10} is the equilibrium distribution function and τ is the characteristic relaxation time. The import of this model is easily appreciated in the absence of spatial gradients and external forces. The Boltzmann Equation (2.21) becomes:

$$\frac{\partial f_1(\vec{V}, t)}{\partial t} = -\frac{1}{\tau}\left(f_1(\vec{V}, t) - f_{10}\right),$$

or

$$f_1(\vec{V}, t) = \left(f_1(\vec{V}, 0) - f_{10}\right)e^{-t/\tau} + f_{10}, \tag{2.28}$$

where $f_1(\vec{V}, 0)$ is the distribution function at $t = 0$ and f_{10} is the distribution function at $t = \infty$. The difference between the initial and the equilibrium values of the distribution function decays exponentially.

2.7. The Boltzmann Collision Model

In this model, the collision terms $\left.\dfrac{\delta f_1}{\delta t}\right|_c$ are determined by restricting the interactions among particles to only binary collisions. The particles are modeled as hard spheres undergoing two-body elastic collisions. The Boltzmann model is applicable when (i) the particle density is low so that triple and higher order interactions are reasonably neglected; (ii) the particles experience only short range forces which are negligible for large interparticle distances; (iii) within the range of forces, the short range force dominates over any external force; (iv) the spatial variation of the joint probability f_2 is significant only over distances shorter than or equal to the range of the force; (v) the temporal variation of the joint probability f_2 is negligible during the time of binary collision and (vi) there are no correlations $(f_c = 0)$.

The evolution equation of the joint probability f_2 can be found by integrating Equation (2.17) over all the coordinates except $(\vec{q}, \vec{V}_1, \vec{q}_2, \vec{V}_2, t)$. We get

$$\frac{\partial f_2(\vec{q}_1, \vec{V}_1, \vec{q}_2, \vec{V}_2, t)}{\partial t} + \sum_{i=1}^{2} \vec{V}_i \cdot \frac{\partial f_2(\vec{q}_1, \vec{V}_1, \vec{q}_2, \vec{V}_2, t)}{\partial \vec{q}_i}$$

$$+ \sum_{i=1}^{2} \frac{\vec{F}_{ext}}{m} \cdot \frac{\partial f_2(\vec{q}_1, \vec{V}_1, \vec{q}_2, \vec{V}_2, t)}{\partial \vec{V}_i} + \frac{\vec{F}_{12}}{m} \cdot \frac{\partial f_2(\vec{q}_1, \vec{V}_1, \vec{q}_2, \vec{V}_2, t)}{\partial \vec{V}_1}$$

$$+ \frac{\vec{F}_{21}}{m} \cdot \frac{\partial f_2(\vec{q}_1, \vec{V}_1, \vec{q}_2, \vec{V}_2, t)}{\partial \vec{V}_2} = 0, \tag{2.29}$$

where we have put $f_3(q_1, \vec{V}_1, q_2, \vec{V}_2, q_3, \vec{V}_3, t) = 0$ and $\vec{F}_{12}$ and $\vec{F}_{21}$ are the two-body internal forces. We introduce a relative coordinate $\vec{q} = \vec{q}_2 - \vec{q}_1$; integrate over $d\vec{q}d\vec{V}_2$ and use assumptions (i) to (v) to reduce Equation (2.29) to:

$$\int \left[(\vec{V}_2 - \vec{V}_1) \cdot \frac{\partial f_2}{\partial \vec{q}} + \frac{\vec{F}_{12}}{m} \cdot \frac{\partial f_2}{\partial \vec{V}_1} \right] d\vec{q}d\vec{V}_2 = 0, \tag{2.30}$$

so that the collision term $\left. \frac{\delta f_1}{\delta t} \right|_c$ becomes:

$$\left. \frac{\delta f_1}{\delta t} \right|_c = \int (\vec{V}_2 - \vec{V}_1) \cdot \frac{\partial f_2}{\partial \vec{q}} d\vec{q}d\vec{V}_2. \tag{2.31}$$

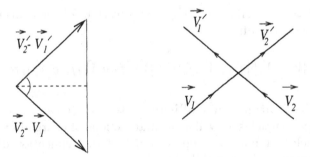

Figure 2.2.   The Magnitude of the Relative Velocity $(\vec{V}_2 - \vec{V}_1)$ Remains Invariant in an Elastic Binary Collision.

In a binary collision, let $\vec{V}_1$ and $\vec{V}_2$ be the particle velocities before the collision and $\vec{V}_1'$ and $\vec{V}_2'$ be the velocities after the collision. From the conservation of momentum and energy, it can be easily shown that the centre of mass and the magnitude of the relative velocities remain unchanged and the collision only rotates the relative velocity vector $(\vec{V}_2 - \vec{V}_1)$ into $(\vec{V}_2' - \vec{V}_1')$

(Figure 2.2). In Equation (2.31), the volume integral over the interaction region $\vec{q} = \vec{q}_1 - \vec{q}_2$ can be converted into a surface integral, so that

$$\left.\frac{\delta f_1}{\delta t}\right|_c = \int (\vec{V}_2 - \vec{V}_1) \cdot d\vec{S} f_2 d\vec{V}_2, \tag{2.32}$$

where $d\vec{S}$ is the surface element. We have to now determine the joint probability f_2, which in the absence of correlations is just the product of the individual probabilities $f_1(\vec{q}_1, \vec{V}_1) f_1(\vec{q}_2, \vec{V}_2)$. There are two types of collisions, one that increases the density of the particles at a given phase space point by bringing in particles from other phase space locations and the other that reduces the density of particles by taking particles away from this point to other phase space locations. Thus, the collisions $\vec{V}_1' + \vec{V}_2' \rightarrow \vec{V}_1 + \vec{V}_2$ enhance the particle density. The net gain in particle density at $\vec{V}_1$ can therefore be expressed in terms of the probabilities for the two processes as:

$$f_2 = f_1(\vec{q}_1, \vec{V}_1') f_2(\vec{q}_2, \vec{V}_2') - f_1(\vec{q}_1, \vec{V}_1) f_1(\vec{q}_2, \vec{V}_2). \tag{2.33}$$

As discussed earlier, the surface integral extends over the interaction sphere whose radius is the range of the two-body forces. Therefore, if $\sigma(\Omega)$ is the differential cross section for scattering into the solid angle $d\Omega$, where Ω denotes the angles between the initial and final relative velocities, the surface element normal to the relative velocity vector can be written as

$$dS = \sigma(\Omega) d\Omega. \tag{2.34}$$

Since the relative velocity $(\vec{V}_2 - \vec{V}_1)$ is an invariant of the collision process, Equation (2.32) simplifies to

$$\left.\frac{\delta f_1}{\delta t}\right|_c = \int |\vec{V}_2 - \vec{V}_1| \left[f_1(\vec{q}, \vec{V}_1') f_2(\vec{q}, \vec{V}_2') - f_1(\vec{q}, \vec{V}_1) f_1(\vec{q}, \vec{V}_2) \right] \sigma(\Omega) d\Omega d\vec{V}_2. \tag{2.35}$$

This is the **Boltzmann Collision Integral**. It can be used for low density neutral particles as well as for interactions of charged particles with neutral particles. It must be appreciated that the truncation of $BBGKY$ hierarchy introduces irreversibility in the system.

2.8. The Fokker - Planck Collision Model

The Fokker-Planck collision model is derived from the Boltzmann collision integral (Equation 2.35) under the assumption that the change in the velocity of a particle due to a collision is rather small i.e. $\vec{V}_1 \sim \vec{V}_1'$. Thus the Fokker-Planck model accounts for grazing collisions during which the colliding particles undergo small and continuous changes in their trajectories due to the long range nature of the force between them; an example is stars in their mutual gravitational force in a galaxy.

Let us multiply the Boltzmann collision term (Equation 2.35) by a function $P(\vec{V}_1)$ and integrate over $\vec{V}_1$ to get:

$$\int P(\vec{V}_1)\frac{\delta f_1}{\delta t}\bigg|_c d\vec{V}_1 = \int \left[f_1(\vec{q},\vec{V}_1')f_1(\vec{q},\vec{V}_2') - f_1(\vec{q},\vec{V}_1)f_1(\vec{q},\vec{V}_2)\right]$$
$$|\vec{V}_2 - \vec{V}_1|P(\vec{V}_1)\sigma(\Omega)d\Omega d\vec{V}_1 d\vec{V}_2. \tag{2.36}$$

Noting that:

$$\int f_1(\vec{q},\vec{V}_1')f_1(\vec{q},\vec{V}_2')|\vec{V}_2 - \vec{V}_1|P(\vec{V}_1)\sigma(\Omega)d\Omega d\vec{V}_1 d\vec{V}_2 = \int f_1(\vec{q},\vec{V}_1)f_1(\vec{q},\vec{V}_2)$$
$$|\vec{V}_2 - \vec{V}_1|P(\vec{V}_1')\sigma(\Omega)d\Omega d\vec{V}_1 d\vec{V}_2. \tag{2.37}$$

We can rewrite Equation (2.36) as:

$$\int P(\vec{V}_1)\frac{\delta f_1}{\delta t}\bigg|_c d\vec{V}_1 = \int \left[P(\vec{V}_1') - P(\vec{V}_1)\right] f_1(\vec{q},\vec{V}_1)f_1(\vec{q},\vec{V}_2)$$
$$|\vec{V}_2 - \vec{V}_1|\sigma(\Omega)d\Omega d\vec{V}_1 d\vec{V}_2. \tag{2.38}$$

The right hand side of Equation (2.37) is obtained by exchanging primed and unprimed velocities and using the identity $d\vec{V}_1'd\vec{V}_2' = d\vec{V}_1 d\vec{V}_2$.

Problem 2.5: Prove that $d\vec{V}_1'd\vec{V}_2' = d\vec{V}_1 d\vec{V}_2$.

Now assuming that the change in velocity of a particle $\vec{V}_1' - \vec{V}_1 = \Delta\vec{V} <<$ $\vec{V}_1$ or $\vec{V}_1'$, we can expand the function $P(\vec{V}_1')$ about $\vec{V}_1$, to get:

$$P(\vec{V}_1') = P(\vec{V}_1) + \sum_i \frac{\partial P(\vec{V}_1)}{\partial V_{1i}}\Delta V_{1i} + \sum_{i,j} \frac{1}{2}\frac{\partial^2 P(\vec{V}_1)}{\partial V_{1i}\partial V_{1j}}\Delta V_i \Delta V_j \cdots \tag{2.39}$$

We substitute Equation (2.39) into Equation (2.38) and perform partial integration over $\vec{V}_1$. The function P and integration over $\vec{V}_1$ being common to both sides of Equation (2.38) get canceled and we obtain:

$$\frac{\delta f_1}{\delta t}\bigg|_c = -\frac{\partial}{\partial V_{1i}}\left[f_1(\vec{V}_1)\langle\Delta V_i\rangle\right] + \frac{1}{2}\frac{\partial^2}{\partial V_{1i}\partial V_{1j}}\left[f_1(\vec{V}_1)\langle\Delta V_i\Delta V_j\rangle\right] + \cdots, \tag{2.40}$$

where

$$\langle\Delta V_i\rangle = \int f_1(\vec{V}_2)\Delta V_i|\vec{V}_2 - \vec{V}_1|\sigma(\Omega)d\Omega d\vec{V}_2, \tag{2.41}$$

and

$$\langle\Delta V_i\Delta V_j\rangle = \int f_1(\vec{V}_2)\Delta V_i\Delta V_j|\vec{V}_2 - \vec{V}_1|\sigma(\Omega)d\Omega d\vec{V}_2. \tag{2.42}$$

Equation (2.40) represents the **Fokker-Planck Collision Model**; $\langle \Delta V_i \rangle$ is known as the **Fokker-Planck Coefficient of Dynamical Friction** and $\langle \Delta V_i \Delta V_j \rangle$ is known as the **Fokker Planck Coefficient of Diffusion** in velocity space.

Problem 2.6. Show that $\langle \Delta V_i \rangle$ is the frictional force per unit mass due to collisions.

It has been found that the two terms of dynamical friction and diffusion in velocity space provide a fairly good representation of the binary collision process in system of particles colliding via long range forces like coulomb and gravitational forces.

2.9. The Kinetic Description

We have seen how a large system can be studied through the evolution of its distribution functions. The time evolution of the one particle distribution function f_1, given by Equation (2.24) represents what is known as the **Kinetic Description,** since it depends on the detailed behaviour of the distribution in the phase space. We will first discuss the moderately coupled systems for which the collisions and correlations included in f_c may be neglected (valid, e.g. for a high temperature plasma). The evolution equation of the distribution function called the **Collisionless Boltzmann Equation** is given by

$$\frac{\partial f(\vec{r}, \vec{V}, t)}{\partial t} + \vec{V} \cdot \frac{\partial f(\vec{r}, \vec{V}, t)}{\partial \vec{r}} + \frac{\vec{F}}{m} \cdot \frac{\partial f(\vec{r}, \vec{V}, t)}{\partial \vec{V}} = 0. \tag{2.43}$$

Here, f is the one particle distribution function and the coordinate $\vec{q}$ has been replaced by $\vec{r}$, $\vec{F}$ is the total force equal to the sum of the external $\vec{F}_{ext}$ and the self-consistent force $\vec{F}_s$ acting on a particle of mass m.

2.10. Stellar Systems

The collisionless Boltzmann equation can be applied to stellar systems since stars rarely collide with each other. They move under their mutual gravitational forces. The total force $\vec{F}$, in this case is given by

$$\vec{F} = -m\vec{\nabla}\varphi_g, \tag{2.44}$$

where φ_g is the gravitational potential. The number density of particles at a position $\vec{r}$ and time t with velocity lying between $\vec{V}$ and $\vec{V} + d\vec{V}$ is given by $f(\vec{r}, \vec{V}, t)d\vec{V}$. The total number density $n(\vec{r}, t)$ is then determined as :

$$n(\vec{r}, t) = \int f(\vec{r}, \vec{V}, t)d\vec{V} \tag{2.45}$$

and the total number of particles N is given by

$$N(t) = \int n(\vec{r}, t) d\vec{r}. \tag{2.46}$$

The Poisson equation for a stellar system can then be written as :

$$\nabla^2 \varphi_g = 4\pi G \rho_m(\vec{r}, t), \tag{2.47}$$

where the mass density ρ_m can be expressed in terms of the distribution function $f(\vec{r}, \vec{v}, t)$ as

$$\rho_m(\vec{r}, t) = m \int f(\vec{r}, \vec{V}, t) d\vec{V}, \tag{2.48}$$

with m as the mass of a star. The system of Equations (2.43), (2.47) and (2.48) are used to study the equilibrium and the stability of a stellar system.

2.11. Collisionless Plasmas

A hot plasma consisting of electrons and protons is another system that is nearly collisionless. The coulomb collisions between these charged particles become very rare at high temperatures. This is due to the fact that at high thermal speeds the particles experience the Coulomb force for an extremely short time interval and therefore the impulse or the momentum transfer is rather small. Thus Equation (2.43) can furnish a fairly accurate evolution of the distribution function of the plasma particles. The mutual forces among the plasma particles are, however, electromagnetic: the Lorentz force. Each particle moves in the Coulomb fields of the other particles and in turn produces its own field. Therefore, again, a self-consistent force is experienced by each particle. If $\vec{E}$ and $\vec{B}$ are the total (external + the self-consistent) electric and magnetic fields, the total force $\vec{F}$ including the force due to gravity is now given as :

$$\vec{F} = Q \left[\vec{E} + \frac{\vec{V} \times \vec{B}}{c} \right] - m \vec{\nabla} \varphi_g, \tag{2.49}$$

and the resulting time evolution of the distribution functions of the plasma particles is given by

$$\frac{\partial f_i}{\partial t}(\vec{r}, \vec{V}, t) + \vec{V} \cdot \frac{\partial f_i}{\partial \vec{r}}(\vec{r}, \vec{V}, t) + \left[-\vec{\nabla}\varphi_g + \frac{Q_i}{m_i} \left(\vec{E} + \frac{\vec{V} \times \vec{B}}{c} \right) \right] \cdot \frac{\partial f_i}{\partial \vec{V}}(\vec{r}, \vec{V}, t)$$

$$= 0. \tag{2.50}$$

This is known as the **Vlasov Equation**. Here, i stands for the particle species, i.e. $i = e$ for electrons and $= p$ for protons or ions; Q_i is the electric charge of particle and m_i is the mass. There is one Vlasov Equation for each species of particles. The electromagnetic fields $\vec{E}$ and $\vec{B}$ are governed by Maxwell's equations :

$$\vec{\nabla} \cdot \vec{E} = \sum_i 4\pi Q_i n_i, \tag{2.51}$$

$$\vec{\nabla} \cdot \vec{B} = 0, \tag{2.52}$$

$$\vec{\nabla} \times \vec{E} = \frac{-1}{c} \frac{\partial \vec{B}}{\partial t}, \tag{2.53}$$

and

$$\vec{\nabla} \times \vec{B} = \frac{1}{c} \frac{\partial \vec{E}}{\partial t} + \frac{4\pi}{c} \sum_i \vec{J_i}, \tag{2.54}$$

where the charge density is

$$Q_i n_i = Q_i \int f_i(\vec{r}, \vec{V}, t) d\vec{V}, \tag{2.55}$$

and the current density is

$$\vec{J_i} = Q_i \int \vec{V} f_i(\vec{r}, \vec{V}, t) \ d\vec{V}. \tag{2.56}$$

Here, Equation (2.51) is the Poisson equation, the differential form of Coulomb's law for the electric field due to a distribution of electric charges; Equation (2.52) describes the divergence free property of the magnetic field i.e. there are no free magnetic charges or monopoles; Equation (2.53) is Faraday's induction law relating the induced electric field and the time varying magnetic flux; Equation (2.54) is Ampere's law modified by the addition of the displacement current introduced by J.C.Maxwell. Thus, the Vlasov Equation (2.50) along with Maxwell's equations provides the so called **Kinetic Description** of a plasma. This description is used to study the microscopic stability or otherwise of a plasma. The cause of instability lies in the particle distribution function. The relative motion between plasma particles, an anisotropy, due to, for example, the presence of a magnetic field, and the ability of a plasma to have several different temperatures, are some of the features that would manifest themselves in the distribution functions and which then become the carriers of free energy. Such a system tries to attain equilibrium by shedding or redistributing this free energy, giving rise to waves and instabilities. The energy associated with these waves and instabilities is spent either in heating the plasma or in the production of electromagnetic radiation, or, most often, both. Thus nonequilibrium plasmas are good sources of electromagnetic radiation.

Problem 2.7 : Using Maxwell's equations in vacuum, show that the electric and magnetic fields $\vec{E}$ and $\vec{B}$ obey a wave equation. What are the group velocity, the phase velocity and the polarization of these waves?

Problem 2.8 : Show that in the absence of forces, $f(\vec{r} - \vec{V}t, \vec{V})$ is a solution of the Vlasov equation.

2.12. The Fluid Description

A system with a large number of particles, under certain situations, can also be described as a continuum − a fluid. All the particles of such a fluid move with the same velocity. Each species of particles corresponds to a fluid. For example in an electron-proton plasma, there are two fluids, one corresponding to the electrons and the other to the protons. These fluids follow a set of laws ensuring the conservation of mass, momentum and energy. The circumstances of the applicability for kinetic or the fluid description will be discussed later. Here, we establish the mathematical framework for the fluid description. The conservation laws can be derived by taking the moments of the Boltzmann or the Vlasov equation in the velocity space. As we go along, it will become evident that the exercise of taking the moments yields meaningful results only for distribution functions possessing certain properties.

The Vlasov equation for one species of particles including the collisions and correlations can be written as:

$$\frac{\partial f(\vec{r}, \vec{V}, t)}{\partial t} + \vec{V} \cdot \frac{\partial f(\vec{r}, \vec{V}, t)}{\partial \vec{r}} + \left[-\vec{\nabla}\varphi_g + \frac{Q}{m} \left(\vec{E} + \frac{\vec{V} \times \vec{B}}{c} \right) \right] \cdot \frac{\partial f(\vec{r}, \vec{V}, t)}{\partial \vec{V}}$$

$$= \left. \frac{\delta f}{\delta t} \right|_c . \tag{2.57}$$

The first velocity moment is taken by integrating Equation (2.57) over the velocity as :

$$\int \frac{\partial f(\vec{r}, \vec{V}, t)}{\partial t} d\vec{V} + \int \vec{V} \cdot \frac{\partial f(\vec{r}, \vec{V}, t)}{\partial \vec{r}} d\vec{V} + \int \left[-\vec{\nabla}\varphi_g + \frac{Q}{m} \left(\vec{E} + \frac{\vec{V} \times \vec{B}}{c} \right) \right]$$

$$\cdot \frac{\partial f(\vec{r}, \vec{V}, t)}{\partial \vec{V}} d\vec{V} = \int \left. \frac{\delta f}{\delta t} \right|_c d\vec{V} . \tag{2.58}$$

The first term in Equation (2.58) can be rewritten as :

$$\frac{\partial}{\partial t} \int f(\vec{r}, \vec{V}, t) d\vec{V} = \frac{\partial}{\partial t} n(\vec{r}, t), \tag{2.59}$$

where, as defined earlier, $n(\vec{r}, t)$ is the particle density.

The second term in Equation (2.58) can be simplified as :

$$
\begin{aligned}
\int \vec{V} \cdot \frac{\partial f(\vec{r}, \vec{V}, t)}{\partial \vec{r}} d\vec{V} &= \int \frac{\partial}{\partial \vec{r}} \cdot \left[\vec{V} f(\vec{r}, \vec{V}, t) \right] d\vec{V}, \\
&= \frac{\partial}{\partial \vec{r}} \cdot \left[n(\vec{r}, t) \vec{U} \right],
\end{aligned}
\tag{2.60}
$$

where the average velocity $\vec{U}$ with which all the particles move in a fluid is defined as :

$$\vec{U} \equiv \overline{\vec{V}} = \frac{\int \vec{V} f(\vec{r}, \vec{V}, t) d\vec{V}}{\int f(\vec{r}, \vec{V}, t) d\vec{V}}. \tag{2.61}$$

While writing Equation (2.60), the fact that $\vec{V}$ is an independent variable and is not acted upon by the differential operator $\frac{\partial}{\partial \vec{r}} \equiv \vec{\nabla}$ has been used. Here, $\vec{U}$ is known as the **Fluid Velocity.**

The third term in Equation (2.58) is:

$$
\begin{aligned}
\int & \left(\frac{Q}{m} \vec{E} - \vec{\nabla} \varphi_g \right) \cdot \frac{\partial f(\vec{r}, \vec{V}, t)}{\partial \vec{V}} d\vec{V} \\
&= \int \frac{\partial}{\partial \vec{V}} \cdot \left[\left(\frac{Q}{m} \vec{E} - \vec{\nabla} \varphi_g \right) f(\vec{r}, \vec{V}, t) \right] d\vec{V} \\
&= \int f(\vec{r}, \vec{V}, t) \left(\frac{Q}{m} \vec{E} - \vec{\nabla} \varphi_g \right) \cdot d\vec{S}_v,
\end{aligned}
\tag{2.62}
$$

where the volume integral over velocity has been converted to a surface integral and dS_v is the surface area element in the velocity space. Now, the value of $\int \left(\frac{Q}{m} \vec{E} - \vec{\nabla} \phi_g \right) f(\vec{r}, \vec{V}, t) dS_v$ on the surface at $V = \infty$ must vanish; otherwise, it will be infinitely large for any distribution function representing a physical system with finite energy. Thus, since $f dS_v \simeq f V^2$, f must drop faster than $\frac{1}{V^2}$ as $V \to \infty$. Under these conditions, the third term vanishes. Note that we have now put a constraint on the functional dependence of the distribution function f on velocity V. The fourth term, as before, can also be converted into a surface integral :

$$\frac{Q}{m}\int\frac{(\vec{V}\times\vec{B})}{c}\cdot\frac{\partial f(\vec{r},\vec{V},t)}{\partial\vec{V}}d\vec{V} = \frac{Q}{m}\int\frac{\partial}{\partial\vec{V}}\cdot\left[f(\vec{r},\vec{V},t)\frac{\vec{V}\times\vec{B}}{c}\right]d\vec{V}$$

$$-\frac{Q}{m}\int f\frac{\partial}{\partial\vec{V}}\cdot\frac{[\vec{V}\times\vec{B}]}{c}\cdot d\vec{V}$$

$$= \frac{Q}{m}\int\left(f(\vec{r},\vec{V},t)\frac{(\vec{V}\times\vec{B})}{c}\right)\cdot d\vec{S}_v. \tag{2.63}$$

Why does this last equality hold?

This time, for the surface integral to vanish, the distribution function f must fall faster than $1/V^3$ as $V \to$ infinity. This is a more stringent constraint on $f(\vec{r},\vec{V},t)$. We recall that the Maxwellian distribution of velocities satisfies these constraints. The three dimensional Maxwellian distribution function for a homogeneous and stationary system is defined as :

$$f(\vec{V}) = n\left(\frac{m}{2\pi K_B T}\right)^{\frac{3}{2}}\exp\left[-\frac{V^2}{(2K_B T/m)}\right], \tag{2.64}$$

where n is the particle density, T is the temperature, m is the mass of a particle and K_B is the Boltzmann constant; $f(\vec{V})$ given by Equation (2.64) falls faster than any power of V for large values of V. Thus, the fluid description is valid for particles obeying Maxwell - Boltzmann distribution of velocities. This distribution is characterized by a single characteristic - the temperature of the system.

Problem 2.9: Show that $\int f(\vec{V})d\vec{V} = n$, where $f(\vec{V})$ is given by Equation (2.64). Determine the fluid velocity $\vec{U}$, the most probable speed U_p and the root mean square velocity U_{rms} for this distribution.

The fifth term of Equation (2.58) describes the change in particle number density due to collisions and correlations.

The collisional processes could cause recombination, ionization and diffusion of the particles. The correlations can cause enhancement or depletion at certain regions. In the absence of these processes, this term vanishes, i.e.

$$\int\left.\frac{\delta f}{\delta t}\right|_c d\vec{V} = 0. \tag{2.65}$$

Now, collecting all the five terms of Equation (2.58), we find that its first velocity moment gives :

$$\frac{\partial}{\partial t}n(\vec{r},t) + \nabla\cdot\left[n(\vec{r},t)\vec{U}\right] = 0. \tag{2.66}$$

Equation (2.66) is known as the **Continuity Equation** and describes particle conservation. This is the first conservation law, that a fluid, during its flow, must satisfy.

Problem 2.10: Find the radial variation of particle density n for a spherically symmetric system in the steady state, for (i) constant fluid velocity and (ii) for velocity of Keplerian motion.

By multiplying Equation (2.66) by the mass m of a particle, the mass conservation law can be rewritten as :

$$\frac{\partial \rho_m}{\partial t} + \nabla \cdot (\rho_m \vec{U}) = 0, \tag{2.67}$$

where $\rho_m = mn(\vec{r}, t)$ is the mass density. The second velocity moment of equation (2.57) is taken by multiplying it by the momentum component mV_i and integrating it over the three dimensional velocity as :

$$m \int V_i \frac{\partial f(\vec{r}, \vec{V}, t)}{\partial t} d\vec{V} + m \sum_j \int V_i \left[V_j \frac{\partial f(\vec{r}, V, t)}{\partial x_j} \right] d\vec{V}$$

$$+ \int V_i \left[Q \left(\vec{E} + \frac{(\vec{V} \times \vec{B})}{c} \right) - m\vec{\nabla}\varphi_g \right]_j \frac{\partial f(\vec{r}, \vec{V}, t)}{\partial V_j} d\vec{V}$$

$$= m \int V_i \frac{\delta f(\vec{r}, \vec{V}, t)}{\delta t} \bigg|_c d\vec{V}. \tag{2.68}$$

It is simple to evaluate the first term which becomes :

$$\frac{\partial}{\partial t} \left[mn(\vec{r}, t) U_i \right]. \tag{2.69}$$

The second term of Equation (2.68) can be written as :

$$\int mV_i \sum_j V_j \frac{\partial f(\vec{r}, \vec{V}, t)}{\partial x_j} d\vec{V} = m \int \sum_j \frac{\partial}{\partial x_j} \left[V_i V_j f(\vec{r}, \vec{V}, t) \right] d\vec{V}$$

$$= \sum_j \frac{\partial}{\partial x_j} \left[\rho_m(\vec{r}, t) \overline{V_i V_j} \right], \tag{2.70}$$

where $\overline{V_i V_j}$ is the average of $V_i V_j$.

In order to understand the content of this term, let us split $\vec{V}$ into two parts such that

$$\vec{V} = \vec{U} + \vec{u}', \tag{2.71}$$

where $\vec{U}$ is the already defined fluid velocity and $\vec{u}'$ is the fluctuating random part with zero average value, i.e.

$$\int \vec{u}' f(\vec{r}, \vec{V}, t) d\vec{V} = 0. \tag{2.72}$$

Substituting Equation (2.71) into Equation (2.70), we get:

$$\frac{\partial}{\partial x_j}[\rho_m(\vec{r}, t)\overline{V_i V_j}] = \frac{\partial}{\partial x_j}[\rho_m(\vec{r}, t)U_i U_j] + \frac{\partial}{\partial x_j}\left[\rho_m(\vec{r}, t)\overline{u_i' u_j'}\right]. \tag{2.73}$$

The quantity

$$\Pi_{ij} \equiv \rho_m(\vec{r}, t)\overline{u_i' u_j'}, \tag{2.74}$$

is called the **Stress Tensor**. It is a symmetric tensor. The diagonal components are $\rho_m\overline{(u_x')^2}, \rho_m\overline{(u_y')^2}$ and $\rho_m\overline{(u_z')^2}$ and represent the **Pressure** in a system. The off-diagonal elements of Π_{ij} are known as **Shear Stresses**.

In the presence of shear, the direction of motion of a fluid element differs from the direction of transfer of momentum. In many situations the shear stresses play a dominant role in the transfer of the mean fluid momentum density $(\rho_m\vec{U})$. A plasma in the presence of a magnetic field becomes anisotropic. The motion of the particles along the magnetic field is different from that in a direction perpendicular to the magnetic field. As a result, a plasma can support different temperatures and pressures in the two directions. In this case the stress tensor is still diagonal but with different values of the diagonal elements $\Pi_\perp$ and $\Pi_\parallel$, perpendicular and parallel to the magnetic field respectively. We shall get another chance to discuss shear stresses.

The quantity $\frac{1}{2}\rho\overline{\vec{V} \cdot \vec{V}}$ is the energy density of total motion and it is equal to the sum of the energy density of random motion $((\rho_m\overline{\vec{u}' \cdot \vec{u}'}/2)$ and the energy density of convective motion $\left(\frac{1}{2}\rho_m\overline{\vec{U}\vec{U}}\right)$.

Problem 2.11: Show that

$$\text{Trace } \Pi = \int m(V^2)f(\vec{r}, \vec{V}, t)d\vec{V} - \rho_m(\vec{r}, t)U^2,$$

i.e. the trace of the stress tensor is equal to twice the total kinetic energy density minus twice the convective energy density.

Let us get back to Equation (2.73); the first term on the RHS can be rewritten as:

$$\begin{aligned}
\sum_j \frac{\partial}{\partial x_j}[\rho_m(\vec{r}, t)U_i U_j] &= \sum_j U_i \frac{\partial}{\partial x_j}(\rho_m U_j) + \rho_m U_j \frac{\partial}{\partial x_j}U_i \\
&= U_i\left(-\frac{\partial \rho_m}{\partial t}\right) + \rho_m(\vec{U} \cdot \vec{\nabla})U_i, \tag{2.75}
\end{aligned}$$

where the continuity equation (2.67) has been used.

The third term of Equation (2.68) can be evaluated as:

$$\sum_j \int V_i \left[Q(\vec{E} + \frac{\vec{V} \times \vec{B}}{c}) - m\vec{\nabla}\varphi_g \right]_j \frac{\partial f(\vec{r}, \vec{V}, t)}{\partial V_j} d\vec{V}$$

$$= \sum_j \int \frac{\partial}{\partial V_j} \cdot \left[f(\vec{r}, \vec{V}, t) V_i \left\{ Q\left(\vec{E} + \frac{\vec{V} \times \vec{B}}{c}\right) - m\vec{\nabla}\varphi_g \right\}_j \right] d\vec{V}$$

$$- \int f(\vec{r}, \vec{V}, t) \left\{ Q\left(\vec{E} + \frac{\vec{V} \times \vec{B}}{c}\right) - m\vec{\nabla}\varphi_g \right\}_j \frac{\partial V_i}{\partial V_j} d\vec{V}$$

$$= -\frac{Q\rho_m(\vec{r}, t)}{m} \left[\vec{E} + \frac{\vec{U} \times \vec{B}}{c} \right]_i - \{\rho_m(\vec{r}, t)\vec{\nabla}\varphi_g\}_i. \qquad (2.76)$$

Do you see how?

The fourth term of Equation (2.68) is:

$$\int m V_i \frac{\delta f(\vec{r}, \vec{V}, t)}{\delta t} \bigg|_c d\vec{V} \equiv \sum_s {}'\Gamma_i^{ss'}, \qquad (2.77)$$

where $\Gamma_i^{ss'}$ is the rate of change of momentum due to collisions between different species of particles s and s' or between different fluids. The stresses between different parts of the same fluid are contained in the off-diagonal elements of the stress tensor Π_{ij} (Equation 2.74).

Collecting the terms in Equations (2.69), (2.73), (2.74), (2.75), (2.76) and (2.77), the second velocity moment written in the vector form for species s of particles becomes :

$$\frac{\partial}{\partial t}(\rho_s \vec{U}_s) + \vec{U}_s(-\frac{\partial \rho_s}{\partial t}) + \rho_s(\vec{U}_s \cdot \vec{\nabla})\vec{U}_s + \vec{\nabla} \cdot \Pi_s + \rho_s(\vec{r}, t)\vec{\nabla}\varphi_g$$

$$-\frac{Q_s}{m_s}\rho_s(\vec{r}, t)[\vec{E} + \frac{\vec{U}_s \times \vec{B}}{c}] = \sum_{s'} \vec{\Gamma}^{ss'},$$

or

$$\rho_s \left[\frac{\partial \vec{U}_s}{\partial t} + (\vec{U}_s \cdot \vec{\nabla})\vec{U}_s \right] = \frac{Q_s \rho_s(\vec{r}, t)}{m_s} \left[\vec{E} + \frac{\vec{U}_s \times \vec{B}}{c} \right] - \rho_s(\vec{r}, t)\vec{\nabla}\varphi_g$$

$$-\vec{\nabla} \cdot \Pi_s + \sum_{s'} \vec{\Gamma}^{ss'}. \qquad (2.78)$$

Equation (2.78) describes **The Fluid Flow in the Presence of Electromagnetic, Gravitational and Collisional Forces.** It expresses conservation of momentum under the action of different forces.

The third velocity moment of equation (2.57) gives the energy equation. Let us first derive the heat transport equation in order to examine the time evolution of kinetic pressure by taking the third moment with the fluctuating velocity $\vec{u}' = \vec{V} - \vec{U}$ instead of the total velocity $\vec{V}$.

We shall multiply Equation (2.57) by $(m/2)(\vec{V} - \vec{U})_i(\vec{V} - \vec{U})_j$ and integrate over $\vec{V}$.

The first term is :

$$\int \frac{\partial f(\vec{r}, \vec{V}, t)}{\partial t} \frac{m}{2} (\vec{V} - \vec{U})_i (\vec{V} - \vec{U})_j d\vec{V} = \frac{\partial}{\partial t} \int f(\vec{r}, \vec{V}, t) \frac{m}{2} (\vec{V} - \vec{U})_i (\vec{V} - \vec{U})_j d\vec{V}$$

$$- \int f(\vec{r}, \vec{V}, t) \frac{\partial}{\partial t} \frac{m}{2} \left\{ (\vec{V} - \vec{U})_i (\vec{V} - \vec{U})_j \right\} d\vec{V},$$

$$= \frac{\partial}{\partial t} \left[\frac{mn(\vec{r}, t)}{2} \overline{(V - U)_i (V - U)_j} \right] - \frac{mn(\vec{r}, t)}{2} \overline{\frac{\partial}{\partial t} (\vec{V} - \vec{U})_i (\vec{V} - \vec{U})_j},$$

$$= \frac{\partial}{\partial t} \left[\frac{\Pi_{ij}}{2} \right] - 0; \quad \text{why?} \quad \left(\text{hint:} \quad \frac{\partial U}{\partial t} = -\frac{\partial u'}{\partial t} \right). \tag{2.79}$$

The second term is :

$$\int \vec{V} \cdot \frac{\partial f(\vec{r}, \vec{V}, t)}{\partial \vec{r}} \frac{m}{2} (\vec{V} - \vec{U})_i (\vec{V} - \vec{U})_j d\vec{V}$$

$$= \sum_k \int V_k \frac{\partial f(\vec{r}, \vec{V}, t)}{\partial x_k} \frac{m}{2} (\vec{V} - \vec{U})_i (\vec{V} - \vec{U})_j d\vec{V},$$

$$= \sum_k \frac{\partial}{\partial x_k} \left[\int \frac{m}{2} V_k (\vec{V} - \vec{U})_i (\vec{V} - \vec{U})_j f(\vec{r}, \vec{V}, t) d\vec{V} \right]$$

$$- \sum_k \int \frac{m}{2} f(\vec{r}, \vec{V}, t) \frac{\partial}{\partial x_k} \left(V_k (\vec{V} - \vec{U})_i (\vec{V} - \vec{U})_j \right) d\vec{V},$$

$$= \sum_k \frac{\partial}{\partial x_k} \left[\frac{m}{2} n(\vec{r}, t) \overline{V_k (\vec{V} - \vec{U})_i (\vec{V} - \vec{U})_j} \right]$$

$$- \sum_k \frac{m}{2} \ n(\vec{r}, t) \overline{V_k \frac{\partial}{\partial x_k} \left[(\vec{V} - \vec{U})_i (\vec{V} - \vec{U})_j \right]},$$

$$= \sum_k \frac{\partial}{\partial x_k} \left(\frac{H_{kij}}{2} + U_k \frac{\Pi_{ij}}{2} \right) + \frac{\partial U_i}{\partial x_k} \frac{\Pi_{jk}}{2} + \frac{\partial U_j}{\partial x_k} \frac{\Pi_{ik}}{2}. \tag{2.80}$$

The tensor H, defined as :

$$H_{kij} = m \int f(\vec{r}, \vec{V}, t)(\vec{V} - \vec{U})_k (\vec{V} - \vec{U})_i (\vec{V} - \vec{U})_j d\vec{V}, \tag{2.81}$$

is the thermal energy flux density (since $(\vec{V} - \vec{U}) = \vec{u}'$ is the random part of the velocity).

Problem 2.12: show that

$$H_{kij} = m \int f(\vec{r}, \vec{V}, t) V_k V_i V_j d\vec{V} - \rho_m U_k U_i U_j - U_k \Pi_{ij}. \qquad (2.82)$$

From Equation (2.82), we observe that H is a measure of the difference between the flux density of total motion and the convective motion. Or the total energy flux density is a sum of the thermal energy flux density H and the convective energy flux density $[(\rho_m UUU/2) + (U, \Pi)/2]$ just as the total energy density is a sum of the thermal energy density $\Pi/2$ and the convective energy density $(\rho_m UU)/2$

The third term is

$$\int \left(\frac{Q}{m} \vec{E} - \vec{\nabla} \varphi_g \right) \cdot \frac{\partial f(\vec{r}, \vec{V}, t)}{\partial \vec{V}} \frac{m}{2} (\vec{V} - \vec{U})_i (\vec{V} - \vec{U})_j d\vec{V}$$

$$= \sum_k \int \left(\frac{Q}{m} \vec{E} - \vec{\nabla} \varphi_g \right)_k \frac{\partial f(\vec{r}, \vec{V}, t)}{\partial V_k} \frac{m}{2} (\vec{V} - \vec{U})_i (\vec{V} - \vec{U})_j d\vec{V},$$

$$= \sum_k \int \frac{\partial}{\partial V_k} \left[\left(\frac{Q}{m} \vec{E} - \vec{\nabla} \varphi_g \right)_k f(\vec{r}, \vec{V}, t) \frac{m}{2} (\vec{V} - \vec{U})_i (\vec{V} - \vec{U})_j \right] d\vec{V}$$

$$- \sum_k \int \left(\frac{Q}{m} \vec{E} - \vec{\nabla} \varphi_g \right)_k f(\vec{r}, \vec{V}, t) \frac{m}{2} \frac{\partial}{\partial V_k} \left((\vec{V} - \vec{U})_i (\vec{V} - \vec{U})_j \right) d\vec{V},$$

$$= - \sum_k \left(\frac{Q}{m} \vec{E} - \vec{\nabla} \varphi_g \right)_k \frac{mn(\vec{r}, t)}{2} \overline{\frac{\partial}{\partial V_k} (\vec{V} - \vec{U})_i (\vec{V} - \vec{U})_j}, \qquad (2.83)$$

$$= \ 0, \qquad \text{why?}$$

The fourth term is :

$$\int \frac{Q}{m} \left\{ \frac{\vec{V} \times \vec{B}}{c} \right\} \cdot \frac{\partial f(r, V, t)}{\partial \vec{V}} \frac{m}{2} (\vec{V} - \vec{U})_i (\vec{V} - \vec{U})_j d\vec{V}$$

$$= \sum_k \int \frac{Q}{m} \left(\frac{\vec{V} \times \vec{B}}{c} \right)_k \frac{\partial f(\vec{r}, \vec{V}, t)}{\partial V_k} \frac{m}{2} (\vec{V} - \vec{U})_i (\vec{V} - \vec{U})_j d\vec{V},$$

$$= \sum_k \int \frac{\partial}{\partial V_k} \left[\frac{Q}{m} \frac{(\vec{V} \times \vec{B})_k}{c} f(\vec{r}, \vec{V}, t) \frac{m}{2} (\vec{V} - \vec{U})_i (\vec{V} - \vec{U})_j \right] d\vec{V}$$

$$- \sum_k \int \frac{m}{2} f(\vec{r}, \vec{V}, t) \frac{\partial}{\partial V_k} \left[\frac{Q}{m} \frac{(\vec{V} \times \vec{B})_k}{c} (\vec{V} - \vec{U})_i (\vec{V} - \vec{U})_j \right] d\vec{V},$$

$$= (\Pi \times \vec{\Omega}_B)_{ij} + (\Pi \times \vec{\Omega}_B)_{ji}, \tag{2.84}$$

where $\vec{\Omega}_B = \dfrac{Q\vec{B}}{mc}$ is the cyclotron frequency.

The collision term is

$$\int \left.\frac{\delta f}{\delta t}\right|_c \frac{m}{2}(\vec{V} - \vec{U})_i(\vec{V} - \vec{U})_j d\vec{V} \equiv R'_{ij}/2. \tag{2.85}$$

Collecting all the terms in Equations (2.79), (2.80), (2.83), (2.84) and (2.85) we get the **Heat Transport Equation:**

$$\frac{\partial}{\partial t}\Pi_{ij} + \sum_k \frac{\partial}{\partial x_k}(H_{kij} + U_k\Pi_{ij}) + \frac{\partial U_i}{\partial x_k}\Pi_{jk} + \frac{\partial U_j}{\partial x_k}\Pi_{ik}$$

$$+(\Pi \times \vec{\Omega}_B)_{ij} + (\Pi \times \vec{\Omega}_B)_{ji} = R'_{ij}. \tag{2.86}$$

The term R' contains all the losses due to radiative, dissipative and viscous processes.

2.13. Heat Diffusion Equation

The heat transport Equation (2.86) acquires a simple form - the heat diffusion equation when the fluid is unmagnetized ($\vec{B} = 0$), at rest ($\vec{U} = 0$) and lossless ($R' = 0$), we get

$$\frac{\partial}{\partial t}\Pi_{ij} + \sum_k \frac{\partial}{\partial x_k}(H_{kij}) = 0. \tag{2.87}$$

The stress tensor $\Pi_{ij} = \overline{\rho u'_i u'_j}$ is a measure of the thermal energy density which can be written as:

$$\frac{\Pi}{2} = \frac{1}{2}\sum_{ij}\Pi_{ij} = \frac{1}{2}\sum_i \Pi_{ii} = \frac{\rho_m}{2}\sum_i u_i'^2 = \frac{3}{2}\frac{\rho_m K_B T}{m} = \frac{3}{2} nK_B T = \frac{3}{2}p. \tag{2.88}$$

in the absence of shear, i.e., when the off-diagonal terms of the stress tensor vanish. Here, T is the temperature, p is the pressure and K_B is the Boltzmann constant. If we write the heat flux $\vec{H} \propto (-\vec{\nabla}T)$, as the experiments tell us, we get the **Heat Diffusion Equation :**

$$\frac{\partial T}{\partial t} = \vec{\nabla} \cdot [\kappa_c \vec{\nabla}T], \tag{2.89}$$

where κ_c is the thermal conductivity which may also vary from place to place in a medium.

2.14. Adiabatic Energy Equation

Again in the absence of viscosity or shear ($\Pi_{ij} = 0, i \neq j$), magnetic field ($\vec{B} = 0$), thermal conductivity or heat flux ($\nabla \cdot H = 0$) and collisional processes ($R' = 0$), we get the adiabatic equation of state from Equation (2.86). From the third, fourth and fifth terms of Equation (2.86), we get

$$\sum_{ij} \sum_{k} \frac{\partial}{\partial x_k} (U_k \Pi_{ij}) + \frac{\partial U_i}{\partial x_k} \Pi_{jk} + \frac{\partial U_j}{\partial x_k} \Pi_{ik}.$$

On separating the terms for $j = i$ and $j \neq i$ the above expression becomes

$$\sum_{i,k} \frac{\partial}{\partial x_k} (U_k \Pi_{ii}) + \frac{\partial U_i}{\partial x_k} \Pi_{ik} + \frac{\partial U_i}{\partial x_k} \Pi_{ik} + \sum_{k} \sum_{i \neq j} \cdots .$$

Using $\sum_i \Pi_{ii} = 3p$ and separating terms for $k = i$ and $k \neq i$, we get:

$$\vec{\nabla} \cdot (3\vec{U}p) + \sum_{i} 2\frac{\partial U_i}{\partial x_i} \Pi_{ii} + \sum_{k \neq i} \cdots + \sum_{k} \sum_{i \neq j} \cdots$$

$$= 3\vec{U} \cdot \vec{\nabla}p + 5p\vec{\nabla} \cdot \vec{U} + \sum_{k \neq i} \cdots + \sum_{k} \sum_{i \neq j} \cdots .$$

We rewrite Equation (2.86) as

$$\frac{\partial}{\partial t} (3p) + 5p\vec{\nabla} \cdot \vec{U} + 3\vec{U} \cdot \vec{\nabla}p = 0, \tag{2.90}$$

and using the mass conservation law (Equation 2.67) we find that

$$\frac{d}{dt}(p\rho_m^{-\frac{5}{3}}) = 0 \quad \text{or} \quad p\rho_m^{-5/3} = \text{Constant} . \tag{2.91}$$

This is the familiar **Adiabatic Energy Equation.** It describes changes in pressure or volume under varying temperature conditions. Here the fraction (5/3) is related to the number of degrees of freedom d by $\gamma = d + 2/d = 5/3$ for $d = 3$ for a monatomic gas; γ is also the ratio (C_p/C_v) where C_p and C_v are the specific heats at constant pressure and constant volume respectively (Tanenbaum 1967).

Problem 2.13: Why did we get the energy equation (Equation 2.91) for a monatomic gas?

From the adiabatic energy equation $p\rho_m^{-\gamma} = $ constant, we find:

$$\frac{dp}{d\rho_m} = \frac{\gamma p}{\rho_m} = c_s^2, \tag{2.92}$$

where c_s is the adiabatic sound speed of a fluid.

Problem 2.14: Derive an expression for the isothermal sound speed.

In both adiabatic and isothermal cases, the changes in pressure and density are simultaneous and proportional. However, the full energy equation has to be used if the changes in pressure are too fast to affect the corresponding changes in density.

Problem 2.15: By taking the third moment with the total velocity, derive the energy equation :

$$\frac{\partial}{\partial t}\left[\frac{\rho_m}{2}U_iU_j + \frac{\Pi_{ij}}{2}\right] + \sum_k \frac{\partial}{\partial x_k}\left[U_k\left(\frac{\rho_m}{2}U_iU_j + \frac{\Pi_{ij}}{2}\right) + \frac{U_i\Pi_{jk}}{2} + \frac{U_j\Pi_{ki}}{2}\right.$$

$$\left. + \frac{H_{ijk}}{2}\right] - \frac{1}{2}(\vec{E} - \frac{m}{Q}\vec{\nabla}\phi_g)_iJ_j - \frac{1}{2c}(\vec{J}\times\vec{B})_iU_j - \frac{1}{2c}(\vec{J}\times\vec{B})_jU_i$$

$$+ \frac{(\Pi\times\Omega_B)_{ij}}{2} + \frac{(\Pi\times\Omega_B)_{ji}}{2} = \frac{R_{ij}}{2}, \tag{2.93}$$

where $\vec{J}(\vec{r},t) = Qn(\vec{r},t)\vec{U}$ is the current density. Equation (2.93) is the conservation law for total energy which is a sum of kinetic energy and potential energy including magnetic energy and as before

$$R_{ij} \equiv \int \left.\frac{\delta f}{\delta t}\right|_c V_iV_j d\vec{V}, \tag{2.94}$$

represents the collisional processes.
Equations (2.67), (2.78), (2.86) and (2.93) describe the conservation of mass, momentum, kinetic pressure and total energy density for each species of particles in a system.

2.15. Correlation Functions

The joint probability distributions contain information on the correlated behaviour of a system, i.e. when the position of one part of a system in the phase space depends upon the distribution of the other parts. We can consider the distribution of galaxies as an example. It is by now well established that, more often than not, galaxies come in pairs, groups, clusters and superclusters (Figure 2.3). In other words, the positions of galaxies are strongly correlated. The clustering can be described in terms of the multi-point correlation functions. Thus the two point correlation function or the pair correlation function f_c defined in Equation (2.26) can be used to describe the clustering to the lowest order; f_c would be zero if the phase space positions of the two galaxies in a pair were random and independent of each

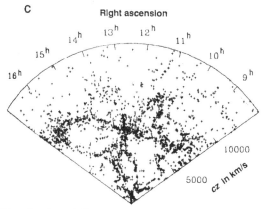

Figure 2.3. Distribution of Galaxies Exhibiting the Tendency for Clustering

other. Further, the pair correlation function f_c could be averaged over both velocities to get the pair correlation function $\xi(R)$ in the coordinate space or f_c could be averaged over the two coordinates to get the pair correlation function ϕ in the velocity space. Mixed pair correlation functions can also be used. The radial autocorrelation function $\xi(\vec{R})$ for a homogeneous and isotropic system is related to f_c as

$$\xi(\vec{R}) = \frac{1}{4\pi\bar{n}^2} \int f_c(\vec{q}_1, \vec{V}_1, \vec{q}_1 + \vec{R}, \vec{V}_2, t) d\vec{q}_1 \ d\vec{V}_1 d\vec{V}_2 d^2\Omega, \qquad (2.95)$$

where $d^2\Omega$ is the differential solid angle in the direction of the separation $\vec{R}$ of the two galaxies, and

$$n(\vec{q}_1, t) = \int f_1(\vec{q}_1, \vec{V}_1, t) d\vec{V}_1. \qquad (2.96)$$

For a uniform and time independent equilibrium distribution, $n(\vec{q}, t)$ $= \bar{n} = $ a constant, is the spatial particle density. In the standard model, the galaxies are assumed to be distributed uniformly to the zeroth order and the correlations are a small first order effect. From Equation (2.26), it is clear that the correlations i.e., a nonzero f_c, enhance the density over the average density $\bar{n}$ in a region. The quantity $\xi(\vec{R})$ is the fractional excess in the number of pairs due to the spatial correlations. Thus, the average number of particles $\Delta N(R)$ in a small volume ΔV at a distance R from a given particle are found to be :

$$\Delta N(R) = [1 + \xi(R)]\bar{n}\Delta V. \qquad (2.97)$$

But, what about the correlations in velocities of the particles, if any? This information can also be extracted by taking the velocity moments of the two particle distribution function f_2. One can define the first relative velocity moment as :

$$\langle V_{R12}(R) \rangle = \int (\vec{V}_2 - \vec{V}_1) \cdot \hat{q} f_2(\vec{q}, \vec{V}_1, \vec{q} + \vec{R}, \vec{V}_2) d\vec{q} d\vec{V}_1 d\vec{V}_2 d^2\Omega, \qquad (2.98)$$

and the second velocity moments as :

$$\left\langle V_{12}^2 \right\rangle (R) = \int (\vec{V}_2 - \vec{V}_1)^2 f_2(\vec{q}_1, \vec{V}_1, \vec{q}_1 + \vec{R}, \vec{V}_2) d\vec{q}_1 d\vec{V}_1 d\vec{V}_2 d^2\Omega,$$

and

$$\left\langle V_{R12}^2 \right\rangle (R) = \int \left[(\vec{V}_2 - \vec{V}_1) \cdot \hat{q}_1 \right]^2 f_2(\vec{q}_1, \vec{V}_1, \vec{q}_1 + \vec{R}, \vec{V}_2) d\vec{q}_1 d\vec{V}_1 d\vec{V}_2 d^2\Omega,$$

$$(2.99)$$

where $\hat{q}_1$ is the unit vector. Here $\langle V_{R12} \rangle (R)$ is the average radial velocity of a pair of galaxies, $< V_{12}^2 > (R)$ is the mean square velocity dispersion of the pairs and $< V_{R12}^2 > (R)$ is the mean square radial velocity dispersion.

According to the Hubble's law of the expansion of the universe, the galaxies are receding away from each other and the speed of recession is directly proportional to the distance of a galaxy from another galaxy. The constant of proportionality is known as **Hubble's Constant H.** The recessional motion produces a redshift in the spectrum of a galaxy. The measurement of this redshift, therefore, provides an estimate of the distance of a galaxy, which when combined with the angular position, gives the three dimensional position of a galaxy. The three dimensional galaxy-galaxy correlation function ξ (R) (Equation 2.95) is found to be (Porter 1988) :

$$\xi(R) = \left(\frac{Rh}{5.4Mpc} \right)^{-1.74}, \qquad (2.100)$$

for 100 h^{-1} kpc $< R < 10h^{-1}$ Mpc. Here, the Hubble constant $H = 100$ h km/sec/Mpc has been used. The corresponding two point correlation function at the scales of clusters of galaxies can be determined similarly and is found to follow a power law

$$\xi(R) \sim 360 \left(\frac{Rh}{Mpc} \right)^{-1.8}, \qquad (2.101)$$

in the range R$< 150h^{-1}$ Mpc. It is intriguing that clusters of galaxies are more strongly clustered than the galaxies. One can define a correlation length for any system. This is the value of R for which $\xi(R) = 1$.

Problem 2.16: Compare the correlation lengths of galaxies and clusters of galaxies.

2.16. Determining f_c

In order to determine the correlation function f_c, we have to determine $f_2(\vec{q}_1, \vec{V}_1, \vec{q}_2, \vec{V}_2, t)$, the joint distribution function for two particles. This can be done by following the steps taken to derive the Boltzmann Equation for f_1. Thus, integrating Equation (2.17) over all the coordinates except $(\vec{q}_1, \vec{V}_1)$ and $(\vec{q}_2, \vec{V}_2)$ and in the absence of external forces (only to keep simplicity), we find :

$$\frac{\partial f_2}{\partial t} + \vec{V}_1 \cdot \frac{\partial f_2}{\partial \vec{q}_1} + \vec{V}_2 \cdot \frac{\partial f_2}{\partial \vec{q}_2} + \frac{\vec{F}_{12}}{m} \cdot \frac{\partial f_2}{\partial \vec{V}_1} + \frac{\vec{F}_{21}}{m} \cdot \frac{\partial f_2}{\partial \vec{V}_2}$$

$$+ \int \sum_{l=3}^{N} \left[\vec{F}_{1l} \cdot \frac{\partial f_3}{\partial \vec{V}_1} + \vec{F}_{2l} \cdot \frac{\partial f_3}{\partial \vec{V}_2} \right] d\vec{q}_l d\vec{V}_l = 0, \qquad (2.102)$$

where $\vec{F}_{ll'}$ is the internal force on particle l due to another particle l' and $f_3(\vec{q}_1, \vec{V}_1, \vec{q}_2, \vec{V}_2, \vec{q}_3, \vec{V}_3, t)$ is the three particle distribution function. Let us recall Equation (2.21) for f_1, neglecting external forces.

$$\frac{\partial f_1}{\partial t} + \vec{V}_1 \cdot \frac{\partial f_1}{\partial \vec{q}_1} + \int \sum_{l=2}^{N} \frac{\vec{F}_{1l}}{m} \cdot \frac{\partial f_2}{\partial \vec{V}_1} d\vec{q}_l d\vec{V}_l = 0. \qquad (2.103)$$

The two particle distribution function f_2 can be expressed in terms of single particle distribution f_1 and the correlation function f_c using Equation (2.26). Is f_3 related to f_1 and f_2? An expression for f_3 can be derived again through the solution of BBGKY Equations by using the standard expansion procedure. However, we can also arrive at this expression by invoking all possible combinations of probabilities and thus write :

$$f_3(1,2,3) = f_2(1,2)f_1(3) + f_2(1,3)f_1(2) + f_2(2,3)f_1(1) - 2f_1(1)f_1(2)f_1(3),$$
$$(2.104)$$

which satisfies the criterion that in the absence of correlations f_3 must equal to $f_1(1) f_1(2) f_1(3)$. Can you verify this? Thus, a formal equation for f_c can be set up by using Equations (2.26), (2.102), (2.103) and (2.104), but finding its solution is a formidable task. Several simplifying assumptions have to be made before it even begins to resemble something we could dare to bare!

2.17. The Single Fluid Description

Under certain circumstances, which we would discuss in detail later, it is not essential to treat the various species of particles individually. For example, a plasma consisting of electrons and ions can be described by a single fluid with a mass density $\rho(\vec{r}, t)$, a velocity $\vec{U}(\vec{r}, t)$ instead of mass densities ρ_e and ρ_i and velocities $\vec{U}_e$ and $\vec{U}_i$. Let us begin with the two fluid equations. The equation of motion of an electron fluid is (Equation 2.78) :

$$\rho_e[\frac{\partial \vec{U}_e}{\partial t} + (\vec{U}_e \cdot \vec{\nabla})\vec{U}_e] = -\frac{e\rho_e}{m_e}\left[\vec{E} + \frac{\vec{U}_e \times \vec{B}}{c}\right] - \rho_e\vec{\nabla}\phi_g - \vec{\nabla}\cdot\Pi_e + \vec{\Gamma}_{ei}, \quad (2.105)$$

where the species index e is for electrons and i is for ions of charge ze. The equation of motion of an ion fluid is,

$$\rho_i[\frac{\partial \vec{U}_i}{\partial t} + (\vec{U}_i \cdot \vec{\nabla})\vec{U}_i] = \frac{ze\rho_i}{m_i}\left[\vec{E} + \frac{\vec{U}_i \times \vec{B}}{c}\right] - \rho_i\vec{\nabla}\phi_g - \vec{\nabla}\cdot\Pi_i + \vec{\Gamma}_{ie}. \quad (2.106)$$

Now, define the single fluid quantities $\rho_m(\vec{r}, t), \vec{U}(r, t)$ and the current density $\vec{J}(\vec{r}, t)$ as :

$$\rho_m = \rho_e + \rho_i = n(m_e + m_i), \quad\quad\quad (2.107)$$

$$\vec{U} = \frac{m_e\vec{U}_e + m_i\vec{U}_i}{m_e + m_i}, \quad\quad\quad (2.108)$$

$$\vec{J}(\vec{r}, t) = ne[z\vec{U}_i - \vec{U}_e], \quad\quad\quad (2.109)$$

where we have assumed $n_e = n_i$. The addition of Equations (2.105) and (2.106) gives

$$n\frac{\partial}{\partial t}[m_e\vec{U}_e + m_i\vec{U}_i] + n[m_e(\vec{U}_e \cdot \vec{\nabla})\vec{U}_e + m_i(\vec{U}_i \cdot \vec{\nabla})\vec{U}_i]$$

$$= en(1 - z)\vec{E} + en\left[(z\vec{U}_i - \vec{U}_e) \times \frac{\vec{B}}{c}\right] - \vec{\nabla} \cdot (\Pi_e + \Pi_i) - (\rho_e + \rho_i)\vec{\nabla}\phi_g,$$

$$(2.110)$$

where $\vec{\Gamma}_{ie} = -\vec{\Gamma}_{ei}$ has been used. This is true, when the rate of momentum density transfer between the two fluids is proportional to their relative velocity, i.e., $\Gamma_{ei} = -\nu_{ei}(\vec{U}_e - \vec{U}_i)\rho_e$ where ν_{ei} is the collision frequency for momentum transfer for electrons colliding with ions. Since momentum is conserved during interparticle collisions, $\vec{\Gamma}_{ei} + \vec{\Gamma}_{ie} = 0$, and the collision frequencies are related as :

$$\rho_e \nu_{ei} = \rho_i \nu_{ie}. \tag{2.111}$$

We observe that in the absence of all other except the collisional forces and for a homogeneous system,

$$\frac{\partial \vec{U}_e}{\partial t} = -\nu_{ei}(\vec{U}_e - \vec{U}_i), \tag{2.112}$$

which says that in equilibrium $\left(\partial \vec{U}_e/\partial t = 0\right)$, the two fluids electrons and ions flow with a common velocity $(\vec{U}_e = \vec{U}_i)$. This result is a consequence of our model for the collisional force.

By using Equations (2.107), (2.108) and (2.109), Equation (2.110) can be written in terms of a single fluid quantities as :

$$\rho_m \left[\frac{\partial \vec{U}}{\partial t} + (\vec{U} \cdot \vec{\nabla})\vec{U}\right] = -\rho_m \vec{\nabla}\phi_g + \frac{\vec{J} \times \vec{B}}{c} - \vec{\nabla} \cdot \Pi - \frac{m_e m_i \rho_m}{e^2 n^2}(\vec{J} \cdot \vec{\nabla})\vec{J}$$

$$-\left[\frac{m_e m_i (m_e + m_i)(1 - z)}{e(m_e z + m_i)^2}(\vec{J} \cdot \vec{\nabla})\vec{U} + (\vec{U} \cdot \vec{\nabla})\vec{J}) - en(1 - z)\vec{E}\right]. \tag{2.113}$$

Here, we notice that the electric force term and the current density advection terms vanish for $z = 1$, i.e., for a plasma with electrons and singly charged positive ions, for example for an electron − proton plasma. The term $(\vec{J} \cdot \vec{\nabla})\vec{J}$ is proportional to the square of the relative velocity of the electrons and ions. This term must be small compared to the mean motion term $(\vec{U} \cdot \vec{\nabla})\vec{U}$ if the plasma behaves dominantly as a single fluid.

Problem 2.17: Compare the terms $(\vec{J} \cdot \vec{\nabla})\vec{J}$ and $(\vec{U} \cdot \vec{\nabla})\vec{U}$ in Equation (2.113) and find the limiting value of the relative velocity of electron and ion for their single fluid description.

The continuity equation for a single fluid can be derived by adding the continuity equations for electron and ion fluids. We get:

$$\frac{\partial \rho_m}{\partial t} + \vec{\nabla} \cdot \left[\frac{m_e + m_i}{z m_e + m_i}\left\{\rho_m z + \rho_i(1 - z)\right\}\vec{U}\right] = 0, \tag{2.114}$$

which, for $z = 1$, reduces to the more familiar form

$$\frac{\partial \rho_m}{\partial t} + \vec{\nabla} \cdot \left[\rho_m \vec{U}\right] = 0. \tag{2.115}$$

Ohm's law for a single conducting fluid can be derived by subtracting the equation of motion (2.105) of the electron fluid multiplied by the mass of the ion from the equation of motion (2.106) of the ion fluid multiplied by the mass of an electron. We find :

$$\left(\frac{m_e + m_i}{m_e z + m_i}\right)(1-z)\frac{\partial \vec{U}}{\partial t} + \frac{(m_e + m_i)}{ne(m_e z + m_i)}\frac{\partial \vec{J}}{\partial t} + \left(\frac{m_e + m_i}{m_e z + m_i}\right)^2 (1-z^2)(\vec{U}\cdot\vec{\nabla})\vec{U}$$

$$+ \frac{(m_e + m_i)(m_e + m_i z)}{ne(m_e z + m_i)^2}[(\vec{U}\cdot\vec{\nabla})\vec{J} + (\vec{J}\cdot\vec{\nabla})\vec{U}] + \frac{m_e^2 - m_i^2}{n^2 e^2 (m_e z + m_i)^2}(\vec{J}\cdot\vec{\nabla})\vec{J}$$

$$= \frac{(m_e z + m_i)e}{m_e m_i}\vec{E} + \frac{(m_e + m_i)^2 ze}{m_e m_i (m_e z + m_i)}\frac{\vec{U}\times\vec{B}}{c} + \frac{m_e^2 z - m_i^2}{n(m_e z + m_i)m_e m_i}\frac{\vec{J}\times\vec{B}}{c}$$

$$- \frac{1}{n}\vec{\nabla}\cdot\left[\frac{\Pi_i}{m_i} - \frac{\Pi_e}{m_e}\right] + \frac{(m_e + m_i)}{nm_e m_i}\vec{\Gamma}_{ie}. \qquad (2.116)$$

Well, whatever happened to the short and cute Ohm's law "V = IR", we learnt about in school or its little more matured version "$\vec{J} = \sigma\vec{E}$," we learnt in undergraduate classes? Well, this is a graduate level course! Equation (2.116) is called the **Generalized Ohm's Law.** The additional terms have arisen from: (i) the mass motion of the medium or the fluid with velocity $\vec{U}$; (ii) the medium is inhomogeneous, i.e., it supports pressure gradients and shear included in the stress tensors Π; (iii) the nonsteady nature, i.e., the medium varies with time; (iv) the usually small terms the nonlinear term $(\vec{J}\cdot\vec{\nabla})\vec{J}$, which is quadratic in the relative velocity between electrons and ions, and (v) the magnetic field $\vec{B}$. Suppose we neglect all these five effects and assume ions to be singly charged i.e. $z = 1$, we get

$$\vec{E} = -\frac{1}{ne}\vec{\Gamma}_{ie}. \qquad (2.117)$$

It still does not look like the familiar forms. We need to express $\vec{\Gamma}_{ie}$ which is the rate of change of momentum density of the ions due to their collisions with electrons. Using the collision model discussed earlier, we write :

$$\vec{\Gamma}_{ie} = -\rho_i(\vec{U}_i - \vec{U}_e)\nu_{ie}, \qquad (2.118)$$

where ν_{ie} is the ion - electron collision frequency and $\vec{U}_i$ and $\vec{U}_e$ are the velocities of the ion and electron fluids respectively. Expressing the velocity difference in terms of the current density $\vec{J}$, we get :

$$\vec{E} = \frac{m_i \nu_{ie}}{ne^2}\vec{J} = \frac{\vec{J}}{\sigma} = \eta\vec{J}, \qquad (2.119)$$

where σ can be identified with the conductivity and η with the resistivity of the single conducting fluid. So, this is how one recovers the familiar Ohm's law from the generalized Ohm's law.

Problem 2.18: Derive the charge conservation law :

$$\frac{\partial \rho_c}{\partial t} + \vec{\nabla} \cdot \vec{J} = 0, \qquad (2.120)$$

where ρ_c is the charge density.

The energy equations for each fluid can also be added to find the energy equation for a single fluid with $\Pi = \Pi_e + \Pi_i$ as the total stress tensor and similarly adding the heat fluxes and the collisional contributions.

Problem 2.19: Using the energy Equation (2.91) for electrons and ions, show that in the single fluid description, the corresponding energy equation is $\frac{d}{dt}\left(p\rho_m^{-5/3}\right) = 0$, where p and ρ_m are the total pressure and density.

2.18. Summary

In this Chapter, beginning with notions of phase space, ensemble, probability density and distribution function, we have presented three descriptions of a many body system. In the **Kinetic Description**, the motion of each particle in the phase space is studied through an N-particle distribution function f_N which remains an invariant in the phase space in the absence of collisions. The temporal change in f_N due to collisions is described by Equation (2.16). By integrating over all except the phase space coordinates of one particle, **Boltzmann Equation** for one particle distribution function is derived (Equation 2.21). This provides a description of a system in which each particle moves under the joint influence of external and internal forces. Three models of the collision term namely the **Krook Model**, the **Boltzmann Model** and the **Fokker Planck Model** have been discussed.

The Kinetic Description of a system with long range internal forces, e.g. a plasma is included in a set of equations called **Vlasov Equations** as given below for each species of particles :

VLASOV EQUATIONS

$$\frac{\partial f_s(\vec{r},\vec{V},t)}{\partial t} + \vec{V}\cdot\frac{\partial f_s(\vec{r},\vec{V},t)}{\partial \vec{r}} + \left[-\vec{\nabla}\phi_g + \frac{Q_s}{m_s}\left(\vec{E} + \frac{\vec{V}\times\vec{B}}{c}\right)\right]\cdot\frac{\partial f_s(\vec{r},\vec{V},t)}{\partial \vec{V}} = 0,$$

$$\rho_c = \sum_s Q_s \int f_s(\vec{r},\vec{V},t)d\vec{r}d\vec{V},$$

$$\vec{J} = \sum_s Q_s \int \vec{V} f_s(\vec{r}, \vec{V}, t) d\vec{V}. \qquad (2.121)$$

The fluid description of each species of particles is quite handy to investigate a certain class of problems. The Multifluid description is included in the following, mass, momentum and kinetic pressure conservation laws.

MULTIFLUID EQUATIONS

$$\frac{\partial \rho_s}{\partial t} + \nabla \cdot [\rho_s \vec{U}_s] = 0,$$

$$\rho_s \left[\frac{\partial \vec{U}_s}{\partial t} + (\vec{U}_s \cdot \vec{\nabla}) \vec{U}_s \right] = \frac{Q_s}{m_s} \rho_s \left[\vec{E} + \frac{\vec{U}_s \times \vec{B}}{c} \right] - \rho_s \vec{\nabla} \phi_g - \vec{\nabla} \cdot \Pi_s + \vec{\Gamma}^{ss'},$$

$$(2.122)$$

$$\frac{\partial}{\partial t} \Pi_{ij}^s + \sum_k \frac{\partial}{\partial x_k} (H_{kij}^s + U_k^s \Pi_{ij}^s) + \frac{\partial U_i^s}{\partial x_k} \Pi_{jk}^s + \frac{\partial U_j^s}{\partial x_k} \Pi_{ik}^s$$

$$+ (\Pi \times \vec{\Omega}_B)_{ij} + (\vec{\Pi} \times \vec{\Omega}_B)_{ji} = R_{ij}'.$$

We have indicated briefly how one proceeds to determine the pair correlation function f_c.

Finally, a single fluid description of a system containing electrons and ions of charge ze and mass m_i is derived and is represented in the following set of equations.

SINGLE FLUID EQUATIONS

$$\frac{\partial \rho_m}{\partial t} + \vec{\nabla} \cdot \left[\frac{m_e + m_i}{zm_e + m_i} \{ \rho_m z + \rho_i (1 - z) \} \vec{U} \right] = 0,$$

$$\rho_m \left[\frac{\partial \vec{U}}{\partial t} + (\vec{U} \cdot \vec{\nabla}) \vec{U} \right] = -\rho_m \vec{\nabla} \phi_g + \frac{\vec{J} \times \vec{B}}{c} - \nabla \cdot \Pi$$

$$- \left[\frac{m_e m_i (m_e + m_i)(1 - z)}{e(m_e z + m_i)^2} \left\{ (\vec{J} \cdot \vec{\nabla}) \vec{U} + (\vec{U} \cdot \vec{\nabla}) \vec{J} \right\} \right.$$

$$- en(1 - z) \vec{E} - \frac{m_e m_i \rho_m}{e^2 n^2} (\vec{J} \cdot \vec{\nabla}) \vec{J} \right]. \qquad (2.123)$$

GENERALIZED OHM'S LAW FOR $z = 1$.

$$\frac{1}{ne}\frac{\partial \vec{J}}{\partial t} + \left[(\vec{U}\cdot\vec{\nabla})\vec{J} + (\vec{J}\cdot\vec{\nabla})\vec{U}\right] + \frac{m_e^2 - m_i^2}{n^2 e^2 (m_e + m_i)^2}(\vec{J}\cdot\vec{\nabla})\vec{J}$$

$$= \frac{(m_e + m_i)e}{m_e m_i}\vec{E} + \frac{(m_e + m_i)e}{m_e m_i}\frac{\vec{U}\times\vec{B}}{c} + \frac{m_e - m_i}{n(m_e + m_i)m_e m_i}\frac{\vec{J}\times\vec{B}}{c}$$

$$-\frac{1}{n}\vec{\nabla}\cdot\left[\frac{\Pi_i}{m_i} - \frac{\Pi_e}{m_e}\right] + \frac{(m_e + m_i)}{nm_e m_i}\vec{\Gamma}_{ie}. \tag{2.124}$$

And, finally

MAXWELL'S EQUATIONS

$$\vec{\nabla}\cdot\vec{E} = 4\pi\rho_c,$$

$$\vec{\nabla}\cdot\vec{B} = 0,$$

$$\vec{\nabla}\times\vec{E} = -\frac{1}{c}\frac{\partial\vec{B}}{\partial t},$$

$$\vec{\nabla}\times\vec{B} = \frac{4\pi}{c}\vec{J} + \frac{1}{c}\frac{\partial\vec{E}}{\partial t}. \tag{2.125}$$

We believe, we have most of the mathematical tools required to study a single particle system, a kinetic many body system, a multifluid system and a single fluid in their conducting and nonconducting forms. Some of the physical quantities like stress tensors and collisional forces have been written in a formal way. Their explicit forms for particular systems will be obtained when we study these systems in some detail under various simplifying and revealing assumptions.

References

Bogolioubov, N.N., 1946, Problemi Dynamitcheskij Theorie V Statisticcheskey Phisike, OGIS, Moscow, English translation by E.K. Gora, Problems of a Dynamical Theory in Statistical Physics, ASTI A document No. AD213317

Born, M. and Green, H.S., 1946, Proc. Roy. Soc. (London), A188, 10; 1947, A189, 103; 1947, A190, 455; 1947, A191, 168.

Goldstein, H., 1950, "Classical Mechanics", Addison Wesley Publishing Company, Inc. U.S.A.

Kirkwood, J.G., 1946, J. Chem. Phys., 14, 180; 1947, 15, 72.

Nicholson, D.R., 1983, Introduction to Plasma Theory, John Wiley & Sons, New York.

Porter, D.H., 1988, Statistical Measures of Galaxy Clustering in the Minnesota Lectures on Clusters of Galaxies and Large Scale Structure, Ed: J.M. Dickey.

Prigogine, I., 1962, Non-Equilibrium Statistical Mechanics, Vol 1, Interscience Publishers.

Tanenbaum, B.S., 1967, Plasma Physics, McGraw-Hill Book Company.

Yvon, J., 1935, La Thoerie Statistique des Fluids et l'equation d'Etat, Hermann, Paris.

Yvon, J., 1937, Fluctuations en Densite, Hermann, Paris.

Appendix

1. The quantity $(\vec{U}, \Pi)$ stands for the permutational product defined as:

$$(\vec{U}, \Pi)_{xyz} = U_x \Pi_{yz} + U_y \Pi_{zx} + U_z \Pi_{xy}.$$

2. The vector product of a vector $\vec{\Omega}$ with a second rank tensor Π_{ij} is defined as:

$$\left(\vec{\Omega} \times \Pi\right)_{ij} = -\sum_{l,m} \epsilon_{lmi} \Omega_m \Pi_{lj},$$

where ϵ_{lmi} is the antisymmetric third rank tensor whose value is plus one when l, m and i are in cyclic order, minus one when l, m and i are not in cyclic order and zero when two out of the three indices are equal. The expansion of $\left(\vec{\Omega} \times \Pi\right)$ is:

$$\vec{\Omega} \times \Pi = \begin{pmatrix} \Omega_y \Pi_{zx} - \Omega_z \Pi_{yx} & \Omega_y \Pi_{zy} - \Omega_z \Pi_{yy} & \Omega_y \Pi_{zz} - \Omega_z \Pi_{yz} \\ \Omega_z \Pi_{xx} - \Omega_x \Pi_{zx} & \Omega_z \Pi_{xy} - \Omega_x \Pi_{zy} & \Omega_z \Pi_{xz} - \Omega_x \Pi_{zz} \\ \Omega_x \Pi_{yx} - \Omega_y \Pi_{xx} & \Omega_x \Pi_{yy} - \Omega_y \Pi_{xy} & \Omega_x \Pi_{yz} - \Omega_y \Pi_{xz} \end{pmatrix}.$$

3. The scalar product of a vector $\vec{\Omega}$ with a second rank tensor Π is written as:

$$\left(\vec{\Omega} \cdot \Pi\right)_i = \sum_j \Omega_j \Pi_{ji},$$

and

$$\left(\Pi \cdot \vec{\Omega}\right)_i = \sum_j \Pi_{ij} \Omega_j.$$

4. The gradient $\vec{\nabla}\vec{\Omega}$ of a vector $\vec{\Omega}$ is a tensor written as:

$$\vec{\nabla}\vec{\Omega} = \begin{pmatrix} \dfrac{\partial \Omega_x}{\partial x} & \dfrac{\partial \Omega_y}{\partial x} & \dfrac{\partial \Omega_z}{\partial x} \\ \dfrac{\partial \Omega_x}{\partial y} & \dfrac{\partial \Omega_y}{\partial y} & \dfrac{\partial \Omega_z}{\partial y} \\ \dfrac{\Omega}{R} & 0 & \dfrac{\partial \Omega_z}{\partial z} \end{pmatrix}.$$

The sum of the diagonal terms is the divergence $\vec{\nabla} \cdot \vec{\Omega}$. The third column represents the vector $\vec{\nabla}\Omega^2/2\Omega$. The third row represents the vector $\dfrac{1}{\Omega}\vec{\Omega} \cdot \vec{\nabla}\vec{\Omega}$.

Here, R is the radius of curvature of the field line of $\vec{\Omega}$, passing through the origin and defined as:

$$\frac{\partial \Omega_x}{\partial z} = \frac{\Omega}{R},$$

where $\partial \Omega_y / \partial z = 0$.

5. Scalar product of a vector $\vec{\Omega}$ and a third-order tensor Q is a second order tensor, written as:

$$\left(\vec{\Omega} \cdot Q\right)_{ij} = \sum_l \Omega_l Q_{lij}.$$

CHAPTER 3

PARTICLE AND FLUID MOTIONS IN GRAVITATIONAL AND ELECTROMAGNETIC FIELDS

3.1. Back to Single Particle Motion

In Chapter 2, we presented the kinetic and the fluid descriptions of an N-particle system. We can recover the familiar single particle equations by ignoring the many body effects like collisions and stresses and the extensive quantities like pressure and temperature. For an important class of problems, the knowledge of the motion of a single particle under the action of various forces provides us with great insight at a modest effort. Under certain circumstances, the entire fluid does what each particle does. We shall study these drifts which are common to both a single particle as well as a fluid. But there are additional drifts which originate entirely due to the fluid properties such as pressure forces. The two types of drifts in the presence of space and time dependent fields will be studied in this chapter.

3.2. Purpose of Studying Single Particle Motion

Yes, we must learn to walk before we can run! Identification of a simpler unit of a complex whole is an essential first step in any endeavour. For example, instead of the frustrating prospect of taking into account the gravitational forces of all the members of the Solar System, we can delineate the major component of the motion of any planet by assuming that it moves under the gravitational force of the sun alone. The finer details can be worked out later depending upon the purpose for which they are required.

The study of motion of a particle in the presence of electromagnetic and gravitational fields is inherently a nonlinear problem since the fields have to be evaluated at the instantaneous position of the particle, which is determined by the action of these very fields. The nonlinearity, however, can be broken by using perturbation methods if the spatial and temporal variations of the fields are slower than the spatial and time scales of the phenomenon under investigation. We can then determine the velocity and the trajectory of a particle. If the velocity is a function of charge and or mass of a particle, the relative motion between different species of particles like electrons and ions can result in a current flowing through the system. This current, if exceeds a certain threshold, can make the system unstable. Thus, the determination of particle drifts forms an important part of the study of the stability of a plasma. The concepts of adiabatic invariance are also found to be quite useful in analyzing the motion in complex field configurations and thereby in studying the transfer of energy from one degree of freedom to another (as we will see in a magnetic mirror geometry).

Further, the motion of a star in a galaxy, of an asteroid or a comet in the solar system, or of an atom in the ionizing electromagnetic field is being studied with a new awareness of the sensitive role of the initial conditions. This constitutes chaotic dynamics which has sprung many surprises in a seemingly well determined classical system.

The action of a strong electromagnetic wave on a particle can be described in terms of a nonlinear potential called the **Ponderomotive Potential**. In the fluid limit, then, all the particles feel this force, which is akin to a pressure gradient force. This technique comes quite handy for understanding some of the nonlinear plasma phenomena. In the following sections, we will study the variety of motions that a particle and a fluid undergo under the action of gravitational and electromagnetic forces, acting singly or jointly.

3.3. Equation of Motion of a Single Particle

Ignoring the terms containing many-body effects i.e. the stress tensor Π and the collisional force Γ in Equation (2.78), we find the equation of a single particle of species s:

$$\frac{d\vec{U}_s}{dt} = \frac{Q_s}{m_s}\left[\vec{E} + \frac{\vec{U}_s \times \vec{B}}{c}\right] - \vec{\nabla}\varphi_g, \tag{3.1}$$

where $\dfrac{d}{dt} \equiv \left[\dfrac{\partial}{\partial t} + \vec{U}_s \cdot \vec{\nabla}\right]$ is the total time derivative. It is convenient to use the form in Equation (3.1) for point particles. Here, the time derivative is determined in a frame moving with the particle. Fluids are studied by using either of the forms given in Equation (3.1) or Equation (2.78), called the Lagrangian and the Eulerian description, respectively. We will have more to say on this issue when we study fluids in a later chapter.

Problem 3.1: Find the acceleration, the velocity and the trajectory of an electron in a uniform and time independent electric field.

3.4. Motion of a Charged Particle in a Uniform Magnetic Field

The equation of motion of a charged particle in a uniform magnetic field can be written as (from Equation 3.1):

$$\frac{d\vec{U}_s}{dt} = \frac{Q_s}{m_s} \frac{\vec{U}_s \times \vec{B}}{c}. \tag{3.2}$$

Problem 3.2: Prove that the uniform magnetic field does no work on a charged particle i.e. $\frac{d}{dt}(\frac{1}{2}m_s U_s^2) = 0$.

We can split Equation (3.2) into two, one describing the motion in a direction parallel to the magnetic field and the other describing the motion in a direction perpendicular to the magnetic field. The motion in the parallel direction can be studied by taking the scalar product of Equation (3.2) with the magnetic field $\vec{B}$. This gives:

$$\vec{B} \cdot \frac{d\vec{U}_s}{dt} = 0. \tag{3.3}$$

Thus, the acceleration parallel to the magnetic field is zero, or the magnetic field exerts no force along its direction and therefore the particle moves with a constant velocity $\vec{U}_{z0}$ along $\vec{B}$. The motion in a plane perpendicular to $\vec{B}$ can be analyzed by writing Equation (3.2) in component form as:

$$\frac{dU_x}{dt} = \frac{Q}{mc}(U_y B_z), \tag{3.4}$$

$$\frac{dU_y}{dt} = \frac{Q}{mc}(-U_x B_z),$$

where we have taken $\vec{B} \equiv (0, 0, B_z)$ to be along the z-axis and the species index s has been deleted. Observe that parallel and perpendicular motions are completely independent of each other. The solution of Equation (3.4) is found to be:

$$U_x = U_{\perp 0} \cos \Omega_B t; \quad U_y = -U_{\perp 0} \sin \Omega_B t; \quad U_z = \text{Constant} = U_{z0},$$

$$x = x_0 + \frac{U_{\perp 0}}{\Omega_B} \sin \Omega_B t; \quad y = y_0 + \frac{U_{\perp 0}}{\Omega_B} \cos \Omega_B t; \quad z = z_0 + U_{z0} t. \tag{3.5}$$

Here, $\Omega_B = \frac{QB_z}{mc}$ is the **Cyclotron Frequency** and $(U_{\perp 0}, U_{z0}, x_0, y_0, z_0)$ are the initial values of the velocity and position of the particle. From Equation (3.5), we see that in the perpendicular plane (x, y), the particle moves along a circle of radius R_B with centre at (x_0, y_0) determined from the trajectory as:

$$(x - x_0)^2 + (y - y_0)^2 = \frac{U_{\perp 0}^2}{\Omega_B^2} \equiv R_B^2 \tag{3.6}$$

This circular motion combined with a motion with constant velocity U_{z_0} along the field $\vec{B}$ becomes a screw type motion or a helical motion with the axis of the helix along $\vec{B}$, as shown in Figure (3.1); R_B is called the **Cyclotron Radius** and is given by the ratio of the total perpendicular speed i.e. $(U_x^2 + U_y^2)^{\frac{1}{2}}$ and cyclotron frequency Ω_B.

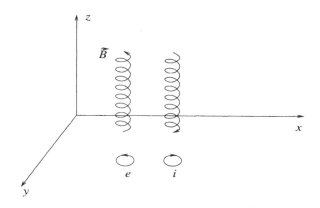

Figure 3.1. Helical Motion of Electrons and Ions in a Uniform Magnetic Field.

Problem 3.3: Show that $U_x^2 + U_y^2 =$ constant.

The helical motion of a charged particle can also be described in terms of the cylindrical coordinates (r, θ, z). We see that in a uniform magnetic field $r = R_B$ is a constant and the angular coordinate $\theta = \tan^{-1} \frac{y - y_0}{x - x_0} = \frac{\pi}{2} - \Omega_B t$. From this we conclude that the circular motion of a particle with positive charge $(Q > 0)$, e.g., for a proton, is in the clockwise direction and that for a negatively charged particle $(Q < 0)$ e.g. for an electron is in the anti-clockwise direction.

In strong magnetic fields such as might exist in hard x-ray emitting binaries like Her X-1 or in pulsars, the cyclotron radius R_B, also known as the **Larmor Radius** may become comparable to the **de Broglie Wavelength** of the particle. Under these circumstances the motion of a particle must be treated quantum mechanically. The energy of the particle gets quantized and is given by $E_l = (l + \frac{1}{2})\hbar\Omega_B$, where $\hbar = h/2\pi$ and h is the Planck constant and l is an integer ranging from 0 to infinity. Thus, a strongly magnetized particle behaves like an atom with discrete eigen-frequencies in its motion perpendicular to the magnetic field but as a free particle along the direction of the field.

Problem 3.4a: For an electron with speed c, the speed of light, find the critical value of the magnetic field for which the quantum effects become important. What is the critical field for protons?

Problem 3.4b: Determine the motion of a relativistic charged particle that obeys the equation $\dfrac{d\vec{p}}{dt} = \dfrac{Q}{c}\vec{U} \times \vec{B}$ where $\vec{p} = \gamma m \vec{U}$ and $\gamma = (1 - U^2/c^2)^{-1/2}$.

Problem 3.4c: A charged particle executing circular motion in a uniform magnetic field constitutes a current loop. Show that this current loop produces a magnetic field in a direction opposite to the original field.

3.5. Motion of a Charged Particle in Uniform Magnetic and Electric Fields

Neglecting the gravitational force in Equation (3.1), we recover the equation of motion of a charge in uniform magnetic and electric fields:

$$\frac{d\vec{U}}{dt} = \frac{Q}{m}\left[\vec{E} + \frac{\vec{U} \times \vec{B}}{c}\right].\tag{3.7}$$

We can again analyze this motion by taking the scalar and the vector product of Equation (3.7) with $\vec{B}$. Taking the scalar product with $\vec{B}$ gives:

$$\vec{B} \cdot \frac{d\vec{U}}{dt} = \frac{Q}{m}\vec{E} \cdot \vec{B}.\tag{3.8}$$

If the electric field $\vec{E}$ is perpendicular to $\vec{B}$ the particle has no acceleration along $\vec{B}$. If, on the other hand, $\vec{E}\|\vec{B}$, the particle is accelerated along $\vec{B}$ forever. Of course, in a real situation, there will be other particles which will produce a retarding force due to collisions and some other effects.

An example where a strong electric field parallel to a magnetic field is believed to exist is a neutron star or a pulsar. A fast rotating magnetized neutron star generates a large parallel electric field $E_\|$ of the order of $R\Omega B/c$ at the surface of the star. Here R is the radius of the star, Ω is its spinning rate and B is the surface magnetic field.

Problem 3.5 Compare the electric and the gravitational forces on a proton at the surface of a neutron star with typical values $R = 10^6$ cm, $\Omega \sim 1$ sec^{-1} and $B = 10^{12}$ Gauss.

This electric force is believed to be responsible for pulling charged particles, electrons and protons, from the surface of a neutron star thereby creating a plasma and giving the star a magnetosphere, which has turned out to be a challenge for the imagination of the pulsar plasma physicists!

We can find a steady state solution of Equation (3.7) by putting $\dfrac{d\vec{U}}{dt} = 0$, $\vec{U} = \vec{U}_E$ and taking a cross product with $\vec{B}$. This gives:

$$\vec{E} \times \vec{B} + \frac{1}{c}(\vec{U}_E \times \vec{B}) \times \vec{B} = 0,$$

or

$$\vec{U}_E = c\frac{\vec{E} \times \vec{B}}{B^2}. \tag{3.9}$$

Thus, in a steady state, a particle moves in a direction perpendicular to both $\vec{E}$ and $\vec{B}$. This is known as the $\mathbf{E} \times \mathbf{B}$ **Drift**. It is independent of the charge, mass and energy of the particle. The electrons and protons, treated as single particles will move together with a common velocity under the crossed fields $\vec{E}$ and $\vec{B}$. Equation (3.9) can be inverted to show that

$$\vec{E} = -\frac{\vec{U}_E \times \vec{B}}{c}, \tag{3.10}$$

for $\vec{E} \cdot \vec{B} = 0$.

This implies that the electric field vanishes in a frame of reference moving with the particles. This has an important consequence for a plasma, since in the absence of pressure and collisional forces, all the particles will move with velocity given by Equation (3.9). If a perpendicular electric field is applied to a magnetized plasma, it will always move with the $\mathbf{E} \times \mathbf{B}$ **Drift** so that the particles do not see any electric field as they move. Plasmas always adjust their motion so as to nullify the effects of any imposed electric fields. We shall come across this idiosyncracy of plasmas again.

We can find the solution of full Equation (3.7) by using the $\vec{E} \times \vec{B}$ drift, which is a constant as $\vec{E}$ and $\vec{B}$ are space and time independent. Let us write:

$$\vec{U} = \vec{U}_E + \vec{U}_B, \tag{3.11}$$

and substitute in Equation (3.7). We find

$$\frac{d\vec{U}_B}{dt} = \frac{Q}{mc}\vec{U}_B \times \vec{B}. \tag{3.12}$$

Equation (3.12) is identical to Equation (3.2) for the motion of a charged particle in a uniform magnetic field and its solution is given in Equation (3.5). Thus, the total motion of a charged particle in crossed electric and magnetic fields is composed of two components: a steady drift $\vec{U}_E$ imposed

on the helical motion. The trajectory of a particle in crossed fields is found
to be

$$x = (x_0 + U_{E_x}t) + \frac{U_{\perp 0}}{\Omega_B}\sin\Omega_B t,$$

$$y = (y_0 + U_{E_y}t) + \frac{U_{\perp 0}}{\Omega_B}\cos\Omega_B t, \tag{3.13}$$

$$z = (z_0 + U_{z_0}t).$$

This is as if the centre of the helical motion (x_0, y_0) has been replaced by
$x_0(t) = x_0 + U_{E_x}t$ and $y_0(t) = y_0 + U_{E_y}t$, i.e. it is the centre that moves with
the $\vec{E} \times \vec{B}$ velocity. This is known as the **Guiding Centre Motion**. The
constants (x_0, y_0, z_0) and $(U_{\perp 0}, U_{z_0})$ are determined from initial conditions.

Problem 3.6 By taking an average over the cyclotron time period $T_c = 2\pi/\Omega_B$, show that a particle is left only with a rectilinear motion.

This rectilinear motion is just the drift of the gyration centre (x_0, y_0).
Thus, depending upon the time scale of interest, we can either consider the
complete motion Equation (3.13) or only the drift motion that we obtain
after taking an average over the circular motion.

3.6. Motion of a Charged Particle in Uniform Magnetic and Gravitational Fields

The equation of motion is:

$$\frac{d\vec{U}}{dt} = \frac{Q}{m}\frac{\vec{U} \times \vec{B}}{c} + \vec{g}, \tag{3.14}$$

where $\vec{g}$ is the acceleration due to gravity. It is easy to see that this is
equivalent to Equation (3.7) provided we replace the electric force $Q\vec{E}$ by
the gravitational force $m\vec{g}$. With this replacement we find a gravitational
drift $\vec{U}_g$ in a magnetic field, called the $\vec{g} \times \vec{B}$ **Drift** given by:

$$\vec{U}_g = \frac{mc}{Q}\frac{\vec{g} \times \vec{B}}{B^2}. \tag{3.15}$$

Compare it with the $\vec{E} \times \vec{B}$ **Drift** (Equation 3.9), to see that $\vec{U}_g$ depends
upon the mass and charge of a particle and therefore it has different values
for electrons and protons. The relative motion of electrons and protons sets
up a current density $\vec{J}_g$ in an electron-proton plasma, (of electron density
n_e and proton density n_p) which can be determined as:

$$\vec{J}_g = -n_e e\vec{U}_{g_e} + n_p e\vec{U}_{g_p}. \tag{3.16}$$

It is important to estimate the magnitude of any current flowing in a plasma since under large currents, a plasma becomes unstable. The stability is restored by getting rid of this current, say, by an ohmic or a plasma dissipation mechanism.

The $\vec{g} \times \vec{B}$ drift will be significant in regions of high $\vec{g}$ and low $\vec{B}$. Neutron stars are known to be extremely compact objects as their central mass densities are comparable to nuclear densities. A typical neutron star has a mass of the order of (1.4) times the mass of the sun $M_\odot$, whereas the radius of the sun $R_\odot$ is $\sim (7 \times 10^4)$ times the radius of a neutron star. A neutron star has a magnetic field $B \sim 10^{12}$ Gauss, whereas the average magnetic field on the sun is about 1 Gauss. You may wonder if a black hole is not more compact than a neutron star. The radius of a black hole is identified with its event horizon, and is known as the **Schwarzschild radius** R_s (for a non-rotating black hole). It is the radius at which a particle acquires a velocity equal to the velocity of light c under the influence of the gravitational field of a black hole and can be estimated by equating the kinetic energy and the potential energy of the particle which gives $R_s = 2GM/c^2$ where M is the mass of a black hole. A black hole can have a magnetic field only if it is charged and rotating.

Problem 3.7: Estimate $\vec{g} \times \vec{B}$ drift for a proton and an electron on the surface of (1) sun and (2) a neutron star.

It is an interesting fact that in a magnetized plasma, an electric field cannot produce charge separation whereas a gravitational field can.

3.7. Motion of a Charged Particle in an Inhomogeneous Magnetic Field

Magnetic fields are always inhomogeneous. The field lines are always curved as exhibited by iron filings around a bar magnet. It is only far from the source of the magnetic field that it may be approximated to be constant, uniform and straight. Magnetic fields acquire special and often complex configurations due to the non-existence of magnetic monopoles. The simplest form of a magnetic field has to satisfy the divergence-free condition:

$$\vec{\nabla} \cdot \vec{B} = 0, \tag{3.17}$$

and in addition, the curl-free condition if there are no currents i.e..

$$\vec{\nabla} \times \vec{B} = 0. \tag{3.18}$$

This means that

$$\vec{B} = -\vec{\nabla}\varphi_B, \tag{3.19}$$

and

$$\nabla^2 \varphi_B = 0, \tag{3.20}$$

where φ_B, a scalar, is called the **Magnetic Potential**. The solution of the Laplace Equation (3.20) can be found for given boundary conditions in different coordinate systems. We know that most of the celestial bodies, whether planets or pulsars, have a magnetic field with a dominant dipole component. A dipole magnetic field obtains at large distances from a circular current loop. We give here the expressions for a dipole field in the spherical polar coordinates $(B_r, B_\theta, B_\varphi)$ as well as in Cartesian coordinates (B_x, B_y, B_z). In the spherical polar coordinates the magnetic potential at a distance $\vec{r}$ from the dipole is given by:

$$\varphi_B = \frac{\vec{m}_B \cdot \vec{r}}{r^3},$$

and

$$\begin{aligned}
B_r &= \frac{2m_B}{r^3} \cos\theta, \tag{3.21} \\
B_\theta &= -\frac{m_B}{r^3} \sin\theta, \\
B_\varphi &= 0.
\end{aligned}$$

where $\vec{m}_B$ is the magnetic dipole moment and is given by the product of the current and the area of the loop divided by the speed of light c. In the Cartesian coordinates:

$$B_x = \frac{3xzm_B}{r^5}; \quad B_y = \frac{3yzm_B}{r^5}; \quad B_z = (3z^2 - r^2)\frac{m_B}{r^5}; \tag{3.22}$$

where $r = (x^2 + y^2 + z^2)^{1/2}$.

We see that even in this rather basic form of a magnetic field, determining the trajectory of a charged particle is a nontrivial task. The components of the magnetic field given by Equations (3.21) and (3.22) possess gradients along as well as transverse to them. The magnitude of $\vec{B}$ has all possible gradients; and the vector field $\vec{B}$ has curvature. All these field characteristics add complexities to the simple circular motion of a charged particle in a uniform magnetic field. One way of handling this situation is to use perturbation methods. These methods can be applied if we can split the total magnetic field into its uniform and nonuniform parts and then treat the nonuniform part as small compared to the uniform part. Then, the trajectory of a charged particle obtained in the uniform part of the magnetic field (Equation 3.5) is substituted into the nonuniform part and the new trajectory of the particle is determined. By averaging over the circular motion, we get the additional motion in the form of drifts of the centre of gyration as in the cases with electric and gravitational fields discussed earlier. The underlying assumption is that the magnetic field does not vary very much over a distance of the order of Larmor radius R_B and in a time of the order of cyclotron or Larmor period $T_B = 2\pi/\Omega_B$. Such a field can be Taylor expanded about the origin $(x_0 = 0, y_0 = 0)$ and, here, we write one simple form with which all the effects due to inhomogeneities in the field can be studied. We write:

$$
\begin{aligned}
B_x &= B_0 + z\left(\frac{\partial B_x}{\partial z}\right)_0, \\
B_y &= 0, \\
B_z &= B_0 + x\left(\frac{\partial B_z}{\partial x}\right)_0.
\end{aligned}
\tag{3.23}
$$

This field satisfies the divergence free condition and the curl free condition if $(\partial B_x/\partial z)_0 = (\partial B_z/\partial x)_0$.

The equations of motion of a charged particle in the field given by Equation (3.23) are:

$$
\begin{aligned}
\frac{dU_x}{dt} - \Omega_B U_y &= -\frac{U_{\perp 0}^2}{B_0}\left(\frac{\partial B_z}{\partial x}\right)_0 \sin^2 \Omega_B t, \\
\frac{dU_y}{dt} + \Omega_B U_x &= \left[\Omega_B U_{z0}^2 - \frac{U_{\perp 0}^2}{2}\sin 2\Omega_B t\right]\frac{1}{B_0}\left(\frac{\partial B_z}{\partial x}\right)_0, \\
\frac{dU_z}{dt} &= \Omega_B U_{\perp 0} U_{z0} t \sin \Omega_B t \frac{1}{B_0}\left(\frac{\partial B_z}{\partial x}\right)_0.
\end{aligned}
\tag{3.24}
$$

We observe that in the absence of magnetic field gradients, the right hand sides in Equation (3.24) vanish and we recover Equation (3.4) for the homogeneous case. The system of Equations (3.24) represent a simple inhomogeneous harmonic oscillator and can be easily solved by adding a particular solution to the general homogeneous solution (Tanenbaum 1967). We find:

$$
\frac{d^2 U_x}{dt^2} + \Omega_B^2 U_x = \frac{1}{B_0}\left(\frac{\partial B_z}{\partial x}\right)_0 \Omega_B^2 \left[U_{z0}^2 t - \frac{3}{2}\frac{U_{\perp 0}^2}{\Omega_B}\sin 2\Omega_B t\right],
\tag{3.25}
$$

$$
\begin{aligned}
U_x = &\left[U_{\perp 0} - \frac{1}{B_0}\left(\frac{\partial B_z}{\partial x}\right)_0 \frac{U_{z0}^2}{\Omega_B}\right]\cos \Omega_B t + \frac{1}{B_0}\left(\frac{\partial B_z}{\partial x}\right)_0 \\
&\times \left[\frac{U_{\perp 0}^2}{2\Omega_B}\sin 2\Omega_B t + U_{z0}^2 t\right],
\end{aligned}
$$

$$U_y = -\left[U_{\perp 0} - \frac{1}{B_0}\left(\frac{\partial B_z}{\partial x}\right)_0 \frac{U_{z0}^2}{\Omega_B}\right]\sin\Omega_B t + \frac{1}{B_0}\left(\frac{\partial B_z}{\partial x}\right)_0$$

$$\times \left[\frac{U_{\perp 0}^2}{\Omega_B}\cos 2\Omega_B t + \frac{U_{z0}^2}{\Omega_B} + \frac{U_{\perp 0}^2}{\Omega_B}\sin^2\Omega_B t\right], \qquad (3.26)$$

and

$$U_z = U_{z0} + \frac{1}{B_0}\left(\frac{\partial B_z}{\partial x}\right)_0\left(\frac{U_{\perp 0}U_{z0}}{\Omega_B}\right)[\sin\Omega_B t - \Omega_B t\cos\Omega_B t].$$

We see that there is an additional oscillatory motion at twice the cyclotron frequency. An integration of Equation (3.26) gives the position of the particle. One must take care to preserve the initial conditions at $t = 0$, for which the motion in the uniform field was obtained. In order to find drift speeds, we average over the cyclotron motion and find:

$$\overline{U}_x = \frac{1}{B_0}\left(\frac{\partial B_z}{\partial x}\right)_0 U_{z0}^2 t,$$

$$\overline{U}_y = \frac{1}{B_0}\left(\frac{\partial B_z}{\partial x}\right)_0\left[U_{z0}^2 + \frac{U_{\perp 0}^2}{2}\right]\frac{1}{\Omega_B},$$

$$\overline{U}_z = U_{z0}. \qquad (3.27)$$

For the field given by Equation (3.23) it can be shown that the field lines, near the origin, approximately describe a circle with centre at $x = [(1/B_0)(\partial B_z/\partial x)_0]^{-1}$, $z = 0$ and the radius of the circle is $[1/B_0 (\partial B_z/\partial x)_0]^{-1}$ Therefore, we can define a radius of curvature $\vec{R}_c$ of the field as:

$$\vec{R}_c = -\left[\frac{1}{B_0}\left(\frac{\partial B_z}{\partial x}\right)_0\right]^{-1}\hat{x}, \qquad (3.28)$$

where $\hat{x}$ is the unit vector in the x-direction. We can now express the drift speed U_y as:

$$\vec{U}_{GC} = \frac{1}{\Omega_B}\left(\frac{U_{\perp 0}^2}{2} + U_{z0}^2\right)\frac{\vec{R}_c \times \vec{B}}{R_c^2 B}, \qquad (3.29)$$

where only linear terms in field gradients have been retained. In Equation (3.29) the term $U_{\perp 0}^2$ is known as the **Gradient Drift** and the term U_{z0}^2 is called the **Curvature Drift** and the two always occur together. The gradient drift is proportional to the square of the velocity component perpendicular to the ambient field, whereas the curvature drift is proportional to the square of the velocity component parallel to the ambient magnetic field. As shown in Figure (3.2), at the origin, the magnetic field is in the z direction. Away from the origin, it curves and has an x component, which crossed with U_{z0}, produces a Lorentz force in the y direction and causes the curvature drift, which takes the particle out of the plane of the Figure. Another way of looking at the situation is that when a particle moving in a straight line enters a curved path, it experiences a centrifugal force. This force crossed with the magnetic field produces the curvature drift. Particles can cross the field due to this drift motion and populate the planetary radiation belts. Further, we see that these drifts are different for electrons and protons and thus set up an electric current in a system of electrons and protons.

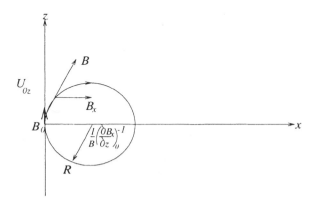

Figure 3.2. Origin of Curvature Drift in a Magnetic Field of Equation (3.23) with Near Circular Field Lines, Near the Origin.

Problem 3.8: Derive an expression for the curvature drift by determining the force on a charged particle moving along a curved field line.

Problem 3.9: Show that the field lines of the dipole field given by Equation (3.21) satisfy a relationship of the type $r \propto \sin^2 \theta$. Determine the gradient and curvature drifts in this field.

Problem 3.10: Discuss the motion of an electron in a magnetic field given by $B_r = 0, B_\theta = B_0 J_1(\alpha r)$ and $B_z = B_0 J_0(\alpha r)$ in cylindrical coordinates (r, θ, z); the J's are the Bessel Functions.

3.8. Motion of a Charged Particle in a $\vec{\nabla} B \| \vec{B}$ Field

Such a field has a gradient in its own direction i.e.. if B_z is the major component of the magnetic field then $\partial B_z / \partial z \neq 0$. In order to satisfy the divergence free condition, we must have (using cylindrical coordinates):

$$\frac{1}{r} \frac{\partial}{\partial r}(r B_r) + \frac{\partial B_z}{\partial z} = 0, \qquad (3.30)$$

or

$$B_r = -\frac{r}{2}\left(\frac{\partial B_z}{\partial z}\right)_0,$$

where $\partial B_z/\partial z$ has been assumed to vary very little with r and its value at the origin($r = 0$) has been taken. It means that the expression for B_z is valid only for small values of r. We can, as before, find, first, the solution of the equation of motion of a charged particle in only the uniform B_z component and then substitute this solution in the additional force arising due to nonzero B_r. We find, after averaging over a Larmor period that

$$U_z = U_{z0} - \frac{V_{r0}^2}{2}\frac{1}{B_z}\left(\frac{\partial B_z}{\partial z}\right)_0 t, \tag{3.31}$$

where V_{r0} is the radial velocity at $t = 0$. Equation (3.31) shows that the particle slows down as it enters a region of increasing field i.e.. a field with $(\partial B_z/\partial z)_0' > 0$, and will come to a standstill at some later time. At this instant, the particle reverses its direction and is now moving in a region of decreasing field strength and undergoes a continuous enhancement in its speed. We say that the particle has been reflected by high field region. However, the increase in the velocity is limited by the conservation of total energy (Problem 3.2). Thus, at any instant,

$$U_{\perp 0}^2 + U_{z0}^2 = U_{\perp 0}^2(t) + U_z^2(t), \tag{3.32}$$

where $U_{\perp 0}$ is the component of the velocity perpendicular to B_z. We notice that at the moment of reflection, the entire kinetic energy is in the perpendicular motion. And the maximum of $U_z(t)$ occurs when $U_{\perp 0}(t) = 0$ in the region of the weakest field.

A configuration of the magnetic field with two maxima and a minimum in between is called a **Magnetic Mirror**, since a charged particle suffers reflections in such a configuration, it remains trapped in it until other effects take over. Planetary and stellar fields with dominant dipole magnetic fields possess magnetic mirror configurations since the field is strong at the poles and weak at the equator.

Problem 3.11: Show that the force on a particle along $\vec{B}$ is given by $\vec{F}_{\parallel} = -\mu\vec{\nabla}_{\parallel}B$, where μ is the magnetic moment of the particle.

3.9. Van Allen Planetary Radiation Belts

We are bombarded by cosmic rays all the time from all directions. Satellite studies revealed that our earth is surrounded by belts of high energy charged particles, called **Van Allen Radiation Belts** after J.A. Van Allen who designed the magnetometers flown over the first Explorer satellites which

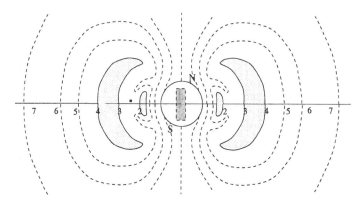

Figure 3.3. The Van Allen Radiation Belts in the Earth's Magnetic Field at ~ 1.5 and ~ 3 Earth Radii.

discovered these radiation belts. The Geiger counters on satellites registered sharp, 10,000 fold, increases in the intensity of cosmic rays at distances of 1.5 and 3 earth radii from the centre of the earth (Figure 3.3). These are known as the inner and the outer belts, respectively. The inner belt is populated by β decay products of the neutrons which are produced through the interaction of cosmic rays with terrestrial oxygen and nitrogen nuclei. The outer belt consists of mainly solar wind particles entering the earth's atmosphere during geomagnetic disturbances. Isotropic intensities of $\sim 2 \times 10^4 \mathrm{cm}^2/\mathrm{sec}$ of protons of energy $\sim 30 \mathrm{MeV}$ exist in the intense regions of the inner belt. Moon, Venus and Mars have either very weak or no dipole magnetic field to trap radiation. However, Mercury has been seen to wear a belt around it. Jupiter, on the other hand with ten times the earth's magnetic field, has radiation belts with protons and electrons of ten's of MeV's (Artsimovich and Lukyanov 1980). Neptune and Uranus are also expected to support radiation belts, but probably close to the polar planes, since their magnetic dipoles lie at large angles to the rotation axis unlike the case of the earth. In the following sections, we shall study more about the conditions of trapping of particles in magnetic field configurations by using a method based on the invariance properties of motion of a charged particle in an inhomogeneous magnetic field.

3.10. Adiabatic Invariants of Motion of a Charged Particle in Slowly Varying Magnetic Fields

We have seen that inhomogeneities in a magnetic field can make the equations of motion difficult to solve unless we resort to perturbation methods, which can be deployed under slow variations of the field. It was proved a long time ago, that certain quantities called the **Action Integrals** (Cowan

1984), which are invariants of a system under homogeneous and time independent fields, remain invariant to the first order in the parameters of the slowly varying fields. This is known as the **Principle of Adiabatic Invariance of Action Integrals**. Many of the characteristics of a system under slowly varying fields can be discussed with the use of adiabatic invariants, without having to solve the equations of motion. The principle of adiabatic invariance has found applications in fields like plasma physics, accelerator physics and galactic astronomy.

The action integral J is defined as

$$J = \int pdq,\tag{3.33}$$

where p is the canonical momentum and q is the corresponding canonical conjugate coordinate. We may recall, here, the definition of the canonical momentum:

$$p_i = \frac{\partial L}{\partial \dot{q}_i},\tag{3.34}$$

where L is the Lagrangian of the system.

Problem 3.12: The Lagrangian of a charged particle in an electromagnetic field is given by $L = \frac{1}{2}m\dot{\vec{r}}^2 - Q\phi(\vec{r},t) + \sum_j \frac{Q}{c}\dot{r}_j \cdot A_j(\vec{r},t)$, where ϕ and $\vec{A}$ are the scalar and the vector potentials respectively. Show that the canonical momentum is $\vec{p} = m\dot{\vec{r}} + (Q/c)\vec{A}$. Can you show that $L = \frac{m}{2}(\dot{\vec{r}})^2 + \vec{\mu} \cdot \vec{B}$, where $\vec{\mu} = \frac{Q}{2mc}(\vec{r} \times m\dot{\vec{r}})$ is the magnetic moment of the particle?

Let us write the Lagrangian of a particle in a magnetic field in cylindrical coordinates (r, θ, z) as:

$$L = \frac{m}{2}(\dot{r}^2 + r^2\dot{\theta}^2 + \dot{z}^2) + \frac{QBr^2\dot{\theta}}{2c},\tag{3.35}$$

where $\vec{B} = (0, 0, B)$ has been assumed. Observe that L is independent of θ and therefore θ is the cyclic coordinate. The corresponding canonical momentum

$$p_\theta = \frac{\partial L}{\partial \dot{\theta}} = mr^2\dot{\theta} + \frac{QB}{2c}r^2,\tag{3.36}$$

is a constant of the motion.

Problem 3.13: Derive the equations of motion from the Lagrangian (Equation 3.35) and show that their solution is given by $r = constant \equiv R_B$ and $\dot{\theta} = -\frac{QB}{mc} = -\Omega_B$.

Using the results obtained above, we find

$$J = -\frac{Q\pi R_B^2 B}{c} = -\frac{2\pi mc}{Q}\frac{mU_{\perp0}^2}{2B} = -\frac{2\pi mc}{Q}\mu = \text{constant}, \qquad (3.37)$$

according to the principle of adiabatic invariance. Thus, the **Magnetic Moment** $\mu = (mU_{\perp0}^2/2B)$ or the **Magnetic Flux** (BR_B^2) are the **Adiabatic Invariants** of motion of a particle in a slowly varying field.

Problem 3.14 By including the electric field associated with a time varying magnetic field in the Lorentz force, prove the invariance of the magnetic moment μ.

3.11. Magnetic Mirror Revisited

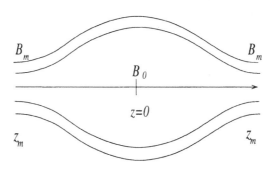

Figure 3.4.   A Magnetic Mirror

We have seen (Equation 3.32) that at any instant, the total energy of a particle moving in a magnetic field remains constant. This is also true at any instantaneous position of the particle. Thus,

$$U_\perp^2(z) + U_z^2(z) = U_{\perp0}^2 + U_{z0}^2 \equiv U^2. \qquad (3.38)$$

Since we now know that the magnetic moment μ is an adiabatic invariant, we have

$$\frac{U_\perp^2(z)}{B_z(z)} = \frac{U_{\perp0}^2}{B_0}, \qquad (3.39)$$

where $U_{\perp0}$ and B_0 are the values at $z = 0$. From Equations (3.38) and (3.39), we get

$$U_z^2(z) = U^2 - \frac{B_z(z)}{B_0}U_{\perp0}^2. \qquad (3.40)$$

We observe that as $B_z(z)$ increases along the z direction, $U_z(z)$ decreases, which implies that the perpendicular component $U_\perp(z)$ increases; the radius of gyration $R_B^2 \propto \left(U_\perp^2(z)/B_z^2(z)\right)$ decreases and the particle spirals in a continuously narrowing helical path. It also means that there is a transfer of energy from one degree of freedom to another – from parallel to perpendicular motion and vice-versa in a region of decreasing $B_z(z)$. Thus, in a magnetic mirror (Figure 3.4) a particle suffers reflection at the two ends where $B_z(z)$ is a maximum and can remain trapped. The condition for trapping is $U_z(z) \leq 0$. Now $U_z(z) = 0$ at $B_z(z) = B_M(z_M)$ where $B_M(z_M)$ is the maximum value of B_z. At this point, we find

$$U^2 = \frac{B_M(z_M)}{B_0}U_{\perp 0}^2,$$

and the condition for trapping, therefore, becomes:

$$\left(\frac{U_{z0}}{U_{\perp 0}}\right)^2 \leq \frac{B_M(z_M)}{B_0} - 1 \qquad (3.41)$$

or

$$\sin^2\theta \geq \frac{1}{R_M} \equiv \sin^2\theta_M, \qquad (3.42)$$

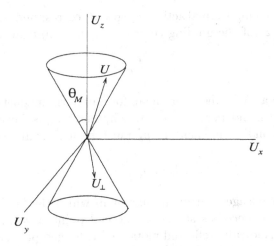

Figure 3.5.   The Loss Cone Velocity Distribution of Particles.

where θ is the angle between the velocity $\vec{U}$ and the magnetic field $\vec{B}$ at $z = 0$ and $R_M = (B_M/B_0)$ is known as the **Mirror Ratio**. It is clear from Equation (3.42) that particles with **Pitch Angles** $\theta > \theta_M$ remain trapped in the magnetic mirror executing oscillations between the two regions of the highest magnetic fields. Whereas particles with pitch angles $\theta < \theta_M$ can leave the mirror. Thus, a magnetic mirror gives rise to a velocity distribution of particles which has no particles with $\theta < \theta_M$. Such a velocity distribution is known as the **Loss Cone Distribution** and is shown in Figure (3.5). The trapped and the untrapped particles are separated by a boundary at $\sin^2 \theta_M = (1/R_M)$. The trapping condition has no dependence upon mass and the charge of a particle. But the presence of collisions among particles can introduce charge and mass dependent differences (Chen 1974).

Problem 3.15: Show that the invariance of the magnetic moment μ can be expressed as $\dfrac{\sin^2 \theta}{B} = \text{constant}$.

Problem 3.16: Taking the typical energy $\sim 1\text{MeV}$ of the particles trapped in the earth's radiation belt, show that the adiabaticity or the slow variation condition is satisfied for both protons and electrons, where $\left| \dfrac{\vec{\nabla} B}{B} \right| \sim$ inverse of the radius of the earth R_e. Estimate the '$\vec{\nabla} B$' drift of the particles and show that these particles take a few hours to traverse the length $2\pi R \sim 2\pi R_e$ of the belt.

3.12. The Longitudinal Adiabatic Invariant

We can define the longitudinal action integral J_z corresponding to the canonical coordinate z (of the guiding centre) and the conjugate momentum p_z such that

$$J_z = \int p_z dz. \tag{3.43}$$

Now, it is clear from the Lagrangian for a uniform magnetic field (Equation 3.35) that it is independent of z, which, therefore, is a cyclic coordinate and the corresponding momentum p_z must be a constant of the motion. Thus

$$J_z = p_z \int dz. \tag{3.44}$$

Now, if we envisage a magnetic mirror in which the field varies slowly in the middle and becomes strong at the end points $z = \pm L$, the spatial period of the periodically reflected motion of a charged particle is $2L$. Thus, according to the principle of adiabatic invariance

$$J_z = p_z \cdot 2L = \text{constant}. \tag{3.45}$$

The invariance of J_z can be used to determine the change in the energy of a particle trapped in a magnetic mirror whose mirror **Separation L Varies with time**. The longitudinal energy W_z is given by

$$W_z^2 = \frac{p_z^2}{2m} = \frac{J_z^2}{8mL^2}, \tag{3.46}$$

from which it follows that W_z increases as L decreases, i.e.. a particle gets accelerated by suffering reflections by approaching magnetic mirrors. This process is known as the **Fermi Acceleration Mechanism** and was first proposed for acceleration of cosmic rays which may suffer reflections by moving magnetized galactic gas clouds. We must appreciate that this is a true acceleration mechanism in contrast to a static magnetic mirror, where there is only a mutual transfer of energy between motions parallel and perpendicular to the magnetic field. A moving magnetized cloud has an associated electric field which accelerates charged particles. The perpendicular energy remains constant due to the invariance of $(mU_{\perp 0}^2/B)$, if we assume B is uniform in the region between the mirrors.

The invariance of J_z can be proved by noting that in analogy with a mechanical reflection at a wall, the gain in a particle momentum $\delta p_{\parallel}$ at each reflection in a magnetic mirror is $2mU_m$ where U_m is the speed of the mirrors situated at $z = \pm L$. The reflections occur at an interval of time $\delta t = \dfrac{2L}{U_m}$. Therefore the average rate of change of momentum is

$$\frac{dp_{\parallel}}{dt} = \frac{\delta p_{\parallel}}{\delta t} = \frac{2mU_m}{(2L/U_m)}. \tag{3.47}$$

But U_m is nothing but the rate of change of the mirror separation, i.e..

$$2U_m = -\frac{d}{dt}(2L), \tag{3.48}$$

since both the mirrors are approaching each other with speed U_m. Combining Equations (3.47) and (3.48), we get:

$$\frac{dp_{\parallel}}{p_{\parallel}} = -\frac{dL}{L} \text{ or } p_{\parallel}L = \text{constant}, \tag{3.49}$$

and hence J_z is an invariant.

Problem 3.17: A simple representation of a magnetic mirror field is $\dfrac{B_z}{B_0} = 1 + (R_M - 1)\left(\dfrac{z^2}{L^2} + \dfrac{z}{L} - \dfrac{z^3}{L^3}\right)$ and the B_y component must be determined to ensure the divergence free condition. Prove invariance of J_z for motion in this field. Determine the period of the longitudinal oscillations. Does it remind you of a particle thrown up with some initial speed?

The invariance of J_z requires that the field must vary little over a period of the longitudinal oscillations.

3.13. Charged Particle in a Nonuniform Electric Field

We know that in a uniform and time independent electric field, a charged particle has an accelerated motion with its trajectory given by

$$\vec{r} = \vec{r}_0 + \vec{U}t + \frac{1}{2}\frac{Q\vec{E}}{m}t^2. \tag{3.50}$$

We can also determine the trajectory of a charged particle, if additionally, a retarding force $(-m\nu\vec{U})$ acts on it.

Problem 3.18 Solve $m(d\vec{U}/dt) = Q\vec{E} - m\nu\vec{U}$, where ν is a constant.

One of the most often encountered forms of a nonuniform electric field is due to the Coulomb force. Now, the Coulomb force is a central force, essentially identical to the gravitational force. Motion of a particle in a central force is well studied and is known as the **Kepler Problem**. A central force is derivable from a potential $\varphi(r)$. The angular momentum of a particle is conserved in it. The trajectory, is given by

$$t(r) = \sqrt{\frac{m}{2}} \int \frac{dr}{[W - \varphi(r) - L_m^2/2mr^2]^{1/2}} \tag{3.51}$$

along with the angular momentum $L_m = mr^2\dot{\theta}_r = $ constant. W is the total energy. The shape of the orbit can be determined from a relation between the radial coordinate r and the angular coordinate θ_r given by:

$$\frac{1}{r} = \frac{1 + e_r\cos\theta_r}{(L_m^2/mk)}, \tag{3.52}$$

where k is the constant appearing in the potential energy $\varphi(r) = -k/r$ corresponding to a central force and

$$e_r = \left[1 + \frac{2WL_m^2}{mk^2}\right]^{1/2},$$

is the eccentricity of the orbit.

It is straight forward to show that equations (3.51) and (3.52) also describe the motion of a two-particle system under the action of their mutual central force like gravitational or Coulomb force. This facilitates the determination of the scattering angle of a particle in a central force produced by another particle. We need this information to find scattering crosssections. The equations of motion for particles with masses m_1 and m_2 and positions

$\vec{r}_1$ and $\vec{r}_2$ in their mutual central force $\vec{F}$ which is a function only of the distance between the particles $R = |\vec{R}| = |\vec{r}_1 - \vec{r}_2|$ are:

$$m_1 \ddot{\vec{r}}_1 = F(R)\frac{\vec{R}}{R} \quad \text{and} \quad m_2 \ddot{\vec{r}}_2 = -F(R)\frac{\vec{R}}{R}, \qquad (3.53)$$

since the force on the two particles must be equal in magnitude and opposite in direction. Equations (3.53) can be rewritten as:

$$m_R \ddot{\vec{R}} = F(R)\frac{\vec{R}}{R}, \qquad (3.54)$$

where

$$m_R = \frac{m_1 m_2}{m_1 + m_2}, \qquad (3.55)$$

is the reduced mass. Equation (3.54) describes the motion of a particle of mass m_R whose position and velocity are the relative position $\vec{R}$ and the relative velocity $\dot{\vec{r}}_1 - \dot{\vec{r}}_2 = \dot{\vec{R}}$. Thus, Equations (3.51) and (3.52) apply equally well to this two-particle system under a force $\vec{F} = -\vec{\nabla}\varphi$ with m replaced by m_R.

In a two-body elastic collision the centre of mass velocity as well as the magnitude of the relative velocity remain constant; only the direction of the velocity vector $\vec{V}_1 - \vec{V}_2$ rotates by an angle θ called the scattering angle. We will now determine the scattering angle as a function of the interparticle force, the impact parameter b and the magnitude of the relative velocity V. We refer to Figure (3.6) where the origin O is at $R = 0$; ABA' is the trajectory of the particle of mass m_1 in the central force excited by the particle of mass m_2, at rest at the origin; b is the impact parameter and would be equal to the minimum distance of approach R_m if there were no force; θ is the scattering angle; θ_R and θ_m are the angular coordinates corresponding to the radial coordinates R and R_m respectively and $\vec{V}_1' - \vec{V}_2'$ is the relative velocity vector after the collision.

The trajectory ABA' is symmetric about the point B, therefore OB makes equal angles with the initial asymptote OZ and the final asymptote OM and this angle is θ_m. The scattering angle θ is then given by:

$$\theta = \pi - 2\theta_m. \qquad (3.56)$$

The total energy $W = \frac{1}{2}m_R V^2$ and the angular momentum $L_m = bm_R V$, where $V = |\vec{V}_1 - \vec{V}_2| = |\vec{V}_1' - \vec{V}_2'|$. The minimum distance of approach R_m and the corresponding angular coordinate θ_m can be determined from the condition:

$$\frac{dR}{d\theta_R} = \frac{\dfrac{dR}{dt}}{\dfrac{d\theta_R}{dt}} = 0 \text{ at } R = R_m \text{ and } \theta_R = \theta_m.$$

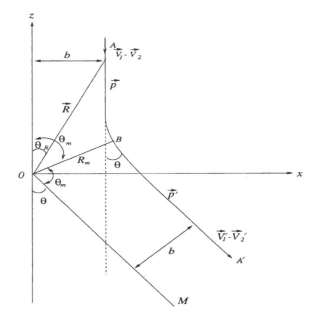

Figure 3.6.   Geometry of a Two-Body Elastic Collision in a Central Force.

Now using Equation (3.51) and the constancy of the angular momentum, we find:

$$\frac{dR}{d\theta_R} = \frac{R^2}{b}\left[1 - \frac{b^2}{R^2} + \frac{2k}{Rm_R V^2}\right]^{1/2},\tag{3.57}$$

and the minimum distance of approach is:

$$R_m = -\frac{k}{m_R V^2}\left[1 \pm \left(1 + \frac{b^2 m_R^2 V^4}{k^2}\right)^{1/2}\right].\tag{3.58}$$

Integration of Equation (3.57) leads to the trajectory in polar coordinates (Equation 3.52). In order to determine θ_m, we integrate Equation (3.57) to get:

$$\theta_R - \theta_m = \pm \int_{Rm}^{R} \frac{b\,dR}{R^2\left(1 - \frac{b^2}{R^2} + \frac{2k}{Rm_R V^2}\right)^{1/2}}.\tag{3.59}$$

We note that initially when $R \to \infty$, $\theta_R \to 0$, therefore

$$\theta_m = \int_{R}^{\infty} \frac{b\,dR}{R^2\left(1 - \frac{b^2}{R^2} + \frac{2k}{Rm_R V^2}\right)^{1/2}},\tag{3.60}$$

where in Equation (3.59) the minus sign is used for $\theta_R < \theta_m$ and the plus sign is used for $\theta_R > \theta_m$. The scattering angle θ is then found to be:

$$\sin\frac{\theta}{2} = \left(1 + \frac{b^2 m_R^2 V^4}{k^2}\right)^{-1/2} \equiv \frac{1}{e_r}. \tag{3.61}$$

Problem 3.19: (i) Show that $\tan\dfrac{\theta}{2} = \dfrac{q_1 q_2 m}{bp^2}$ for scattering of a particle with electric charge q_1, mass m and momentum p by a particle of electric charge q_2 and mass $M \gg m$.
(ii) Show that the change Δp in momentum of the particle is $\Delta p = 2p\sin^2\theta/2$.

Problem 3.20: Assuming the mass-energy equivalence, show that a photon is deflected by an angle $\theta \simeq \dfrac{GM_\odot}{R_\odot c^2}$ at the surface of the sun. Estimate θ; how does it compare with the observed deflection. Such a measurement was first made during the total solar eclipse of 1919, by a team led by the Astronomer Royal Arthur Eddington.

By summing appropriately over a large number of scatterings suffered by a test particle, say an electron, in the presence of scattering centres, say protons, we can determine scattering crosssections, which are needed, for example, to find the electrical resistivity of a plasma (Tanenbaum 1967).

3.14. Charged Particle in a Spatially Periodic Electric Field and Uniform Magnetic Field

Another situation of interest could arise when a charged particle is acted upon by a spatially periodic electric field along with a uniform magnetic field. The spatially periodic electric field could be the field of an electrostatic wave with frequency much smaller than the cyclotron frequency of the particle. Let us represent this field as:

$$\vec{E} = E_0 \cos\frac{2\pi x}{\lambda}\hat{y},$$

where λ is the spatial period. The equation of motion of a charged particle can be written as

$$m\frac{d\vec{U}}{dt} = Q\left[E_0 \cos\frac{2\pi x}{\lambda}\hat{y} + \frac{\vec{U}\times\vec{B}}{c}\right]. \tag{3.62}$$

This equation can be solved by the perturbation method described earlier (Section 3.7).

Problem 3.21: Show that the drift speed of a particle is now given by

$$\vec{U}_E = \frac{c\vec{E} \times \vec{B}}{B^2}\left[1 - \frac{\pi^2 R_B^2}{\lambda^2}\right] \tag{3.63}$$

where $\vec{E}$ is evaluated for the unperturbed pure cyclotron motion.

We see that the nonuniform field has introduced a charge and mass dependence in the drift speed U_E. A proton suffers a much larger reduction in its drift speed than an electron. This is due to the fact that a proton, having a much larger Larmor radius, spends more time in a region of weak E field, as compared to an electron. This modification in the drift speed is called the **Finite Larmor Radius Effect**, and must be included for $\lambda \sim R_B$. We note that finite Larmor radius effects can cause a current in a plasma which, as we shall see later, can make a plasma unstable.

3.15. Motion in Time Varying $\vec{E}$ and $\vec{B}$ Fields

The most common manifestation of time varying $\vec{E}$ and $\vec{B}$ fields is in the electromagnetic radiation in which we bathe every moment of our lives. Electromagnetic radiation exerts a force on charged particles. Before taking up this, a slightly more complicated problem, we can study motion in a uniform and constant $B\hat{z}$ and a time varying $\vec{E}$ field. This time, we represent $\vec{E}$ as

$$\vec{E} = E_0 \cos \omega t \hat{x}, \tag{3.64}$$

and ignore the spatial variation. This is known as the dipole approximation, i.e., we are interested in a phenomenon whose spatial scale is much larger than the wavelength of the electric field. The equation of motion of a charged particle can now be written as:

$$m\frac{d\vec{U}}{dt} = Q\left[E_0 \cos \omega t \hat{x} + \frac{\vec{U} \times \vec{B}}{c}\right]. \tag{3.65}$$

It is easy to see that the solution of this equation, in the limit $\omega^2 \ll \Omega_B^2$, is given by

$$\begin{aligned}
U_x &= U_{\perp 0} \cos \Omega_B t - U_p, \\
U_y &= -U_{\perp 0} \sin \Omega_B t - U_E, \\
U_z &= \text{constant},
\end{aligned} \tag{3.66}$$

where $\vec{U}_p = \dfrac{c}{\Omega_B}\dfrac{1}{B}\dfrac{d\vec{E}}{dt}$ and $\vec{U}_E = c\dfrac{\vec{E} \times \vec{B}}{B^2}$.

So, we see that we have found a new drift velocity $\vec{U}_p$, called the **Polarization Drift**. During the drift motion, the guiding centre changes its direction from along the $\vec{E}$ field to opposite to it, every half cycle. This drift depends on the mass and the charge of a particle and is caused by a time varying electric field; the $\vec{E} \times \vec{B}$ drift $\vec{U}_E$ now also becomes a function of time. The polarization drift originates from a charge separation produced by a time varying electric field, since the response time of the electrons and protons to neutralize the charge separation cannot match the electric field variation in time. In a constant field, electrons and protons due to their motion have all the time to neutralize the charge imbalance created by the field. Thus, polarization drift can also make a plasma unstable by feeding the drift instability arising due to the flow of current.

3.16. Motion in a Time Varying $\vec{B}$ Field

Associated with a time varying $\vec{B}$ field is an $\vec{E}$ field given by Faraday's law of Induction:

$$\vec{\nabla} \times \vec{E} = -\frac{1}{c}\frac{d\vec{B}}{dt}. \tag{3.67}$$

A charged particle in such a $\vec{B}$ field gets energized due to the associated $\vec{E}$ field.

Problem 3.22: Show that the magnetic moment μ of a charged particle remains an invariant in a $\vec{B}$ field varying slowly with time.

We have already seen that the magnetic moment and magnetic flux across a Larmor orbit are the adiabatic invariants. These properties are used in the Betatron acceleration mechanism of charged particles. If B increases with time so does the perpendicular energy $\frac{1}{2}mU_{\perp 0}^2$.

One of the popular models of pulsars is the rotating magnetic dipole model. The rotation produces electric fields of the order of 10^8 volts/cm near the surface of a pulsar. These strong electric fields can pull charged particles from the surface of a pulsar against its gravitational pull which is a billion times weaker than the electric pull (Problem 3.5).

As mentioned earlier, moving magnetized clouds are believed to accelerate cosmic rays again by the induced electric fields, the Fermi acceleration process, a favourite of Astrophysicists!

3.17. Motion in an Electromagnetic Wave - The Ponderomotive Force

You must have appreciated by now that the motion of a charged particle in the presence of the Lorentz force is governed by a nonlinear equation of motion, since the electric and magnetic fields are to be determined at the position of the particle which moves under the action of these fields. We shall again use the perturbation methods to deal with this nonlinearity. We have written the equation of motion of a charged particle under the Lorentz force several times; let us write it once more:

$$m\frac{d\vec{U}}{dt} = Q\left[\vec{E}(\vec{r},t) + \frac{\vec{U} \times \vec{B}(\vec{r},t)}{c}\right], \tag{3.68}$$

where now $\vec{E}(\vec{r},t)$ and $\vec{B}(\vec{r},t)$ are the electric and magnetic fields of an electromagnetic wave of frequency ω. We write

$$\vec{E}(\vec{r},t) = \vec{E}(\vec{r})\cos\omega t,$$

and

$$\vec{B}(\vec{r},t) = -\frac{c}{\omega}\vec{\nabla} \times \vec{E}(\vec{r})\sin\omega t. \tag{3.69}$$

In order to use the perturbation procedure the position vector $\vec{r}$ and the velocity $\vec{U}$ of the particle as well as the electric field $\vec{E}(\vec{r},t)$ are written as:

$$\begin{aligned}
\vec{r} &= \vec{r}_0 + \vec{r}_1 + \vec{r}_2 + \ldots\ldots \\
\vec{U} &= \vec{U}_0 + \vec{U}_1 + \vec{U}_2 + \ldots\ldots \\
\vec{E}(\vec{r},t) &= \vec{E}(\vec{r}_0)\cos\omega t + (\vec{r}_1 \cdot \vec{\nabla})\vec{E}(\vec{r})|_{\vec{r}=\vec{r}_0}\cos\omega t + \ldots\ldots,
\end{aligned} \tag{3.70}$$

where each term is one order higher in smallness than the previous one. To the first order, Equation (3.68) becomes

$$m\frac{d\vec{U}_1}{dt} = Q\vec{E}(\vec{r}_0)\cos\omega t, \tag{3.71}$$

where $\vec{E}(\vec{r}_0)$ is the electric field at the initial position $\vec{r}_0$ of the particle. We find:

$$\vec{U}_1 = \frac{Q\vec{E}(\vec{r}_0)}{m\omega}\sin\omega t,$$

and

$$\vec{r}_1 = -\frac{Q\vec{E}(\vec{r}_0)}{m\omega^2}\cos\omega t. \tag{3.72}$$

To the second order, Equation (3.68) becomes

$$m\frac{d\vec{U}_2}{dt} = Q\left[(\vec{r}_1 \cdot \vec{\nabla})\vec{E}(\vec{r})|_{\vec{r}=\vec{r}_0}\cos\omega t + \frac{\vec{U}_1 \times \vec{B}(\vec{r}_0,t)}{c}\right]. \tag{3.73}$$

We notice that there are two sources of nonlinearity: one, due to the electric field being now determined at the displaced position of the particle and the other due to the magnetic field of the electromagnetic wave. Substituting for $\vec{r}_1, \vec{U}_1$ and $\vec{B}(\vec{r}_0)$, we find

$$m\frac{d\vec{U}_2}{dt} = -\frac{Q^2}{m\omega^2}\left[\vec{E}(\vec{r}_0)\cdot\vec{\nabla}\right]\vec{E}(\vec{r}_0)\cos^2\omega t +$$
$$+\frac{Q^2}{m\omega^2}\left[\vec{E}(\vec{r}_0)\times(\vec{\nabla}\times\vec{E}(\vec{r}_0))\right]\sin^2\omega t. \qquad (3.74)$$

Let us take an average over the period $T = \dfrac{2\pi}{\omega}$, so that

$$\left\langle m\frac{d\vec{U}_2}{dt}\right\rangle = -\frac{Q^2}{4m\omega^2}\vec{\nabla}E^2(\vec{r}_0) = f_{NL} \qquad (3.75)$$

The nonlinear force f_{NL} on a charged particle due to an electromagnetic wave is known as the **Ponderomotive Force**. We can also define a **Ponderomotive Potential** ψ by

$$f_{NL} = -\vec{\nabla}\psi,$$

or

$$\psi = \frac{Q^2}{4m\omega^2}E^2(\vec{r}_0). \qquad (3.76)$$

The force f_{NL} on a charged particle due to electromagnetic radiation arises due to the gradient of energy density E^2. Such a situation arises, for example, in the presence of several different waves. For a constant E_0, this force vanishes. The nature of the ponderomotive force is quite different from the force exerted by radiation due to its absorption in a macroscopic medium. For example, a medium of electrons of density n has an absorbing coefficient $n\sigma_T$ and the force exerted by radiation on the entire fluid of electrons is $n\sigma_T P_R$ where $\sigma_T = (8\pi/3)\left(e^2/m_ec^2\right)^2$ is the **Thomson Scattering Crosssection** of electrons and P_R is the radiation energy density. In the absorption process, each electron, on the average provides an effective absorbing area σ_T.

The ponderomotive force comes very handy while studying the nonlinear interaction of an intense electromagnetic wave with charged particles in a plasma. In the presence of strong electromagnetic fields, a plasma acquires completely novel properties some of which can be attributed to the action of the ponderomotive force.

3.18. Chaotic Motion

So far, we have studied single particle motion using classical mechanics. These types of motion are known as deterministic in the sense that initial

conditions completely determine the system and a small change in the initial
conditions appears as a small change in the resulting motion. We have dealt
with nonlinear motion by using perturbation methods. There are, however
situations, nonlinear, where a system exhibits what is known as **Chaotic
Behaviour**. A simple definition of a chaotic system is that it supports a
large change in its motion for minute changes in its initial conditions. This
feature is known as the 'sensitivity to initial conditions' also picturesquelly
called the **Butterfly Effect**, whereby a flutter of a butterfly in America
causes rains in the plains of India!

Well, there are more manageable cases where a chaotic behaviour can be
seen. We will give here two examples: (1) an hydrogen atom in a microwave
field and (2) stochastic motion of a star in the gravitational field of other
stars in a galaxy.

3.19. Microwave Ionization of Hydrogen

The equation of motion of an electron of charge $(-e)$ and mass m_e in the
electric field of an infinite mass proton of charge e and microwave radiation
field is given by

$$m_e \frac{d\vec{U}}{dt} = -\frac{e^2}{r^2}\hat{r} + \hat{z}F_{max}(t)\cos \omega t, \qquad (3.77)$$

where $(\vec{r}, \vec{p} \equiv m_e\vec{U})$ are the position and momentum of the electron and
$F_{max}(t)\cos \omega t$ is the force due to the microwave field in the $\hat{z}$ direction; $\hat{r}$
and $\hat{z}$ are the unit vectors. In the absence of the microwave field, the electron
describes an orbit around the proton with angular frequency ω_{at} given by

$$\omega_{at} = \frac{E_{at}^{3/2}}{e^2 \sqrt{m_e}}, \qquad (3.78)$$

where $E_{at} = (e^2/a_{at})$ with $a_{at} = \left(n^2\hbar^2/c^2 m_e\right)$, as the semi-major axis of
the orbit and n the principal quantum number.

The force F_{at} exerted by the proton on the electron is

$$F_{at} = m_e a_{at} \omega_{at}^2. \qquad (3.79)$$

Equation (3.77) can be rewritten in the following dimensionless form

$$\ddot{\vec{R}} = -R^{-2}\hat{R} + \hat{z}\frac{F_{max}}{F_{at}}\cos\left(\frac{\omega\tau}{\omega_{at}}\right), \qquad (3.80)$$

with $\vec{R} = \vec{r}/a_{at}$ and $\tau = \omega_{at}t$. Thus, the classical motion of the hydrogen atom in the microwave field depends only on the ratios (ω/ω_{at}) and

(F_{max}/F_{at}). In order to simulate the adiabatic switching on and off of the microwave field, Leopold and Percival (1979) chose;

$$F_{max} \equiv F_{max} A(t),$$

where

$$A(t) = \quad \exp[\lambda(t - t_i)] \quad \text{for} \ 0 \leq t \leq t_i$$
$$= \quad\quad 1 \quad\quad\quad \text{for} \ t_i \leq t \leq T. \quad\quad (3.81)$$

This choice simulates the laboratory experiment conditions and represents a broad-band microwave source as well.

Leopold and Percival (1979) have solved Equations (3.75) numerically using Monte-Carlo trajectory method. It is found that the results can be described in terms of three dimensionless ratios: $(F_{max}/F_{at}), (\omega/\omega_{at})$, and (λ/ω_{at}). The value of (λ/ω_{at}) should be chosen to ensure the adiabatic switching on and off of the microwave field. The ratio (F_{max}/F_{at}) measures the relative importance of microwave force to the atomic force. The ratio (ω/ω_{at}) describes the resonant and nonresonant behaviour. In order to study ionization, Leopold and Percival computed a quantity called the **Compensated Energy** E_c defined as:

$$E_c = E_p + \frac{1}{2m_e} \left[p_x^2 + p_y^2 + \left(p_z - \frac{F_{max}}{\omega} \sin \omega t \right)^2 \right]. \quad\quad (3.82)$$

Problem 3.23: Show that $\left(p_z - \dfrac{F_{max}}{\omega} \sin \omega t \right)$ is a constant of the motion.

The compensated energy E_c remains nearly constant in the presence of an oscillatory microwave field when $E_p \to 0$. Thus, the constancy and the positivity of E_c can be associated with an ionization event when E_p, which is the binding energy, vanishes. We quote results from Jones et. al. (1980)- "For slowly varying fields, Banks and Leopold (1978 a,b,c) made a nonperturbative analysis using classical adiabatic theory and found a minimum value for (F_{max}/F_{at}) below which no ionization should occur. In the dynamic case, however, Leopold and Percival (1979) found considerable ionization even when (F_{max}/F_{at}) was well below this minimum value. The ionization tended to zero in both the high and low frequency limits (provided (F_{max}/F_{at}) was less than the minimum value for adiabatic ionization found by Banks and Leopold (1978a) and had a broad based bell-shaped maximum occurring around $(\omega/\omega_{at}) \simeq 0.5$. The occurrence of maximum ionization below the classical resonance frequency may seem strange at first but can be understood as follows; as soon as we begin to perturb the atom with the field, the system can either gain, or lose energy. If the ionization is to occur, then, on the average, the system must gain more energy than it loses. Suppose the frequency of the electron is initially higher than the frequency of the field. Then as the atom gains energy, its frequency will be reduced, thus bringing it into resonance with the field. Alternatively, if the frequency of the electron is initially lower than the field then as the atom gains energy the electron will move further away from resonance and the ionization will be reduced".

Bayfield and Koch (1974) reported ionization of an hydrogen atom from a state with the principal quantum number $n = 66$ under the action of a microwave field of frequency 9.9 GHz and strength above a critical value of 20 volts/cm. We can check that the transition $n = 66 \rightarrow 67$ has a resonance frequency $=22$GHz, so that $\omega/\omega_{at} \simeq 0.4$. The ionization energy $[13.6eV/(66)^2] \simeq 70$ times the microwave photon energy $(\hbar\omega)$. This means that the ionization process occurs through the absorption of a large number of photons. The multiphoton absorption process depends mainly on the fluence, i.e., on the total energy in the microwave field and not on its detailed time variation. This fluence dependence is attributed to the presence of chaos. The chaotic behaviour of the absorbing nonlinear system makes accessible a large volume of phase space, in a diffusive manner. This is known as **Stochastic Excitation**

The microwave ionization may be operating in some astrophysical situations. Our universe is rich in strong radio sources. We can estimate the physical quantities important for determining micro-wave ionization and excitation of hydrogen in the vicinity of strong radio sources. The theoretical estimates of 50% to 80% ionization for $\gamma_t \equiv [\omega/\omega_{at}][F_{max}/F_{at}]^{-1} \simeq 6$ or 7 have been confirmed by experiments. We find that strong radio sources like 3C273 and 3C345 with luminosity $\sim 10^{42} - 10^{44}$ erg/sec at $\omega \sim 1 - 10$ GHz have $\omega/\omega_{at} \simeq 0.4$ for $n \sim 90$ and $\gamma_t \simeq 6.6$ and can thus ionize hydrogen atom from a state with $n \simeq 90$ upto a distance of a couple of parsecs from the radio source. This process may take place in radio recombination line emission and absorption sites. Shaver (1978) has emphasized the role of stimulated emission of radio recombination lines involving highly excited states of hydrogen atoms in galaxies and quasars.

The stimulated emission can take place only if the radio continuum is strong enough, in which case we cannot discount the possibility of radio excitation leading to ionization (Krishan 1992).

3.20. Motion of a Star in a Galaxy

We have seen that the motion of a particle in a central force can be determined analytically (Equation 3.54). Such a motion has five constants of motion: the total energy, the three components of angular momentum and the direction of the perihelion. This is a completely determined system in the 6 dimensional phase space of position vector and velocity. There is no chaos in this system.

Henon and Heiles (1964) demonstrated the existence of chaos in a low dimensional system and attributed it to the nonexpressibility of at least one constant of motion in an analytical form. They studied the motion of a star in the presence of the gravitational potential due to all the other stars in a galaxy. In cylindrical coordinates (r, θ, z), the equations of motion of the test star are:

$$
\begin{aligned}
\ddot{r} - r\dot{\theta}^2 &= -\frac{\partial \varphi_g}{\partial r}, \\
r\ddot{\theta} + 2\dot{r}\dot{\theta} &= -\frac{1}{r}\frac{\partial \varphi_g}{\partial \theta}, \\
\ddot{z} &= -\frac{\partial \varphi_g}{\partial z},
\end{aligned}
\tag{3.83}
$$

where $\varphi_g(r, \theta, z)$ is the gravitational potential. For an axisymmetric potential, φ_g is independent of θ and the two constants of motion are the total energy

$$
W = \frac{1}{2}m(\dot{r}^2 + r^2\dot{\theta}^2 + \dot{z}^2) + m\varphi_g(r, z),
\tag{3.84}
$$

and the angular momentum

$$
L_z = mr^2\dot{\theta},
\tag{3.85}
$$

which can be used to recast the equations of motion (Equation 3.83) as:

$$
\begin{aligned}
\ddot{r} &= -\frac{\partial \Psi_g(r, z)}{\partial r}, \\
\ddot{z} &= -\frac{\partial \Psi_g(r, z)}{\partial z},
\end{aligned}
\tag{3.86}
$$

where $\Psi_g(r, z) = \varphi_g(r, z) + \dfrac{L_z^2}{2m^2 r^2}$.

Thus, we now have a two dimensional system in four dimensional phase space of $(r, \dot{r}, z, \dot{z})$. Equations (3.86) describe the motion of a star in the plane (r, z) in a potential $\Psi_g(r, z)$ which has no particular symmetry now. Therefore we are left with one constant of motion W. In a four dimensional phase space, there must be three constants of motion. What are the other two? Treating r and z as Cartesian coordinates, we can write the total energy as

$$
W = \frac{1}{2}m(\dot{r}^2 + \dot{z}^2) + m\Psi_g(r, z).
\tag{3.87}
$$

We can use Equation (3.87) to eliminate one of the four phase space coordinates $(r, \dot{r}, z, \dot{z})$, say $\dot{r}$, for a given value of W. The trajectory of the star can now be plotted in three-dimensional space $(r, z, \dot{z})$. The phase space volume is bounded by the condition

$$\dot{r}^2 \geq 0 \text{ or } \Psi_g(r, z) + \frac{1}{2}\dot{z}^2 \leq W/m. \tag{3.88}$$

The trajectory of the star will fill this volume in the absence of any other constraint.

Henon and Heiles chose a fairly general form for the potential Ψ_g, numerically integrated Equations (3.86) and plotted the trajectory in the plane $(z, \dot{z})$ for $r = 0$ for given values of the total energy W. It was found that for low values of W, trajectories for different initial values remain on closed curves in the $(z, \dot{z})$ plane (Figure 3.7). As the value of W was increased, some closed curves remained but the trajectories began to wander in the phase space. Then for W above a critical value, no closed curves were seen, and the trajectories became truly chaotic filling the whole phase space $(z, \dot{z})$ (Figure 3.8). Thus, it would seem that for low energies there must be one more constant of motion due to which the trajectories remain on closed curves and that at high energies, this constant cannot be identified and this causes chaos. A consequence of chaotic motion is that a high energy star may wander in an undeterministic way and may even escape the galaxy. And the energy at which this happens may be less than that corresponding to the escape velocity of the star!

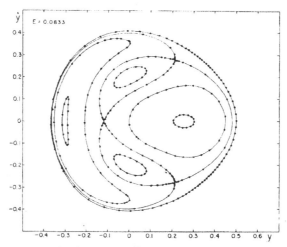

Figure 3.7. Trajectories of a Star in the Gravitational Potential ψ_g for a Given Value of $W = W_0$. Each Set of Points Linked by a Curve Corresponds to One Computed Trajectory ($Y, \dot{Y}$ and E in the Figure Stand for z, $\dot{z}$ and W in the Text Respectively).

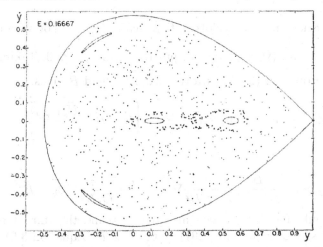

Figure 3.8.   Chaotic Trajectories of a Star in the Gravitational Potential ψ_g for W Larger than a Critical Value (Y, $\dot{Y}$ and E in the Figure Stand for z, $\dot{z}$ and W in the Text Respectively).

We have seen two examples of a chaotic motion with entirely unexpected consequences. The field is rich with possibilities.

3.21. Fluid Drifts

We have determined drifts of single particles under different circumstances of spatially and temporally varying electric and magnetic fields. In all cases, it was the guiding center of a particle that moved with the drift speed. When we treat an entire system of N particles, it acquires new characteristics like pressure, density and transport parameters like thermal and electrical conductivity, as we have seen while deriving fluid equations in Chapter two. There are, thus drifts which are specific to the fluid character of a system and do not exist for single particles (Nicholson 1983).

In the presence of a pressure gradient, a neutral fluid flows from a high pressure region to a low pressure region. However, a new flow is generated when a pressure gradient force acts upon a charged fluid in the presence of a magnetic field. To see this, let us write the equation of motion of charged particles of species s (for example an electron or an ion fluid) (Equation 2.78):

$$\rho_s \left[\frac{\partial \vec{U}_s}{\partial t} + (\vec{U}_s \cdot \vec{\nabla}) \vec{U}_s \right] = -\vec{\nabla} p_s - \rho_s \vec{\nabla} \varphi_g + \frac{\rho_s Q_s}{m_s} \left[\vec{E} + \frac{\vec{U}_s \times \vec{B}}{c} \right], \quad (3.89)$$

where only the diagonal part p_s of the stress tensor Π_s has been retained. There are two time scales in Equation (3.89), the inertial time scale over which $\vec{U}_s$ varies and the cyclotron time scale $(\Omega_B)^{-1}$. For variations of $\vec{U}_s$ much slower than $(\Omega_B)^{-1}$, we can put the term $\dfrac{\partial \vec{U}_s}{\partial t} \simeq 0$. Taking the cross product of Equation (3.89) with the magnetic field $\vec{B}$ gives:

$$\rho_s[(\vec{U}_s \cdot \vec{\nabla})\vec{U}_s] \times \vec{B} = -\vec{\nabla}p_s \times \vec{B} + \frac{\rho_s Q_s}{m_s}\left[\vec{E} \times \vec{B} - \frac{\vec{U}_{s\perp}}{c}B^2\right] - \rho_s \vec{\nabla}\varphi_g \times \vec{B}, \tag{3.90}$$

or

$$\vec{U}_{s\perp} \simeq c\frac{\vec{E} \times \vec{B}}{B^2} - \frac{c}{n_s Q_s B^2}\vec{\nabla}p_s \times \vec{B} - \frac{m_s c}{Q_s B^2}\vec{\nabla}\varphi_g \times \vec{B}. \tag{3.91}$$

You should be able to recognize the first and the third terms on the right–hand side of Equation (3.91) as the $\vec{E} \times \vec{B}$ (Equation 3.9) and $\vec{g} \times \vec{B}$ (Equation 3.15) drifts found earlier for a single particle. The middle term is the **New Fluid Drift - The $\vec{\nabla}p \times \vec{B}$ Drift** that has arisen due to the pressure gradient force in a fluid.

Problem 3.22: Can you show that the left–hand side of Equation (3.90) vanishes for the solution given in equation (3.91)?

The $\vec{\nabla}p_s \times \vec{B}$ drift depends on the charge Q_s and the number density n_s of particles of species s and is known as the **Diamagnetic Drift**. This is akin to the $\vec{\nabla}B \times \vec{B}$ drift since, as we will show later, magnetic field has an associated pressure $(B^2/8\pi)$ with it. The diamagnetic flow is depicted in Figure (3.9) for $\vec{\nabla}n$ in the y direction and $\vec{B}$ in the z direction. There are more electron orbits going up towards x in region A compared to the number of electron orbits coming down in region A'. Why? So, there is a net electron flux $n_e\vec{U}_{e\perp}$ in the x direction. For the same reason, there is a net flux $n_i\vec{U}_{i\perp}$ in the $(-x)$ direction. We must appreciate the fact that particle drifts are the drifts of their guiding centres and fluid drifts are the mass motion of the entire fluid.

We can also determine if there is any drift of a fluid in a direction parallel to the magnetic field. We write the z component of Equation (3.89)

$$\rho_s\left[\frac{\partial U_{sz}}{\partial t} + (\vec{U}_s \cdot \vec{\nabla})U_{sz}\right] = -\frac{\partial p_s}{\partial z} - \rho_s\frac{\partial \varphi_g}{\partial z} + \frac{\rho_s Q_s}{m_s}E_z. \tag{3.92}$$

If we ignore the convective term, we see immediately that the fluid feels an acceleration along the direction z of the magnetic field. Further, neglecting the inertial terms altogether gives us a relation between the particle density, the gravitational potential φ_g and the electric potential φ($E_z = -\partial\varphi/\partial z$):

$$n_s(z) = n_{0s} \exp\left[-\frac{Q_s}{K_B T}\varphi - \frac{m_s}{K_B T}\varphi_g\right], \tag{3.93}$$

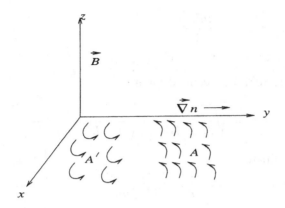

Figure 3.9.   Diamagnetic Flow

where we have expressed $p_s = n_s K_B T$, the isothermal equation of state. Equation (3.93) is the **Boltzmann Relation** for species s of particles and describes the redistribution of particles resulting due to a balance of pressure gradients, gravitational and electric fields. This distribution has important consequences for plasma phenomena as we will find out in the following chapters.

3.22. Some Important Formulae

1. Motion in a uniform magnetic field $\vec{B} = B\hat{z}$

$$x = x_0 + \frac{U_{\perp 0}}{\Omega_B} \sin \Omega_B t, \quad y = y_0 + \frac{U_{\perp 0}}{\Omega_B} \cos \Omega_B t, \quad z = z_0 + U_{z_0} t,$$

$$U_x = U_{\perp 0} \cos \Omega_B t, \quad U_y = -U_{\perp 0} \sin \Omega_B t, \quad U_z = U_{z_0}, \quad R_B = \frac{U_{\perp 0}}{\Omega_B}.$$

2. $\vec{E} \times \vec{B}$ Drift: $\vec{U}_E = c \dfrac{\vec{E} \times \vec{B}}{B^2}.$

3. $\vec{g} \times \vec{B}$ Drift: $\vec{U}_g = \dfrac{mc}{Q} \dfrac{\vec{g} \times \vec{B}}{B^2}.$

4. Gradient and Curvature Drifts:

$$\vec{U}_{GC} = \frac{1}{\Omega_B} \left(\frac{U_{\perp 0}^2}{2} + U_{z_0}^2 \right) \frac{\vec{R}_c \times \vec{B}}{R_c^2 B}.$$

5. Lagrangian of a charged particle in a magnetic field in cylindrical coordinates:

$$L = \frac{m}{2} \left(\dot{r}^2 + r^2 \dot{\theta}^2 + \dot{z}^2 \right) + \frac{QBr^2}{2c} \dot{\theta}.$$

6. Magnetic moment $\mu = \dfrac{mU_{\perp 0}^2}{2B}$.

7. Trapping condition: $\sin^2 \theta \geq \dfrac{1}{R_M}$.

8. Particle trajectory in a central force

$$t(r) = \sqrt{\frac{m}{2}} \int \frac{dr}{(W - \phi(r) - L^2/2mr^2)^{1/2}},$$

in polar coordinates

$$\frac{1}{r} = \frac{1 + e_r \cos \theta_r}{(L^2/mk)}.$$

9. Scattering angle in a central force

$$\sin \theta/2 = \frac{1}{e_r}.$$

10. The eccentricity

$$e_r = \left[1 + \frac{2W L^2}{mk^2}\right]^{1/2}.$$

11. Photon deflection in the gravitational field of the sun

$$\theta \simeq \frac{GM_\odot}{R_\odot c^2}.$$

12. Effect of finite Larmor radius on $\vec{E} \times \vec{B}$ drift:

$$\vec{U}_E = \frac{c\vec{E} \times \vec{B}}{B^2}\left[1 - \frac{\pi^2}{\lambda^2} R_B^2\right].$$

13. Polarization drift: $\vec{U}_p = \dfrac{c}{\Omega_B}\dfrac{1}{B}\dfrac{d\vec{E}}{dt}.$

14. Ponderomotive force: $f_{NL} = -\dfrac{Q^2}{4m\omega^2}\vec{\nabla}(E^2(r_0)).$

15. Thomson scattering crosssection $\sigma_T = \dfrac{8\pi}{3}\left(\dfrac{e^2}{m_e c^2}\right)^2.$

16. Compensated energy:

$$E_c = E_p + \frac{1}{2m_e}\left[p_x^2 + p_y^2 + (p_z - \frac{F_{max}}{\omega}\sin \omega t)^2\right].$$

17. $\vec{\nabla}p \times B$ drift: $\vec{U}_\perp = -\dfrac{c}{nQB^2}\vec{\nabla}p \times \vec{B}.$

18. The Boltzmann relation

$$n(r) = n_0 \exp\left[-\frac{Q_s}{K_B T}\varphi - \frac{m_s}{K_B T}\varphi_g\right].$$

References

Artsimovich, L.A. and Lukyanov, S.Yu. 1980. "Motion of Charged Particles in Electric and Magnetic Fields", Mir Publishers, Moscow

Banks, D. and Leopold, J.G. 1978a., J. Phys. B: Atom. Molec. Phys.,**11**, 37

Banks, D. and Leopold, J.G. 1978b., J. Phys. B: Atom. Molec. Phys.,**11**, L5

Banks, D. and Leopold, J.G. 1978c., J. Phys. B: Atom. Molec. Phys.,**11**,2833

Bayfield, J.E. and Koch, P.M. 1974. Phys. Rev. Lett. **33**, 258

Chen, F.F., 1974., Introduction to Plasma Physics, Plenum Press, New York and London

Cowan, B.P. 1984. Classical Mechanics, Routledge & Kegan Paul, London

Henon, M. and Heiles, C. 1964. Astron. Journal **69**, 73

Jones, D.A., Leopold, J.G. and Percival I.C. 1980., J. Phys. B: Atom. Molec. Phys., **13**, 31

Krishan, V. 1992. Microwave Excitation and Ionization of Hydrogen Around Strong Radio Sources, Preprint

Leopold, J.G. and Percival, I.C. 1979. J. Phys. B: Atom. Molec. Phys., **12**, No.5, 709

Nicholson, D.R. 1983. Introduction to Plasma Theory, John Wiley & Sons, New York

Shaver, P.A. 1978. Astron. Astrophys. **68**, 97

Tanenbaum, B.S. 1967. Plasma Physics, McGraw-Hill Book Company, New York

CHAPTER 4

MAGNETOHYDRODYNAMICS OF CONDUCTING FLUIDS

4.1. Electrically Conducting Fluids

Ionized gases or plasmas and liquid metals such as mercury or liquid sodium are electrically conducting fluids. The outer core of the earth is believed to be molten iron. Magnetospheres of planets and stars, tails of comets, extragalactic jets, accretion disks and many other astrophysical objects are studied by treating them as electrically conducting fluids. The study of magnetohydrodynamics (MHD) draws from two well known branches of physics, electrodynamics and hydrodynamics, along with a provision to include their coupling. The basic laws of electrodynamics described in the form of Maxwell's Equations supplemented by the generalized Ohm's law are sufficient for the purpose. The hydrodynamics of a fluid is expressed in the form of conservation laws of mass, momentum and energy. These laws treat the fluid as a continuum. The continuum description is valid if the mean free path of the constituent particles is much shorter than the spatial scales on which the flow is visualized. Thus, according to this criterion, any substance can be treated as a continuum at some spatial scale. The magnetohydrodynamic phenomena are a consequence of the mutual interaction of the fluid flow and the magnetic field. As is well known, a conductor crossing magnetic field lines gives rise to an induced electric field, which drives an electric current in the conducting fluid. The resulting Lorentz force accelerates the fluid across the magnetic field, which in turn creates another induced electric field and currents which modify the initial magnetic field. Thus, the bulk motion of a conducting fluid and a magnetic field influence each other and must be determined self-consistently.

The interaction of fluid flows and electromagnetic and gravitational fields determines the configurational characteristics like loops, jets, tails and filaments observed on almost all scales in the universe. The configuration of the radiation emitting plasma causes variability in radiation over a host of spatial and temporal scales. The stability or otherwise of the configuration determines the lifetime of the radiating material in a particular mode. The magnetohydrodynamic instabilities help the configuration to relax with an attendant release of kinetic, electromagnetic and gravitational energy.

One of the major results of magnetohydrodynamics is the ability of conducting fluids to amplify magnetic fields, the amplification of magnetic fields being a universal necessity. This aspect of MHD reminds us that more often than not, fluids are turbulent. Turbulent fluids only permit a statistical description. We will, defer the discussion of this topic until the chapter on nonconducting fluids.

4.2. Validity of Magnetohydrodynamics

There is a well defined region of applicability of MHD. Generally MHD addresses the macroscopic, bulk or large spatial scale and large time scale processes occurring in a conducting fluid. More specifically, for example, in an electron-proton fluid, effects associated with fast variations of electric and magnetic fields are neglected. One of these effects is the **Space Charge Effect**. Therefore:

(1) Space Charge Effects are Neglected in MHD. In an electron-proton fluid, electrons and protons are accelerated by the applied electric and magnetic fields and decelerated by the Coulomb collisions between them. Ohm's law $\vec{J} = \sigma \vec{E}$ is a consequence of the balance of these accelerating and retarding forces. In MHD, the mean free path of the particles is very short or the collision frequency is very high. On the other hand, the frequency of the applied fields is low. Under these circumstances, there cannot result any significant charge separation since the large number of collisions can neutralize any charge separation produced by the applied fields. Does it not remind you of section (3.92) where, we showed that a static electric field cannot produce polarization? So, here, applied fields of low frequency may produce a small polarization or net charge density, which is neglected in MHD. The Poisson Equation is therefore never used in MHD to determine electric fields as there is no net charge density. The electric fields are entirely produced due to time varying magnetic fields or charge distributions external to the fluid.

(2) Again, due to the slow time variation of electric and magnetic fields, the **Displacement Current** term in the modified Ampere's law is **Neglected**.

(3) The collision frequency being the highest frequency in MHD phenomena, Maxwellization of velocities of particles is ensured. Please recall that we derived the single-fluid and the two-fluid equations by taking the moments of the Boltzmann equation. In this process, several times, we had to invoke the velocity dependence of the distribution function which would reduce the surface integrals to zero. And we found that the Maxwellian distribution function of velocities fitted the bill very well. Collisions thermalize electrons and protons to a common temperature. Thus, a **Fluid is Characterized by a Single Temperature**.

4.3. Equations of Magnetohydrodynamics

The motion of a conducting fluid in a magnetic field is described by: the usual hydrodynamic variables, mass density, velocity and pressure; Ampere's law without the displacement current; Faraday's induction law; and the generalized Ohm's law. We have derived the single fluid equations in Chapter 3. We use them here to study MHD phenomena. The MHD equations for a fluid consisting of electrons and protons are:

$$\frac{\partial \rho_m}{\partial t} + \vec{\nabla} \cdot [\rho_m \vec{U}] = 0, \tag{4.1}$$

$$\rho_m \left[\frac{\partial \vec{U}}{\partial t} + (\vec{U} \cdot \vec{\nabla}) \vec{U} \right] = -\vec{\nabla} \cdot \vec{\Pi} + \frac{\vec{J} \times \vec{B}}{c} - \rho_m \vec{\nabla} \varphi_g, \tag{4.2}$$

$$\vec{E} + \frac{\vec{U} \times \vec{B}}{c} = \eta \vec{J} + \frac{1}{en} \left[\frac{\vec{J} \times \vec{B}}{c} - \vec{\nabla} \cdot \vec{\Pi}_e \right], \tag{4.3}$$

$$\vec{\nabla} \cdot \vec{J} = 0, \tag{4.4}$$

$$\vec{\nabla} \cdot \vec{B} = 0, \tag{4.5}$$

$$\vec{\nabla} \times \vec{B} = \frac{4\pi}{c} \vec{J}, \tag{4.6}$$

$$\vec{\nabla} \times \vec{E} = -\frac{1}{c} \frac{\partial \vec{B}}{\partial t}, \tag{4.7}$$

and the energy conservation equation, which we shall write after determining a suitable representation for the stress tensor Π. The MHD equations have been written under the assumptions of (i) vanishing electron mass and (ii) neglect of convective terms in the generalized Ohm's law.

The stress tensor Π_{ij}^s for species s of particles was defined in Equation(2.74) and can be split into diagonal and off-diagonal parts as:

$$\Pi_{ij}^s = p_s \delta_{ij} - \Pi_{ijs}', \tag{4.8}$$

such that

$$\text{Trace } \Pi_{ij}^s = 3p_s,$$

and

$$\text{Trace } \Pi_{ijs}' = 0. \tag{4.9}$$

The tensor Π_{ijs}' is the shear stress tensor. Relative displacements of adjacent regions in a fluid are identified with shear. Stress causes strain and shear. A strain is an elongation or compression in the direction of the applied force. Viscous dissipation comes into play in the presence of shear. Experiments tell us that, for small velocities, the viscous force on a moving fluid parcel is proportional to the velocity gradient. The most general way of expressing Π'_{ijs} is therefore

$$\Pi'_{ijs} = \sum_{k,l} C^s_{ijkl} \left(\frac{\partial U_k}{\partial x_l}\right)_s. \tag{4.10}$$

Since Π_{ijs} is symmetric, so must be Π'_{ijs}. The coefficients C_{ijkl} can be expressed using all combinations of Kronecker deltas as

$$C^s_{ijkl} = a^s_1 \delta_{ij}\delta_{kl} + a^s_2\delta_{ik}\delta_{jl} + a^s_3\delta_{il}\delta_{jk}, \tag{4.11}$$

so that,

$$\Pi'_{ijs} = \sum_k a^s_1 \delta_{ij} \left(\frac{\partial U_k}{\partial x_k}\right)_s + a^s_2\left[\left(\frac{\partial U_i}{\partial x_j}\right)_s + \left(\frac{\partial U_j}{\partial x_i}\right)_s\right], \tag{4.12}$$

where due to the symmetry in the i and j indices, we can take $a^s_2 = a^s_3$. The constants a^s_1 and a^s_2 are identified with viscosity coefficients $\mu^s \equiv a^s_2$ and $\xi^s \equiv \frac{2}{3}a^s_2 + a^s_1$. The coefficients μ^s describe the viscous tangential force per unit area per unit velocity gradient and are associated with changes in the geometrical shape without any change in the volume. The coefficient ξ^s is associated with volume changes (compression or expansion) without a change of shape, and is referred to as the second viscosity coefficient.

The divergence of Equation (4.12) gives

$$
\begin{aligned}
(\vec{\nabla} \cdot \Pi'_s)_i &= \sum_j \frac{\partial}{\partial x_j}\Pi'_{ijs} \\
&= \sum_{k,j}(\xi^s - \frac{2}{3}\mu^s)\delta_{ij}\frac{\partial}{\partial x_j}\left(\frac{\partial U_k}{\partial x_k}\right)_s + \mu^s\frac{\partial}{\partial x_j}\left[\left(\frac{\partial U_i}{\partial x_j}\right)_s \right. \\
&\quad \left. + \left(\frac{\partial U_j}{\partial x_i}\right)_s\right] \\
&= (\xi^s + \frac{1}{3}\mu^s)\frac{\partial}{\partial x_i}\vec{\nabla}\cdot\vec{U}^s + \mu^s\nabla^2 U^s_i,
\end{aligned}
\tag{4.13}
$$

We must now sum over the species index s for an electron-proton system and express the species velocities U_s in terms of the single fluid velocity $\vec{U}$ and the current density $\vec{J}$ defined in Chapter 2. In the limit $(m_e/m_i) \to 0$, we find that the single fluid equation (4.2) becomes

$$\rho_m\left[\frac{\partial \vec{U}}{\partial t} + (\vec{U}\cdot\vec{\nabla})\vec{U}\right] = -\vec{\nabla}p + (\xi + \frac{1}{3}\mu)\vec{\nabla}(\vec{\nabla}\cdot\vec{U}) + \mu\nabla^2\vec{U} + \frac{\vec{J}\times\vec{B}}{c}$$

$$-\rho_m\vec{\nabla}\varphi_g, \tag{4.14}$$

where the viscosity coefficients $\xi = \xi_e + \xi_i$, $\mu = \mu_e + \mu_i$ and pressure $p = p_e + p_i$. The generalized Ohm's law Equation (4.3) has also been written under the assumption $(m_e/m_i) \to 0$. By adding the energy equation for each species and expressing the sum in terms of the single fluid variables $\vec{U}$ and $\vec{J}$, the energy equation for an electron-proton fluid can be determined. This is rather a long algebraic exercise. We will here give a simple form of the energy equation by modifying Equation (2.89) with the inclusion of Joule heating rate in the presence of current density $\vec{J}$ and its associated magnetic field $\vec{B}$. The rate of production of heat per unit fluid volume is given by

$$\vec{J} \cdot \vec{E} = \frac{1}{\sigma} J^2 = \frac{c^2}{(4\pi)^2 \sigma} (\vec{\nabla} \times \vec{B})^2.$$

The rate of resulting rise in temperature, ΔT, is given by

$$\rho_m c_p \frac{\Delta T}{\Delta t} = \frac{c^2}{16\pi^2 \sigma} (\vec{\nabla} \times \vec{B})^2. \tag{4.15}$$

The energy equation (2.89), therefore, in the absence of viscous effects, becomes:

$$\frac{\partial T}{\partial t} = \kappa \nabla^2 T + \frac{c^2}{16\pi^2 \sigma c_p \rho_m} (\vec{\nabla} \times \vec{B})^2. \tag{4.16}$$

Equation (4.16) does not include a contribution from the collisional term R. This term tends to equalize the temperatures of the various species. Therefore under the MHD approximation, where we assume that the fluid is characterized by a single temperature, the collisional term R can be neglected. We will study magnetohydrodynamics of conducting fluids by using Equations (4.1), (4.3), (4.4), (4.5), (4.6), (4.7), (4.14) and (4.16).

4.4. Ideal Conducting Fluids

An ideal conducting fluid is one with infinite conductivity σ, or zero electrical resistivity η, and zero viscosity coefficients μ and ξ. In the generalized Ohm's law, Equation (4.3), the Hall term $J \times B$ is usually smaller than the $\vec{U} \times \vec{B}$ term and if additionally the pressure forces are zero, the condition of infinite conductivity implies that

$$\vec{E} + \frac{\vec{U} \times \vec{B}}{c} = 0. \tag{4.17}$$

Faraday's induction law Equation (4.7) becomes:

$$\frac{\partial \vec{B}}{\partial t} = \vec{\nabla} \times (\vec{U} \times \vec{B}). \tag{4.18}$$

It can be easily shown that Equation (4.18) is a statement of conservation of magnetic flux, provided that the area enclosing the flux moves with the fluid with velocity $\vec{U}$. The magnetic flux φ_B can change due to (1) a change in magnetic induction $\vec{B}$ and/or (2) a change in the area enclosing the flux as a result of the moving boundaries of the area. Thus,

$$\frac{d\varphi_B}{dt} = \int_S \frac{\partial \vec{B}}{\partial t} \cdot d\vec{S} + \int_S \vec{B} \cdot \frac{\partial}{\partial t} d\vec{S}$$

$$= \int_S \nabla \times (\vec{U} \times \vec{B}) \cdot d\vec{S} + \int_s \vec{B} \cdot \vec{U} \times d\vec{l},$$

since $\vec{U} \times d\vec{l}$ is the rate of change of area (Figure 4.1).

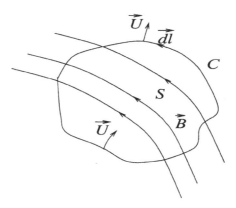

Figure 4.1.   Conservation of Magnetic Flux φ_B in Moving Ideal Fluid.

Problem 4.1: Using Stokes theorem and a well known vector identity, show that

$$\frac{d\varphi_B}{dt} = 0. \tag{4.19}$$

Thus the magnetic flux φ_B remains constant as long as the surface enclosing it moves with the fluid velocity $\vec{U}$. This characteristic of the magnetic induction $\vec{B}$ in a fluid has given rise to the term **Frozen-in Fields**, i.e., an ideal fluid and the magnetic field lines move together with a common velocity. We can easily see that this will not be true for finite conductivity, for, non-ideal fluids suffer viscous dissipation themselves, as well as cause magnetic field dissipation. From Equation (4.17), we see that this common velocity of the field and the fluid is nothing but the familiar $\vec{E} \times \vec{B}$ velocity $\vec{U}_E$. The motion of the fluid parallel to the magnetic field is governed by non-magnetic forces. The case of infinite conductivity is not just a mathematical curiosity. We have seen earlier how induced electric fields generated by a rotating neutron star produce a plasma around the star. We can estimate that the electrical resistivity of a fully ionized plasma is given by (Equation 1.41):

$$\eta \simeq 10^{-7}T^{-3/2}\text{sec,} \qquad (4.20)$$

where T is the temperature of the plasma in Kelvin degrees. Thus for typical temperatures of the order of $10^{8} - 10^{10}K$, $\eta \simeq 10^{-19} - 10^{-22}$sec. Do you know that the resistivity of copper is $\sim 10^{-18}$sec? Many properties of plasmas around neutron stars are studied using ideal magnetohydrodynamics. There are numerous examples of nearly ideal conducting fluids in the universe. Of course, large conductivity or small resistivity is not always an advantage. A phenomenon like solar flares, where large amounts of energy are released in a matter of a few minutes, could not occur in ideal conducting fluid. Several mechanisms to reduce the conductivity and enhance the joule dissipation rate or ohmic heating [Equation 4.16] have been investigated.

4.5. Pragmatic Conducting Fluids

Yes, there are fluids with large electrical conductivities and there are fluids with small conductivities but there are no fluids with infinite conductivities; the large and small are decided by the phenomena we choose to study. Let us see how a normal fluid with finite conductivity behaves. Again neglecting the viscous forces, and Hall current, we substitute for $\vec{J}$ from the generalized Ohm's law in the momentum Equation (4.14) to get

$$\begin{aligned}
\rho_m \frac{\partial \vec{U}}{\partial t} &= \frac{\sigma}{c}\left[\vec{E} + \frac{\vec{U} \times \vec{B}}{c}\right] \times \vec{B} \\
&= \frac{\sigma B^2}{c}\left[\frac{\vec{E} \times \vec{B}}{B^2} - \frac{\vec{U}}{c} + \frac{(\vec{B} \cdot \vec{U})\vec{B}}{cB^2}\right] \\
&= \frac{\sigma B^2}{c^2}\left[\vec{U}_E - \vec{U}_\perp\right].
\end{aligned} \qquad (4.21)$$

The solution of Equation (4.21) is found to be

$$U_\parallel = \text{constant,}$$

and

$$\vec{U}_\perp = \vec{U}_{\perp 0}\,\exp\left[-\frac{\sigma B^2 t}{\rho_m c^2}\right] + \vec{U}_E. \qquad (4.22)$$

That is, if there were an initial perpendicular velocity $\vec{U}_{\perp 0}$ in addition to the $\vec{E} \times \vec{B}$ velocity, it would fall exponentially with time. The e fall time is

$$t_e = \frac{\rho_m c^2}{\sigma B^2}, \qquad (4.23)$$

and ultimately only the $\vec{E} \times \vec{B}$ motion will survive. The magnetic field retards any attempts of the fluid to cross it and offers a kind of **magnetic viscosity**.

Problem 4.2: Estimate t_e for the solar wind with a typical density of 10 protons/cm^3, temperature $T \sim 10^5$K and magnetic field $B \sim 10^{-5}$ Gauss.

What happens to the magnetic field in a non-ideal conducting fluid? Let us substitute the generalized Ohm's law without the viscous and pressure forces and Hall current into Faraday's induction law. We find

$$\frac{\partial \vec{B}}{\partial t} = \vec{\nabla} \times (\vec{U} \times \vec{B}) + \frac{c^2}{4\pi\sigma} \nabla^2 \vec{B}. \qquad (4.24)$$

We have seen the effect of the first term, which established the constancy of magnetic flux in a moving ideally conducting fluid. In order to appreciate the effect of finite conductivity, let us neglect the first term. Now, Equation (4.24) assumes the form of a diffusion equation:

$$\frac{\partial \vec{B}}{\partial t} = \frac{c^2}{4\pi\sigma} \nabla^2 \vec{B}. \qquad (4.25)$$

This equation can be solved by the method of separation of variables. By comparing Equation (4.25) with the momentum Equation (4.14), we see that the quantity $(c^2/4\pi\sigma)$ plays the same role for magnetic field as the kinematic viscosity $\nu = \mu/\rho_m$ plays for the fluid motion. Therefore we can now define the magnetic viscosity ν_m as

$$\nu_m = \frac{c^2}{4\pi\sigma}. \qquad (4.26)$$

From the solution of Equation (4.25), we find that at any given spatial position, the magnetic field decays exponentially with time as

$$B = B_0 \, \exp{-t/t_d}, \qquad (4.27)$$

where the e-fall time t_d is given by

$$t_d = \frac{L^2}{\nu_m} = \frac{4\pi\sigma L^2}{c^2}, \qquad (4.28)$$

for a typical length scale L of the spatial variation of B. Thus, for $\sigma \to \infty$, $\nu_m \to 0$ and $t_d \to \infty$, i.e., there is no decay, a result we have already seen in the conservation of magnetic flux. The relative importance of the two terms in the evolution of magnetic field (Equation 4.24) is decided by their ratio R_m given by

$$R_m = \frac{(UB/L)}{\left(\dfrac{\nu_m B}{L^2}\right)} = \frac{UL}{\nu_m},$$ (4.29)

where we have used dimensional analysis to arrive at Equation (4.29). For small values of ν_m or large values of the ratio R_m, called the **Magnetic Reynolds Number**, the magnetic field suffers very little diffusion and is simply carried away by the fluid. We can also define the **Kinetic Reynolds Number** R_k as the ratio of the convective term $(\vec{U} \cdot \vec{\nabla})\vec{U}$ and the diffusion term $\nu \nabla^2 \vec{U}$ to find

$$R_k = \frac{UL}{\nu}.$$ (4.30)

Ideal conducting fluids have infinitely large values of R_m and R_k. Astrophysical conducting fluids often satisfy the conditions $R_m >> 1$ and $R_k >> 1$, because of their large characteristic spatial scales.

Problem 4.3: Estimate t_d and R_m for the Earth's interior, the Solar Photosphere and the various phases of the interstellar medium of Milky Way.

4.6. Conducting Fluid in Equilibrium

We have seen that a conducting fluid experiences inertial, gravitational, pressure gradient, electromagnetic and viscous forces. A fluid can attain a state of equilibrium if the net force on it vanishes. In the equilibrium state, all physical quantities, including mass density, fluid velocity, pressure, current density and magnetic field are independent of time. From the momentum Equation (4.14), we study the various equilibria by neglecting viscous forces. The equilibrium condition is:

$$\rho_m(\vec{U} \cdot \vec{\nabla})\vec{U} = -\vec{\nabla}p + \frac{\vec{J} \times \vec{B}}{c} - \rho_m \vec{\nabla}\varphi_g.$$ (4.31)

4.7. Hydrostatic Equilibrium

Perhaps the most familiar case is that of **Hydrostatic Equilibrium** for which $\vec{U} = 0$ and $\vec{J} \times \vec{B} = 0$ and the pressure gradient force balances the gravitational force, so that

$$\vec{\nabla}p = -\rho_m \vec{\nabla}\varphi_g.$$ (4.32)

This equation is of great importance for many astrophysical situations such as stars or galactic clouds. The gravitational potential φ_g for an extended mass distribution is given by

$$\varphi_g = -\frac{GM(r)}{r}, \tag{4.33}$$

where $M(r)$ is the spherically distributed mass producing the gravitational force on a fluid element of mass density ρ_m. Equation (4.32) can be recast as:

$$\frac{d}{dr}\left[\frac{r^2}{\rho_m}\frac{dp}{dr}\right] = -4\pi G\rho_m(r)r^2, \tag{4.34}$$

by using

$$\frac{dM(r)}{dr} = 4\pi\rho_m r^2.$$

We can now take a general form of the equation of state

$$p = K_m\rho_m^{1+\frac{1}{n}}, \tag{4.35}$$

where K_m and n are constants and n is known as the **Polytropic Index**. The case $n = \infty$ represents a constant temperature fluid and this equilibrium condition is called the **Isothermal Sphere**. Other values of n give radial variations of temperature. After substituting for pressure from Equation (4.35), Equation (4.34) is studied by recasting it in dimensionless variables defined as:

$$\rho_m = \lambda\theta^n,$$

and

$$r = \beta\xi. \tag{4.36}$$

We get:

$$\frac{1}{\xi^2}\frac{d}{d\xi}\left[\xi^2\frac{d\theta}{d\xi}\right] + \theta^n = 0, \tag{4.37}$$

and

$$\beta = \left[\left(\frac{n+1}{4\pi G}\right)K_m\lambda^{\frac{1-n}{n}}\right]^{1/2}.$$

Equation (4.37) is known as the **Lane-Emden Equation**. Here λ is the central density so that $\theta = 1$ for $\xi = 0$ and $(d\varphi_g/dr) = 0$ at $r = 0$ or $\theta = 1$. The constant $K_m = \frac{1}{3}U_{rms}^2$ for an isotropic velocity dispersion U_{rms}^2. Equation (4.37) can be solved for different values of n. For $n = 0$, we find:

$$\theta = 1 - \frac{\xi^2}{6}, \tag{4.38}$$

from which we learn that the boundary of zero density lies at $\xi = \sqrt{6}$, $\rho_m = \lambda$ and $p = K_m\rho_m = $ constant. This equilibrium consists of a sphere of constant density λ and radius $= (6p/4\pi G)^{1/2}$.

Problem 4.4: By substituting $\theta = \dfrac{f(\xi)}{\xi}$, show that $\theta = \dfrac{\sin\xi}{\xi}$ is the solution of the Lane-Emden Equation for $n = 1$. Find the boundary of zero density. Estimate the total mass contained in this sphere.

The Lane-Emden Equation can also be solved analytically for $n = 5$ by making the substitutions:

$$\xi = \exp[-\psi(z)] \text{ and } \theta = \frac{z}{\sqrt{2}}\exp(\psi(z)/2),$$

and the solution is

$$\rho_m = (1 + \frac{1}{3}\xi^2)^{-5/2}, \qquad (4.39)$$

which varies as r^{-5} for large r. We see that this configuration has an infinite radius but finite mass. For $n > 5$, we find that the total mass becomes infinite.

The case of $n = \infty$ for isothermal spheres has been found to nicely describe some gravitating systems. Here, from Equation (4.35)

$$p = K_m\rho_m,$$

and Equation (4.34) becomes:

$$\frac{1}{r'^2}\frac{d}{dr'}\left[r'^2\frac{dt}{dr'}\right] = \exp(-t), \qquad (4.40)$$

where $r = \beta'r'$, $\beta' = \left(\dfrac{K_m}{4\pi G\lambda}\right)^{1/2}$, $\rho_m = \lambda\exp(-t)$. The boundary conditions are $t = \dfrac{dt}{dr'} = 0$ at $r' = 0$. Equation (4.40) can be solved analytically only for $r' < 1$, since, then t can be expanded in a power series in r'. We find

$$t = \frac{r'^2}{6} - \frac{r'^4}{120}. \qquad (4.41)$$

Isothermal spheres with mass density given by Equation (4.41) for $n = \infty$ have been used to represent mass distribution of spherical galaxies and clusters of galaxies. A more detailed discussion of hydrostatic equilibria of self-gravitating systems, Equation (4.32) can be found in An introduction to the study of Stellar Structure − a book by the Nobel laureate S. Chandrasekhar.

Traditionally, the study of the equilibrium of self-gravitating systems is not included in a section on MHD equilibrium, since these systems are

believed to be mostly neutral hydrogen. However, we see that even a conducting fluid can have these mass configurations, provided the Lorentz force
$\vec{J} \times \vec{B} = 0$, which implies that the magnetic field must be given by:

$$(\vec{\nabla} \times \vec{B}) \times \vec{B} = 0,$$

or

$$\vec{\nabla} \times \vec{B} = \alpha(\vec{r})\vec{B}, \tag{4.42}$$

and $\vec{B} \cdot \vec{\nabla}\alpha(\vec{r}) = 0$. Why?

Such a magnetic field is known as a **Force-Free Magnetic Field** for the
obvious reason that it exerts no force on a fluid. So, we reach the conclusion
that a self-gravitating conducting fluid in a force-free magnetic field can
attain all the configurations that a self-gravitating non-conducting fluid can!

Another type of hydrostatic equilibrium results when a fluid is in a gravitational field of another object; e.g., the earth's atmosphere in the earth's
gravitational field. In such a case, one writes the gravitational potential φ_g
as that due to a point mass M situated at the center of the object of radius
R. The potential at any point at a distance r above the surface of the object
is then

$$\varphi_g = -\frac{GM}{(R+r)}, \tag{4.43}$$

and the hydrostatic balance condition gives

$$\frac{1}{\rho_m}\frac{dp}{dr} = -\frac{GM}{(R+r)^2}. \tag{4.44}$$

Problem 4.5: For an isothermal equation of state $p = K_m\rho_m$, find the scale
height of density in the solar corona.

4.8. Magnetohydrostatic Equilibrium

We now retain a non-zero Lorentz force, but investigate a static equilibrium,
so that $\vec{U} = 0$, and Equation (4.31) then gives:

$$-\vec{\nabla}p + \frac{\vec{J} \times \vec{B}}{c} - \rho_m\vec{\nabla}\varphi_g = 0,$$

or

$$\vec{\nabla}\left[p + \rho_m\varphi_g + \frac{B^2}{8\pi}\right] - \frac{1}{4\pi}(\vec{B} \cdot \vec{\nabla})\vec{B} = 0. \tag{4.45}$$

We see that the Lorentz force contributes to pressure balance in two ways:
(1) through $(B^2/8\pi)$ which acts like pressure and is known as magnetic

hydrostatic pressure and (ii) through $(\vec{B} \cdot \vec{\nabla})\vec{B}$ which acts like tension along the magnetic field lines. If, for a certain magnetic field configuration, the magnetic tension term $(\vec{B} \cdot \vec{\nabla})\vec{B}$ vanishes (when B does not vary in its own direction) and if the gravitational effects can be ignored, we see that Equation (4.45) tells us that the sum of mechanical pressure p and the magnetic pressure $B^2/8\pi$ must be a constant, i.e., space independent, or

$$p + \frac{B^2}{8\pi} = \text{constant.} \qquad (4.46)$$

Equation (4.46) shows us a way of confining a conducting fluid by a magnetic field, which acts like a container for the fluid. A low pressure region should have a high magnetic pressure and vice-versa. The predominance of mechanical pressure over the magnetic pressure can be expressed by a ratio, called the plasma β_p, defined as:

$$\beta_p = \frac{p}{B^2/8\pi}. \qquad (4.47)$$

A fluid is said to be confined by a magnetic field if $\beta_p < 1$. A variety of loop like structures seen in the solar corona have $\beta_p < 1$, whereas in the solar photosphere $\beta_p > 1$.

In cylindrical geometry, for example, an azimuthal current density J_θ crossed with an axial magnetic field B_z can support a radial pressure gradient p'. Such a configuration is known as the θ **Pinch** (Figure 4.2); the

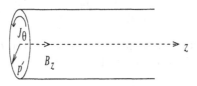

Figure 4.2. The θ Pinch

applied magnetic field limits or pinches the radius of the fluid column. We can estimate the current required to confine a fluid from the following considerations:

$$\vec{\nabla}p \times \vec{B} = \frac{1}{c}(\vec{J} \times \vec{B}) \times \vec{B},$$

or

$$\vec{J}_\perp = c\frac{\vec{\nabla}p \times \vec{B}}{B^2}, \qquad (4.48)$$

where $\vec{J}_\perp$ is called the diamagnetic current, the same diamagnetic current that we discussed in Chapter 3, being associated with the diamagnetic drifts of charged particles. In a direction parallel to the magnetic field, we find

$$\vec{B} \cdot \vec{\nabla} p = 0, \tag{4.49}$$

i.e., the pressure is constant along the magnetic field.

Problem 4.6: Show that for $\beta_p = 1$, the diamagnetic current generates a magnetic field equal and opposite to the externally imposed field $\vec{B}$.

Another equilibrium called the **Z Pinch** (Figure 4.3) obtains when

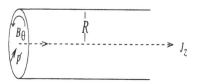

Figure 4.3. The Z Pinch

an axial current density J_z, produces an azimuthal magnetic field B_θ to support a radial pressure gradient p'. This can be seen from the equilibrium condition, Equation (4.45) (with $\varphi_g = 0$):

$$\frac{\partial p}{\partial r} = -\frac{B_\theta^2}{4\pi r} - \frac{1}{8\pi}\frac{\partial B_\theta^2}{\partial r}, \tag{4.50}$$

and

$$B_\theta(r) = \frac{2\pi}{c} J_z r, \tag{4.51}$$

Using the boundary conditions $p = 0$ at $r = R$, the external boundary of the fluid and $p = p_0$ at $r = 0$, we find

$$p(r) = \frac{p_0}{R^2}(R^2 - r^2), \tag{4.52}$$

where

$$p_0 = \frac{\pi J_z^2 R^2}{c^2} = \frac{B_\theta^2(R)}{4\pi}.$$

Thus, for a given central pressure p_0, the radius R of the fluid is determined from

$$R^2 = \frac{p_0^2}{4\pi J_z^2}, \tag{4.53}$$

which shows the pinching effect produced by the current density J_z: the higher the current density J_z, the smaller the radius R.

Figure 4.4.   Solar Coronal Loops (Bray et. al. 1991)

Problem 4.7: Cylindrical plasma loops (Figure 4.4) of typical radius $\sim$ 10^9cm are commonly seen in the solar corona. Estimate the current density J_z for typical values of coronal density and temperature.

The pressure balance condition Equation (4.46) also ensures that a fluid of pressure p_1 with magnetic field B_1 can be in equilibrium with a fluid of pressure p_2 without magnetic field, provided

$$p_1 + \frac{B_1^2}{8\pi} = p_2, \tag{4.54}$$

which, for an isothermal equation of state states that the density ρ_2 of the magnetic field-free fluid must be larger than the density ρ_1 of the magnetized fluid. Such magnetized fluid elements being lighter than the surrounding heavier fluid, experience buoyant forces in a gravitational field and rise up. The appearance of discrete and strongly magnetized structures on the solar surface is attributed to this process. A high density fluid overlying a low density fluid in a gravitational field pointing downwards, can be in equilibrium, but it is an unstable equilibrium. It is the same mechanism due to which we can invert a glass full of water (I haven't tried with wine!) without the water flowing down, but how well we know that a slight carelessness or an air current can destroy the equilibrium.

4.9. Magnetohydrodynamic Equilibrium

We will now discuss MHD equilibrium including flow, i.e., $\dfrac{\partial \vec{U}}{\partial t} = 0$ but $\vec{U} \neq 0$. Neglecting viscous effects, the equilibrium condition becomes:

$$\rho_m \left[(\vec{U} \cdot \vec{\nabla}) \vec{U} \right] = -\vec{\nabla}p + \frac{\vec{J} \times \vec{B}}{c} - \rho_m \vec{\nabla}\varphi_g. \tag{4.55}$$

The inertial force can be balanced either singly or jointly by the pressure gradient, the Lorentz force and the gravitational force. Either in the absence of magnetic field or for a force free magnetic field, the equilibrium (4.55) can be expressed as

$$\vec{\nabla}\left[\frac{\vec{U}^2}{2} + h + \varphi_g\right] = \vec{U} \times (\vec{\nabla} \times \vec{U}), \tag{4.56}$$

where we have written

$$\vec{\nabla}p = \rho_m \vec{\nabla}h. \tag{4.57}$$

For adiabatic variations of pressure and density h is the specific enthalpy since from thermodynamics

$$dh = Tds + \frac{1}{\rho_m}dp, \tag{4.58}$$

and the change in entropy, $ds = 0$ for adiabatic changes. The right hand side of Equation (4.56) vanishes either for **Irrotational Flows**, i.e., when $\vec{\nabla} \times \vec{U} = 0$, or for **Aligned Helical Flows**, i.e., when $\vec{U}$ is parallel to $\vec{\nabla} \times \vec{U} \equiv \vec{\omega}$, where $\vec{\omega}$ is known as the vorticity and $(\vec{U} \cdot \vec{\omega})$ is called the helicity. Thus, for irrotational or fully helical flows, we get the well known Bernoulli's relation:

$$\frac{U^2}{2} + h + \varphi_g = 0. \tag{4.59}$$

We will say more on this equilibrium in a later chapter on Non-conducting Fluids.

Aligned helical flows, similarly to force free magnetic fields, satisfy the following equation

$$\vec{\nabla} \times \vec{V} = \alpha_v \vec{V}. \tag{4.60}$$

Such flows are called **Beltrami Flows**. An equilibrium in which the pressure gradients are balanced by a Beltrami flow has been shown to offer a good description of a class of solar coronal loops, since gravitational effects are negligible and magnetic fields are believed to be nearly force-free in the solar corona (Krishan 1996).

Problem 4.8: Solve equation (4.60) in cylindrical coordinates and determine the radial variation of pressure for constant mass density and no gravitational field.

We can find another equilibrium, if we rewrite Equation (4.55) as:

$$-\vec{\nabla}\left[p + \rho_m \varphi_g + \frac{B^2}{8\pi}\right] = -\frac{1}{4\pi}\left[(\vec{B} \cdot \vec{\nabla})\vec{B} - 4\pi\rho_m(\vec{U} \cdot \vec{\nabla})\vec{U}\right]. \tag{4.61}$$

This shows that

$$p + \rho_m \varphi_g + \frac{B^2}{8\pi} = \text{constant},$$

if

$$\vec{U} = \pm \frac{\vec{B}}{(4\pi\rho_m)^1/2} \equiv \pm\vec{V}_A. \tag{4.62}$$

$\vec{V}_A$ is known as the Alfven velocity and the equilibrium, which is nothing but the magnetostatic equilibrium for a tension free magnetic field, corresponds to what is known as the **Alfvenic State**. The alfvenic state is one in which a conducting fluid flows parallel or anti-parallel to the magnetic field with the Alfven speed. We shall learn more about the Alfven velocity in a later section.

4.10. Magnetohydrodynamic Waves

In order to learn about a system, we must disturb or nudge it and watch how it responds. For example, when we displace a pendulum, a little, from its equilibrium vertical position and release it, the pendulum begins to oscillate. For small displacements the oscillations are harmonic. The pendulum will oscillate forever, if there are no retarding forces due to the surrounding environment. For large displacements, the oscillations are nonlinear, i.e., the amplitude of the oscillations is no longer a constant. The period of oscillations gives us a relation between the characteristics of the system; here, for example, the length of the pendulum, and the forces trying to restore equilibrium, here, for example, the gravitational force. In the same way, when a conducting fluid is disturbed from its equilibrium configuration we see it set into oscillations. The period of the oscillations is related to the characteristics of the conducting fluid such as mass density, pressure, temperature and the restoring forces, which may include pressure gradient, Lorentz and gravitational. These oscillations, also called **Waves** since they propagate in the fluid, have a great diagnostic potential. We can estimate the fluid properties through the detection of these waves. Further, in the presence of dissipative effects like viscous and resistive forces, the amplitude of the waves decreases with time. The energy carried by waves is deposited in the fluid as a result of which it may heat up. Magnetohydrodynamic waves have been considered very favourably for heating the solar corona, which at a temperature of $\sim 10^6 K$, lies outside the solar photosphere with temperature $\sim 6000K$, and therefore needs sources of heat and mechanisms to maintain its temperature.

In the next section, we shall study different types of waves that a conducting fluid exhibits, when disturbed by a small amount from its equilibrium. These waves are called linear waves. We can introduce a small disturbance in the various parameters singly or jointly, depending upon our interest. We may wish to know the response of the conducting fluid to a perturbation in its, say, density. A change in density will produce a change in the gravitational force, a change in pressure, a change in the fluid velocity and a change in magnetic field such that the conservation laws of mass, momentum and energy as well as the Maxwell equations always remain satisfied. In order to see the restoring action of one particular force, we may ignore other forces. Of course, if we include all the forces the problem becomes quite complex, though not intractable. Anyway, it helps if we have some idea of the relative importance of the various forces. In the study of linear waves, we get a dispersion relation which contains everything on propagation characteristics: the phase and group velocities as well as the polarization characteristics. The only property we cannot determine is the amplitude of the wave, for which we must learn to do nonlinear studies. However, for the present, we limit ourselves to linear studies.

4.11. Dispersion Relation of Ideal MHD Waves

Let the equilibrium state of an ideal MHD fluid be described by the space and time independent mass density ρ_0 (for the rest of this chapter, the subscript m will be dropped), the fluid velocity $\vec{U}_0 = 0$, the uniform and time independent magnetic field B_0, the uniform pressure p_0, the current density $\vec{J}_0 = 0$ and the inductive electric field $\vec{E}_0 = 0$. We now perturb this equilibrium such that

$$
\begin{aligned}
\rho &= \rho_0 + \rho_1, \\
\vec{U} &= \vec{U}_1, \\
\vec{B} &= \vec{B}_0 + \vec{B}_1, \\
p &= p_0 + p_1, \\
\vec{E} &= \vec{E}_1, \\
\vec{J} &= \vec{J}_1,
\end{aligned}
\tag{4.63}
$$

where all the quantities with subscript 1 are much smaller than the corresponding equilibrium values (except $\vec{E}_1, \vec{J}_1$ and $\vec{U}_1$). The linearized ideal MHD equations of mass and momentum conservation, neglecting shear and, dissipative effects, the linearized generalized Ohms's law, and the Maxwell equations are:

$$
\frac{\partial \rho_1}{\partial t} + \vec{\nabla} \cdot \left[\rho_0 \vec{U}_1 \right] = 0,
$$

$$
\rho_0 \frac{\partial \vec{U}_1}{\partial t} = -\vec{\nabla} p_1 + \frac{\vec{J}_1 \times \vec{B}_0}{c},
$$

$$\vec{E}_1 = -\frac{\vec{U}_1 \times \vec{B}_0}{c}, \tag{4.64}$$

$$\vec{\nabla} \times \vec{B}_1 = \frac{4\pi}{c} \vec{J}_1,$$

$$\frac{\partial \vec{B}_1}{\partial t} = \vec{\nabla} \times (\vec{U}_1 \times \vec{B}_0).$$

We still have to use the energy conservation law to relate perturbations in density ρ_1 to perturbations in pressure p_1. We recall that p is the sum of pressures due to each species of fluid, i.e.

$$p = \sum_s p_s. \tag{4.65}$$

For an adiabatic energy equation:

$$p \propto \rho^\gamma, \tag{4.66}$$

we find, to the first order,

$$\vec{\nabla} p = \vec{\nabla} p_1 = \sum_s \gamma \frac{p_{0s}}{\rho_{0s}} \vec{\nabla} \rho_{1s}. \tag{4.67}$$

We then assume that the fractional change in density for all species is the same i.e.:

$$\frac{\rho_{1s}}{\rho_{0s}} = \frac{\rho_1}{\rho_0}, \tag{4.68}$$

so that

$$\vec{\nabla} p_1 = \sum_s \gamma p_{0s} \frac{\vec{\nabla} \rho_1}{\rho_0} = \frac{\gamma p_0}{\rho_0} \vec{\nabla} \rho_1 = c_s^2 \vec{\nabla} \rho_1, \tag{4.69}$$

where c_s^2 is the adiabatic sound speed. We can use the general relation,

$$\vec{\nabla} p = c_s^2 \vec{\nabla} \rho, \tag{4.70}$$

for adiabatic or isothermal cases and identify c_s with the corresponding sound speed.

What remains to be done is to eliminate all the first order quantities except one among Equations (4.64) and (4.69). This is easy as we have six first order quantities $(\rho_1, \vec{U}_1, p_1, \vec{E}_1, \vec{B}_1, \vec{J}_1)$ and six linearized equations. The elimination procedure becomes simple when we assume a plane wave type variation for all the first order quantities. We write:

$$\vec{U}_1(\vec{r}, t) = \vec{U}_1' \exp\left[i\vec{k} \cdot \vec{r} - i\omega t\right], \tag{4.71}$$

and similarly for the other five quantities. $\vec{U}'_1$ is the space and time independent amplitude of the oscillating velocity $\vec{U}_1(\vec{r}, t)$. On the completion of the
elimination exercise, we find an equation of the form:

$$D(\text{First order quantity}) = 0. \tag{4.72}$$

Since the first order quantity $\neq 0$, we obtain the dispersion relation $D = 0$,
which is a relation between the wave frequency ω and the wave vector $\vec{k}$.
For the present case, we find:

$$\left[-\omega^2 + \left(\vec{k} \cdot \vec{V}_A\right)^2\right]\vec{U}'_1 + \left[\left(c_s^2 + V_A^2\right)\left(\vec{k} \cdot \vec{U}'_1\right) - \left(\vec{k} \cdot \vec{V}_A\right)\left(\vec{V}_A \cdot \vec{U}'_1\right)\right]\vec{k}$$

$$- \left(\vec{k} \cdot \vec{V}_A\right)\left(\vec{k} \cdot \vec{U}'_1\right)\vec{V}_A = 0. \tag{4.73}$$

Here, $\vec{V}_A = \vec{B}_0(4\pi\rho_0)^{-1/2}$ is the Alfven velocity. In order to find the different
wave motions corresponding to the roots of the dispersion relation, we write
Equation (4.73) in the component form as:

$$\left[-\omega^2 + (kV_A \cos\theta)^2\right]U'_{1x} + \left[(c_s^2 + V_A^2)(k_xU'_{1x} + k_zU'_{1z})\right.$$

$$\left. -kV_A^2 \cos\theta U'_{1z}\right]k_x = 0, \tag{4.74}$$

$$\left[-\omega^2 + (kV_A \cos\theta)^2\right]U'_{1y} = 0, \tag{4.75}$$

and

$$\left[-\omega^2 + (kV_A \cos\theta)^2\right]U'_{1z} + \left[(c_s^2 + V_A^2)(k_xU'_{1x} + k_zU'_{1z})\right.$$

$$\left. -kV_A^2 \cos\theta U'_{1z}\right]k_z - kV_A^2 \cos\theta(k_xU'_{1x} + k_zU'_{1z}) = 0, \tag{4.76}$$

where θ is the angle between the zeroth order magnetic field $\vec{B}_0$ (which
we have taken to be in the z direction), and the wave vector $\vec{k}$, so that
$k_x = k\sin\theta$, $k_z = k\cos\theta$ and $k_y = 0$. We see that with this arbitrary choice
the motion in the y direction is decoupled from the motion in the (x, z)
plane. Equation (4.75) gives a root

$$\omega^2 = k^2V_A^2 \cos^2\theta \quad \text{or} \quad \omega = \pm k_zV_A = \pm\vec{k} \cdot \vec{V}_A, \tag{4.77}$$

for $U'_{1y} \neq 0$. Equation(4.77) is the dispersion relation of the **Alfven Wave**
propagating at an angle θ to the zeroth order magnetic field $\vec{B}_0$. For this
wave

$$\vec{\nabla} \cdot \vec{U}'_1 = 0, \quad \rho_1 = 0, \tag{4.78}$$

i.e., this wave does not produce any density and therefore pressure changes. Such a wave is called **Transverse** and **Noncompressional**. The phase velocity V_{ph} of the Alfven wave is

$$V_{ph} = \frac{\omega}{k} = \pm V_A \cos\theta, \qquad (4.79)$$

and the group velocity $V_g = \dfrac{d\omega}{dk} = \pm V_A \cos\theta$. There is no Alfven wave for $\theta = \pi/2$. The Alfven wave has the maximum phase and group velocity parallel and antiparallel to the field $\vec{B}_0$. The polarization of the Alfven wave i.e., the relative orientations of the electric field $\vec{E}_1$, the magnetic field $\vec{B}_1$ can be determined from Equations (4.64). We find $\vec{B}_1$ is in the y direction and $\vec{E}_1$ lies in the (x, z) plane, the plane containing the wave vector $\vec{k}$ and the magnetic field $\vec{B}_0$, as shown in Figure (4.5). It is clear that in the linear study of waves, we cannot estimate the absolute value of the amplitudes, $\vec{U}_1, \vec{E}_1, \vec{B}_1$ etc. But we can estimate their relative values. Thus we find that the electric energy density $(E_1^2/8\pi)$, the magnetic energy density$(B_1^2/8\pi)$, and the kinetic energy density $(\rho_0 U_1^2/2)$ are in the ratio $1 : c^2/V_A^2 : c^2/V_A^2$.

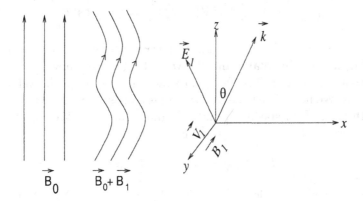

Figure 4.5. The Alfven Wave

The electric energy density is much smaller than the kinetic and magnetic energy densities. Further the kinetic and magnetic energy densities are equal, this is again a reminder of the field frozen to the fluid in the absence of dissipative effects.

The physical mechanism underlying the excitation of Alfven waves is identical to that of the transverse oscillations of a plucked stretched string. A wavy disturbance $\vec{B}_1$ curves the magnetic field lines, but the tension in the curved field tries to straighten the field lines, and the Alfven oscillations set in.

Problem 4.9: (i) Show that the tension of the magnetic field $(\vec{B}_0 + \vec{B}_1)$ provides the restoring force for Alfvenic oscillations. (ii) Show that the fluid moves with the $E \times B$ drift velocity in the Alfvenic wave mode.

Thus, we find that the velocity $\vec{U}_1'$ of the conducting fluid is given by

$$\vec{U}_1' = \frac{c\vec{E}_1 \times \vec{B}_0}{B_0^2}. \tag{4.80}$$

This, combined with Faraday's law of induction in a moving medium, again leads to the conclusion that the fluid and the field remain together until dissipation parts them.

We find two more waves, by eliminating say U_{1x}' between Equations (4.74) and (4.76). The dispersion relations of these waves are:

$$\omega_F^2 = \frac{k_F^2}{2}\left(V_A^2 + c_s^2\right) + \frac{k_F^2}{2}\left[\left(V_A^2 + c_s^2\right)^2 - 4c_s^2 V_A^2 \cos^2\theta\right]^{1/2} \tag{4.81}$$

and

$$\omega_S^2 = \frac{k_S^2}{2}\left(V_A^2 + c_s^2\right) - \frac{k_S^2}{2}\left[\left(V_A^2 + c_s^2\right)^2 - 4c_s^2 V_A^2 \cos^2\theta\right]^{1/2}. \tag{4.82}$$

Here, the wave with frequency ω_F and wave vector k_F is known as the **Fast Magnetosonic Wave** and the wave (ω_S, k_S) is known as the **Slow Magnetosonic Wave**. The fast and slow refer to the phase velocities of these waves. We notice that the fast wave has a phase velocity which is larger than both the Alfven speed V_A and the sound speed c_s. We can determine other properties of these waves now. First, since

$$\vec{k}_F \cdot \vec{U}_1' \neq 0 \quad \text{and} \quad \vec{k}_S \cdot \vec{U}_1' \neq 0, \tag{4.83}$$

both these waves affect density variations, i.e., $\rho_1 \neq 0$. The restoring force for both the waves is provided jointly by the gradient of kinetic and magnetic pressures; that is why they are called Magnetosonic waves. These waves are neither transverse nor longitudinal. They have mixed polarizations. For $\theta = \pi/2$, $\omega_S = 0$, i.e., the slow wave does not exist, whereas the fast wave has the maximum frequency and phase speed and becomes purely longitudinal, i.e., $\vec{k}_F \parallel \vec{U}_1$. In this case the directions of the various fields are $\vec{B}_0 = B_z$, $\vec{B}_1 = B_{1z}$, $\vec{E}_1 = E_{1y}$, $\vec{U}_1 = U_{1x}$ and $\vec{k} = k_x$. The fast wave produces density as well as magnetic field condensations and rarefactions as shown in Figure(4.6). For $\theta = 0$, we find from Equations (4.74), (4.75) and (4.76) that there are two types of waves: (i) a transverse wave with $\omega = \pm kV_A$, $\vec{k} \cdot \vec{U}_1 = 0$; this is the Alfven wave, we have already studied and

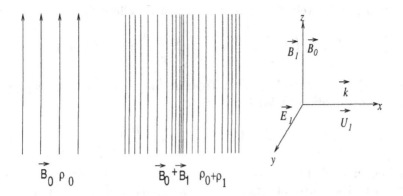

Figure 4.6.   Magnetosonic Waves Produce Condensations and Rarefactions.

(ii) a longitudinal wave with $\omega = \pm k c_s$, $\vec{k} \parallel \vec{U}_1$; this is the ordinary sound wave. Thus, we see that only for oblique propagation, i.e. at an angle to the ambient magnetic field $\vec{B}_0$, do all three waves, the Alfven and the fast and slow magnetosonic waves exist.

Problem 4.10: Find the directions of $\vec{U}_1, \vec{E}_1, \vec{B}_1$ for obliquely propagating fast and slow magnetosonic waves. Find the ratios of kinetic, magnetic and electric energy densities for the two waves.

The fluid and the magnetic field, in reality, do not keep oscillating forever, for there are resistive forces: the fluid is viscous and the magnetic field decays due to the electrical resistivity of the fluid. The magnetohydrodynamic waves suffer damping due to finite viscosity and electrical resistivity. We can study the MHD waves in non-ideal fluids by including the viscous force in the momentum equation, and the resistivity term $\eta \vec{J}$ in Ohm's law. We write the linearized momentum equation (of 4.14), sans gravitational force as:

$$\rho_0 \frac{\partial \vec{U}_1}{\partial t} = -c_s^2 \vec{\nabla} \rho_1 + \frac{\vec{J}_1 \times \vec{B}_0}{c} + \mu \nabla^2 \vec{U}_1 + (\xi + \frac{\mu}{3}) \vec{\nabla}(\vec{\nabla} \cdot \vec{U}_1). \qquad (4.84)$$

Using, Ohm's law with conductivity σ, the linearized Faraday's law becomes (Equation 4.24):

$$\frac{\partial \vec{B}_1}{\partial t} = \vec{\nabla} \times (\vec{U}_1 \times \vec{B}_0) + \nu_m \nabla^2 \vec{B}_1. \qquad (4.85)$$

We can carry out the elimination procedure as before and determine the dispersion relations of the three MHD waves. For Alfven waves, $\vec{\nabla} \cdot \vec{U}_1 = 0$

and the dispersion relation including dissipative effects becomes for $\theta = 0$;

$$\omega^2 + i\omega \left(\nu_m k^2 + \frac{\mu k^2}{\rho_0} \right) - \left(\frac{\nu_m \mu k^4}{\rho_0} + k^2 V_A^2 \right) = 0. \qquad (4.86)$$

This equation has complex roots. Treating the dissipative effects as small, the roots of Equation (4.86) are:

$$\omega \simeq \pm k V_A - \frac{i}{2} k^2 \left(\nu_m + \frac{\mu}{\rho_0} \right). \qquad (4.87)$$

Recalling that all the first order quantities have a time dependence $e^{-i\omega t}$, we see that dissipative effects produce an exponential damping of the wave amplitudes $(\vec{U}_1, \vec{B}_1, \vec{E}_1)$. The damping rate ω_{IA}, equal to the imaginary part of the complex frequency ω, is:

$$\begin{aligned} \omega_{IA} &= -\frac{k^2}{2}(\nu_m + \mu/\rho_0) \\ &\simeq -\frac{\omega^2}{2V_A^2}(\nu_m + \mu/\rho_0). \end{aligned} \qquad (4.88)$$

We see that high frequency or short wavelength waves suffer more damping than do the low frequency waves, for constant values of the magnetic field $\vec{B}_0$ and the mass density ρ_0. The wave intensity decays to e^{-1} of its initial value in a time $(2\omega_{IA})^{-1}$, which is the same as the diffusion time t_d (Equation 4.28) of the magnetic field $\vec{B}_0$, in the absence of the fluid viscosity. The distance L_d travelled by the wave in the time $(2\omega_{IA})^{-1}$ is:

$$L_d = V_A (2\omega_{IA})^{-1} = \frac{V_A^3}{\omega^2 (\nu_m + \mu/\rho_0)}. \qquad (4.89)$$

So, the high frequency waves have short damping lengths. Similarly, we can determine the damping rates for other waves too.

The damping rate of the fast MHD mode propagating perpendicular to the magnetic field $\vec{B}_0 (\theta = \pi/2)$ is found to be:

$$\omega_{IF} \simeq -\frac{k^2}{2} \left[\nu_m \frac{V_A^2}{V_A^2 + c_s^2} + \frac{\mu}{\rho_0} + \frac{(\xi + \mu/3)}{\rho_0} \right], \qquad (4.90)$$

with the real part

$$\omega_{RF}^2 \simeq k^2 (V_A^2 + c_s^2),$$

in which corrections due to the dissipative effects have been ignored. Again, the damping rate increases with frequency. The fast wave, being compressional, has an additional contribution to its damping rate from compressibility of the fluid.

Problem 4.11 Show that the dispersion relation of the MHD waves for oblique propagation including dissipative effects is given by:

$$\left[-\omega^2\left(1+\frac{i\nu_m k^2}{\omega}\right)\left(1+\frac{i\nu_m k^2}{\rho_0 \omega}\right)+\left(\vec{k}\cdot\vec{V}_A\right)^2\right]\vec{U}_1' +$$

$$\left[\left\{c_s^2\left(1+\frac{i\nu_m k^2}{\omega}\right)+V_A^2\right\}\left(\vec{k}\cdot\vec{U}_1'\right)-\left(\vec{k}\cdot\vec{V}_A\right)\left(\vec{V}_A\cdot\vec{U}_1'\right)-\right.$$

$$\left[i\omega\frac{(\xi+\mu/3)}{\rho_0}\left(1+\frac{i\nu_m k^2}{\omega}\right)\left(\vec{k}\cdot\vec{U}_1'\right)\right]\vec{k}-$$

$$\left(\vec{k}\cdot\vec{V}_A\right)\left(\vec{k}\cdot\vec{U}_1'\right)\vec{V}_A=0. \qquad (4.91)$$

It is quite easy to excite MHD waves. All it takes is to shake the magnetized plasma like one shakes a string. One of the most favourable astrophysical sites for excitation of MHD waves is the solar atmosphere. The outer layers of the solar atmosphere - the chromosphere and the corona - are at a much higher temperature than is the photosphere to which we owe our existence. Further, the chromosphere and the corona are highly inhomogeneous media supporting a variety of filamentary structures in the form of arches and loops Figure(4.4). A coronal loop is a bipolar structure whose foot points are anchored in the poles of the sub-photospheric magnetic field. The foot points undergo a continuous turning and twisting due to convective motions in the subphotospheric layers of the sun. This turning and twisting is enough to excite MHD waves in coronal loops. These waves then dissipate and spend their energy in heating the corona. Typically, waves of periods of a few seconds are believed to be excited in the corona. These waves can be detected through the periodic variations in the intensities of the continuum and line radiation as well as through the Doppler shifts of the line radiation. The Alfven waves, which are purely velocity and magnetic field oscillations without any accompanying mass density oscillations, do not produce any changes in the intensity and are observed through the Doppler effect. The magnetosonic waves, which are compressional, show up both as intensity and velocity oscillations. Although the MHD waves have received a lot of attention from the theoreticians for a long time, their unambiguous detection in the solar atmosphere is still awaited.

Problem 4.12: Determine the dissipation lengths of the Alfven and the fast magnetosonic waves of periods ~ 100 sec. in the solar corona with typical values of $\rho_0 \simeq 10^{-14}$ gm/cm^3, $B_0 \simeq 10$ Gauss, $T \sim 10^6$K, electrical conductivity $\sigma \simeq 10^{16}$ sec^{-1} and viscosity $\mu \sim 10^{-1}$gm cm^{-1}sec^{-1}.

At high frequencies, the displacement current begins to contribute and we must use the Ampere law as modified by Maxwell:

$$\vec{\nabla} \times \vec{B} = \frac{4\pi}{c}\vec{J} + \frac{1}{c}\frac{\partial \vec{E}}{\partial t},$$

with

$$\vec{E} = -\frac{\vec{U} \times \vec{B}}{c}. \tag{4.92}$$

The inclusion of the displacement current modifies the dispersion relation of the Alfven wave to

$$\omega = \frac{\vec{k} \cdot \vec{V}_A}{\left(1 + \dfrac{V_A^2}{c^2}\right)^{1/2}}. \tag{4.93}$$

In low density, astrophysical plasmas, the Alfven speed V_A can approach the speed of light c. We can also define the refractive index n_R for Alfven waves as:

$$n_R = \frac{kc}{\omega} = \left(1 + \frac{c^2}{V_A^2}\right)^{1/2}. \tag{4.94}$$

A conducting fluid has an index of refraction n_R for electromagnetic waves of frequency smaller than the electron-ion collision frequency when charge separation effects are negligible.

4.12. Gravitohydrodynamic Waves

Motions in the interior and the atmospheres of planets, stars and galaxies are influenced by their gravitational field in addition to other effects. Gravity produces stratification, as a result of which mass and therefore motion are inhomogeneously distributed. We obtain a case of nonuniform equilibrium in which mass density, pressure and temperature vary with space in the direction of the gravitational force. Waves in such a system acquire new characteristics such as low and high frequency cut-offs. The additional restoring force due to the gravitational field gives rise to new modes of oscillation. The observations of these waves provide information on the structure of the interior as well as the exterior of a gravitating body. Like the earth's seismology, helio-and astro-seismology have proved to be very effective probes of the solar and stellar interiors.

4.13. More on Hydrostatic Equilibrium

In the approximation of a static atmosphere of a gravitating body, the equilibrium is described by the balance of pressure gradient force and the gravitational force as:

$$-\vec{\nabla}p + \rho\vec{g} = 0, \qquad (4.95)$$

where $\vec{g}$ is the acceleration due to gravity and is assumed to be a constant. If $\vec{g} = -g\hat{z}$ then we find

$$\frac{dp}{dz} = -\rho g. \qquad (4.96)$$

Using the equation of state $p = (\rho K_B T/M)$ of a perfect gas of molecular weight M and temperature T, we get

$$\frac{dp}{p} = -\frac{dz}{H},$$

or

$$p = p_0 \exp\left[-\int_0^z \frac{dz}{H}\right], \qquad (4.97)$$

where p_0 is the pressure at $z = 0$ and $H = (K_B T/Mg)$ is known as the pressure scale height.

Problem 4.13: Estimate the pressure scale height in (i) the earth's atmosphere, (ii) the solar photosphere and (iii) the solar corona.

The rise and fall of a parcel of fluid in a stratified atmosphere is an important problem, since it has a bearing on the stability or instability of convective motions. Using the ideal gas law and the adiabatic equation of state, the pressure variation Equation (4.96) can be converted into a temperature variation equation:

$$\left.\frac{dT}{dz}\right|_{adiabatic} = \frac{(1-\gamma)gT}{c_s^2} \equiv -\Gamma_d, \qquad (4.98)$$

where $\left.(dT/dz)\right|_{adiabatic}$ is known as the **Adiabatic Temperature Gradient**. It is ~ 18 K/km for the earth's atmosphere. The stability of a fluid parcel rising vertically in an atmosphere depends on the temperature profile of the atmosphere. Let, in Figure (4.7), d represent the adiabatic lapse rate curve and d' represent the actual temperature variation of an atmosphere. If a fluid parcel rises adiabatically from its initial position at A, it will follow the curve d and reaches the point A' where it will be surrounded by the atmosphere at a temperature higher than its own, i.e., the fluid parcel is colder than the atmosphere in which it is rising and therefore falls back towards the point A. Point A on the curve d', therefore, is said to be stable. On the other hand a parcel beginning its journey from the point B encounters a colder atmosphere and therefore keeps rising. Point B on the curve d' is said to be unstable. Thus, a temperature gradient (dT/dz) represents

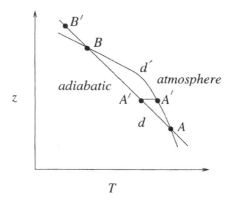

Figure 4.7.　Adiabatic Motion of a Fluid Parcel in an Atmosphere.

stable condition if $(-dT/dz) < \Gamma_d$ and unstable if $(-dT/dz) > \Gamma_d$. A fluid parcel can also execute oscillations in a stratified atmosphere. We shall now study these oscillations.

Problem 4.14: Derive Equation (4.98). Estimate Γ_d for the convective zone of the sun.

4.14. Adiabatic Gravito-acoustic Waves

We know that in an adiabatic process the total energy is conserved and there is no loss or gain of heat. What we mean by adiabatic oscillations is that their excitation and damping are not included while deriving their dispersion characteristics. The term gravito-acoustic implies that these oscillations are the result of the joint action of the gravitational and the pressure gradient restoring forces, in a stratified equilibrium. We study these oscillations in spherical geometry to facilitate their identification in a real system like a planet or a star. The hydrostatic equilibrium is described by the mass density $\rho_0(r)$ the pressure $p_0(r)$, the fluid velocity $\vec{U}_0 = 0$ and the acceleration due to gravity $g\hat{r}$ satisfying the equation:

$$\frac{dp_0}{dr} = -\rho_0 g. \tag{4.99}$$

We perturb the equilibrium quantities and write

$$
\begin{aligned}
p(\vec{r}, t) &= p_0(r) + p_1(\vec{r}, t), \\
\rho_0(\vec{r}, t) &= \rho_0(r) + \rho_1(\vec{r}, t), \\
\vec{U}(\vec{r}, t) &= \vec{U}_1(\vec{r}, t) = \frac{\partial \vec{S}}{\partial t},
\end{aligned}
\tag{4.100}
$$

where p_1, ρ_1, $\vec{U}_1$ and the displacement $\vec{S}$ are the perturbations at a given point $(\vec{r})$ (Eulerian Perturbations). We do not admit perturbations in $\vec{g}$. This is known as the Cowling approximation. The linearized equations of mass, momentum and energy conservation are:

$$\frac{\partial \rho_1}{\partial t} + \vec{\nabla} \cdot \left[\rho_0 \vec{U}_1\right] = 0,$$

or

$$\rho_1 + \vec{\nabla} \cdot \left[\rho_0 \vec{S}\right] = 0, \tag{4.101}$$

$$\rho_0 \frac{\partial^2 \vec{S}}{\partial t^2} = -\vec{\nabla}p_1 + \frac{\rho_1}{\rho_0}\vec{\nabla}p_0, \tag{4.102}$$

$$\frac{1}{\rho_0}\left[\frac{\partial \rho_1}{\partial t} + \vec{U}_1 \cdot \vec{\nabla}\rho_0\right] = \frac{1}{\gamma p_0}\left[\frac{\partial p_1}{\partial t} + \vec{U}_1 \cdot \vec{\nabla}p_0\right],$$

or

$$\frac{\rho_1}{\rho_0} + \frac{\vec{S} \cdot \vec{\nabla}\rho_0}{\rho_0} = \frac{1}{\gamma}\left[\frac{p_1}{p_0} + \frac{\vec{S} \cdot \vec{\nabla}p_0}{p_0}\right]. \tag{4.103}$$

We now express the perturbed quantities in terms of the Spherical Harmonic functions Y_l^m as (Christensen-Dalsgaard and Berthomieu 1991):

$$p_1(\vec{r}, t) = \sqrt{4\pi}\mathrm{Real}\left[p_1'(r)Y_l^m(\theta, \varphi)\exp(-i\omega t)\right]. \tag{4.104}$$

We learn from Equation (4.102) that the displacement $\vec{S}$ can be expressed as:

$$\vec{S}(\vec{r}, t) = \sqrt{4\pi}\mathrm{Real}\left[\left\{S_1(r)Y_l^m\hat{r} + S_2(r)\left(\frac{\partial Y_l^m}{\partial \theta}\hat{\theta}\right.\right.\right.$$
$$\left.\left.\left. + \frac{1}{\sin\theta}\frac{\partial Y_l^m}{\partial \phi}\hat{\phi}\right)\right\}\exp(-i\omega t)\right]. \tag{4.105}$$

After substituting Equations (4.104) and (4.105) in Equations (4.101)-(4.103), we get:

$$\frac{dS_1}{dr} = -\left(\frac{2}{r} + \frac{1}{\gamma}\frac{d\ln p_0}{dr}\right)S_1 + \frac{r\omega^2}{c_s^2}\left(\frac{\omega_L^2}{\omega^2} - 1\right)S_2$$

and

$$\frac{dS_2}{dr} = \frac{1}{r}\left(1 - \frac{\omega_{BV}^2}{\omega^2}\right)S_1 + \left(\frac{\omega_{BV}^2}{g} - \frac{1}{r}\right)S_2, \tag{4.106}$$

where

$$\omega_L = [l(l+1)]^{1/2}\frac{c_s}{r}, \tag{4.107}$$

is known as the **Lamb Frequency** and

$$\omega_{BV} = \pm\sqrt{g}\left[\frac{1}{\gamma}\frac{d\ln p_0}{dr} - \frac{d\ln\rho_0}{dr}\right]^{1/2}, \tag{4.108}$$

is known as the **Brunt-Väisälä Frequency**. We observe from Equation (4.106) that for purely radial perturbations, i.e, for $S_1 \neq 0$ and $S_2 = 0$, we get:

$$\omega^2 = \omega_{BV}^2, \tag{4.109}$$

and

$$S_1 = \text{constant } \exp\left[-\int\left(\frac{2}{r} + \frac{1}{r}\frac{d\ln p_0}{dr}\right)dr\right].$$

This mode is known as the **Internal Gravity Wave**. A fluid parcel of a density in excess of the density of the ambient medium, if displaced from its equilibrium position has a tendency to return to its equilibrium. A fluid parcel of lower density, however, feeling buoyant, will keep rising. For sufficiently slow motions, the fluid parcel maintains pressure equilibrium and undergoes density oscillations with frequency ω_{BV}. The Brunt-Väisälä frequency can be expressed in terms of the difference between the temperature gradient of an atmosphere and the adiabatic temperature gradient by using the equation of hydrostatic equilibrium, the ideal gas law and the adiabatic equation of state, so that:

$$\omega_{BV} = \pm\sqrt{\frac{g}{T_0}}\left[\left.\frac{dT_0}{dr}\right|_{atmosphere} - \left.\frac{dT_0}{dr}\right|_{adiabatic}\right]^{1/2}. \tag{4.110}$$

Problem 4.15 Convert Equation (4.108) into Equation (4.110).

We see that gravity oscillations exist only if the atmospheric temperature decreases with height more slowly than the adiabatic fall. Under the opposite circumstances, ω_{BV} becomes imaginary, the amplitude of the motion of the parcel increases at an exponential rate, the parcel moves farther and farther from its equilibrium location and the system is said to become unstable.

The condition that ω_{BV} is imaginary is known as the **Schwarzschild Criterion for Convective Instability**, since convection in a stratified medium may yield destabilizing circumstances, as for example, in the convection zone of the sun. If the acceleration due to gravity $\vec{g}$ makes an angle θ with the radial displacement S_1, then the gravitational force in the direction of the radial displacement becomes $g\cos\theta$. Consequently ω_{BV} must be replaced by $\omega_{BV}\cos\theta$ and the dispersion relation of the internal gravity waves is modified to:

$$\omega = \omega_{BV}\cos\theta. \tag{4.111}$$

This shows that ω_{BV} is the maximum frequency of the internal gravity waves.

We further observe from Equations (4.106) that for nonradial perturbations giving rise to toroidal oscillations, i.e., for $S_1 = 0$ and $S_2 \neq 0$, we get

$$\omega^2 = \omega_L^2 = k_h^2 c_s^2, \tag{4.112}$$

where $k_h = \dfrac{\sqrt{l(l+1)}}{r} \equiv \dfrac{L_h}{r}$ is known as the local horizontal wave number of the mode. Equation (4.112) describes the dispersion relation of the familiar sound waves. The effect of gravity on sound waves can be more easily deduced by solving Equations (4.101) - (4.103) in Cartesian geometry (Problem 4.16).

Problem 4.16: Assuming a plane wave variation of the type $\vec{U}_1 \propto e^{\alpha z} e^{i(k_x x + k_z z) - i\omega t}$, show that the sound waves satisfy a dispersion relation:

$$k_z^2 c_s^2 = \omega^2 - \left(k_x^2 + \frac{1}{4H^2}\right) c_s^2. \tag{4.113}$$

where H is the density scale height.

Thus, in a stratified atmosphere, there is a minimum frequency $\omega_{ac} = (c_s/2H)$, called the **Acoustic Cutoff**, below which the sound waves cannot propagate.

Of course, the full glory of Gravito-acoustic oscillations in spherical geometry can only be seen by solving Equations (4.106). In order to proceed, we need two boundary conditions, one at the center at $r = 0$ and the second at the surface at $r = R$.

For $r \to 0$, Equations (4.106) reduce to

$$\frac{dS_1}{dr} = -\frac{2S_1}{r} + \frac{L_h^2}{r} S_2, \tag{4.114}$$

and

$$\frac{dS_2}{dr} = \frac{S_1}{r} - \frac{S_2}{r}.$$

We can use the method of series solution for solving Equations (4.114) and we find that

$$S_1 = lS_2; \ S_1 \propto r^{l-1}, \ p_1, \rho_1 \propto r^l \ \text{for} \ r \to 0 \ \text{and} \ l \neq 0. \tag{4.115}$$

So, we have found one boundary condition.

Problem 4.17 Show that $S_1 \propto r$ for $l = 0$.

The second boundary condition is derived by assuming that there are no forces on the surface, which implies that the pressure perturbation vanishes on the perturbed surface, i.e.,

$$\Delta p = p_1(\vec{r}) + \frac{dp_0(r)}{dr} S_1 = 0 \ \text{ at } r = R, \qquad (4.116)$$

where Δp is the Lagrangian perturbation, the one that follows the motion. The elders have arrived at an approximate form of Equation (4.106) by defining a quantity ψ as (Lamb 1932, Deubner and Gough 1984):

$$\psi(r) = \rho_0^{1/2} c_s^2 \vec{\nabla} \cdot \vec{S}, \qquad (4.117)$$

which varies as

$$\frac{d^2\psi}{dr^2} + \frac{1}{c_s^2}\left[\omega^2 - \omega_{ac}^2 - \omega_L^2\left(1 - \frac{\omega_{BV}^2}{\omega^2}\right)\right]\psi = 0, \qquad (4.118)$$

and the acoustic-cutoff frequency for nonconstant density scale height is given by

$$\omega_{ac}^2 = \frac{c_s^2}{4H^2}\left[1 - 2\frac{dH}{dr}\right]. \qquad (4.119)$$

Equation (4.118) shows that ψ will be an oscillating function of r if

$$\omega^2 - \omega_{ac}^2 - \omega_L^2\left(1 - \frac{\omega_{BV}^2}{\omega^2}\right) > 0. \qquad (4.120)$$

Thus, the oscillations are confined to the regions defined by Equation (4.120) and they decay exponentially outside these regions. Each mode of oscillation has its own habitat, known as the trapping region, which in turn is determined by the actual variations of density and temperature in an atmosphere. Equation (4.120) admits two types of trapping regions; one, defined by $\omega^2 > \omega_{ac}^2$ and $\omega^2 > \omega_L^2$, is the high frequency region and is known as the **p-Mode Region** since it supports acoustic waves for which pressure provides the restoring force. The other region defined by $\omega^2 < \omega_{BV}^2$ is the low frequency region and is known as the **g-Mode Region** since it supports internal gravity modes for which gravity provides the restoring force. It is obvious that observations and identification of these modes provide a direct view of the internal structure of a gravitating body like a planet or a star. Due to this, **Helioseismology** has become, perhaps, the most important tool in the hands of the solar astronomers. A staggering number of modes have been seen in the high resolution observations of the solar photosphere. The GONG (Global Oscillation Network Group) project is providing round the clock observations of velocity fields on the solar surface. A large number of scientists are engaged in converting this information into the density and temperature structure of the solar interior. There is additional fine structure in the oscillation modes due to the presence of highly inhomogeneous magnetic fields and the differential rotation of the sun. An accurate separation of these effects promises a good diagnostics of the solar interior. The observed solar p-mode oscillations are shown in Figure (4.8). A model of the solar interior inferred from the observations of global oscillations is shown in Figure (4.9). For rotating bodies, the coriolis force provides one more type of restoring force. Waves excited in response to this restoring force are more typical of non-conducting fluids and will be studied in a later chapter.

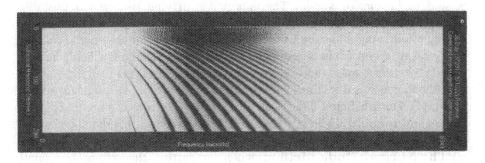

Figure 4.8. Spectrum of Observed Solar p-Mode Oscillations.

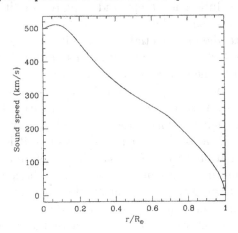

Figure 4.9. Model of the Solar Interior i.e. the Radial Variation of Sound Speed, Inferred From Helioseismological Observations.

4.15. Magnetohydrodynamic Instabilities

We have seen one type of response of a system to a small stimulus. The
system begins to oscillate about its equilibrium and never ventures too
far from it. It could happen, however, that the restoring forces are not
strong enough to bring the system back towards its equilibrium. On the
other hand, there are forces which drive the system farther and farther
from its equilibrium. This is a circumstance of an instability, a glimpse
of which we received for a rather fleeting moment when we arrived at the
Schwarzschild criterion for convective instability while studying internal
gravity waves. The frequency of the waves becomes imaginary because the
magnitude of the temperature gradient of the atmosphere was larger than
the adiabatic temperature gradient. As a result of this the perturbation in
density and velocity of the fluid begins to grow exponentially and the system
is said to become unstable. The energy for driving the system unstable
lies in the temperature gradient of the atmosphere. In the same vein, the
energy for driving a system unstable could be in any of the spatial and/or
temporal gradients of the other characteristics such as magnetic field, current
density, pressure, mass density, gravity or rotation of a magnetofluid. Thus
the inhomogeneous distribution of fluid parameters can lead to macroscopic
or **MHD Instabilities**. The system, in response, reconfigures itself to a
new equilibrium by shedding the excess energy stored in inhomogeneities.
There are two methods for exploring whether a system is stable or unstable.
(i) **The Normal Mode Method**, where we perturb the system by a small
extent, linearize the mass, momentum and energy conservation equations,
find a plane wave solution for them and explore conditions for complex
roots of the frequency; since the complex roots occur in pairs, one of the
roots corresponds to the exponential growth with time of the amplitude of
the perturbation and we have a case of an instability. It is of the utmost
importance to identify the source of energy responsible for the excitation of
the instability. (ii) **The Energy Principle**, where we perturb the system
by a small extent, linearize the relevant equations, calculate the potential
energy of the perturbed system; if the potential energy of the perturbed
system is larger than that of the unperturbed system, it is said to be stable
against this perturbation and vice-versa, since a system always likes to
acquire a state of minimum potential energy. The change in the potential
energy can then be related to the frequency of oscillations of the perturbed
quantities. The normal mode method is used when we wish to include more
complex physical processes like the finite Larmor radius effect or the Hall
current, whereas the energy principle is used when the complexity lies in
the geometrical configuration of fluids. We shall illustrate both the methods
through some examples.

4.16. The Rayleigh-Taylor Instabilities

The simplest example of the class of Rayleigh-Taylor (R-T) instabilities is the inverted glass full of water where the heavy fluid, water, is supported by the light fluid, air, at least for a few uncertain moments. This is a case of an unstable equilibrium since it is easily lost by a small air current. Thus, a fluid with an inverted density gradient, i.e., where the mass density increases in the direction of decreasing gravity is Rayleigh-Taylor unstable. In fact, we have already studied the instability! Let us recall the dispersion relation of the internal gravity waves, Equations (4.108) and (4.109):

$$\omega^2 = \omega_{BV}^2 = g \left[\frac{1}{\gamma} \frac{d\ln p_0}{dr} - \frac{d\ln \rho_0}{dr} \right] \tag{4.121}$$

$$= i^2 g \left[\frac{1}{H_\rho} - \frac{1}{\gamma H_p} \right]$$

$$\equiv i^2 \omega_I^2.$$

We observe that:

(i) The frequency ω becomes purely imaginary if the density scale height $H_\rho = (1/\rho_0)(d\rho_0/dr)$ is smaller than the pressure scale height $\gamma H_p = (1/p_0)(dp_0/dr)$. The perturbations therefore grow with time as $\exp[\omega_I t]$. The imaginary part ω_I of the frequency ω is called the **Growth Rate** of the instability. This is a case of Rayleigh-Taylor Instability for compressible perturbations for which $\vec{\nabla} \cdot \vec{U}_1 \neq 0$. Equation (4.121) can be recast in terms of the entropy $s \equiv p_0/\rho_0^\gamma$, so that the growth rate ω_I becomes

$$\omega_I = \left[\frac{g}{\gamma} \frac{d}{dr} \ln \left(\frac{1}{s} \right) \right]^{1/2}, \tag{4.122}$$

and the **Instability** is driven for negative entropy gradient $(ds/dr) < 0$ even when $(d\rho_0/dr) < 0$.

(ii) For $\gamma H_p \gg H_\rho$, i.e., for the incompressible case for which $\vec{\nabla} \cdot \vec{U}_1 = 0$, we get

$$\omega_I^2 = \frac{g}{H_\rho}. \tag{4.123}$$

We again have a **R-T Instability** if the density scale height H_ρ is positive, i.e., when the density increases in a direction opposite to that of the acceleration due to gravity. The growth rate is, now, $(g/H_\rho)^{1/2}$.

(iii) For an isothermal equation of state, $\gamma = 1$ and $p_0 = \dfrac{\rho_0 K_B T_0}{M}$. The scale height H_T for temperature is related to H_ρ and H_p as

$$\frac{1}{H_\rho} = \frac{1}{H_p} - \frac{1}{H_T}, \tag{4.124}$$

so that Equation (4.121) gives

$$\omega_I^2 = -\frac{g}{H_T}. \tag{4.125}$$

We, again, have R-T Instability for $H_T < 0$, i.e., if the temperature decreases in a direction opposite to $\vec{g}$. The growth rate is, now, $(g/H_T)^{1/2}$.

In conclusion, in the absence of the Lorentz force, either due to the absence of the magnetic field or due to the magnetic field being force free, the Rayleigh-Taylor instability is excited (i) when a heavy fluid lies at the top of a light fluid, (ii) when a cold fluid lies at the top of a hot fluid and (iii) when the upper fluid has lower entropy than the lower fluid. The result of R-T instability is the mixing of either different fluids or different parts of a fluid. The internal gravity waves in their stable and unstable form (R-T instability) are believed to play an important role in the mixing of elements and distribution of angular momentum in the radiative zones of stars. The passage of a shock wave through the stratified distribution of heavy elements in a star during a supernova explosion also creates circumstances of R-T instability with a typical exponentiation time of $\sim 10^4 \sec$.

4.17. Rayleigh-Taylor Instability in Magnetized Fluid

Astrophysical fluids are, more often than not, magnetized. We study the effect of magnetic field on the growth rate of the R-T instability for two cases: (i) when the magnetic field $\vec{B}$ is parallel to the acceleration due to gravity $\vec{g}$ and (ii) when $\vec{B}$ is perpendicular to $\vec{g}$. We shall neglect all nonideal effects. We shall use what is known as the **Boussinesq Approximation** to deal with the density variations. Under this approximation, we neglect all changes in density except where they are coupled with external forces like gravity. It is valid when small changes in temperature lead to small changes in density due to smallness of the coefficient of volume expansion. So, we have $\vec{\nabla} \cdot \vec{U} = 0$ along with $\rho_1 \vec{g} \neq 0$.

(i) $\vec{g} = -g\hat{z}$; $\vec{B}_0 = B_0\hat{z}$ and $\rho_0 = \rho_0(z)$

Assuming a spatial and time dependence of the perturbed quantities of the form:

$$\vec{U}_1 = \vec{U}_1(z)e^{ik_x x + ik_y y + i\omega t}, \tag{4.126}$$

we write the linearized MHD equations including the gravitational forces as:

$$i\omega\rho_0 U_{1x} - \frac{B_0}{4\pi}\left[B_{1x}' - ik_x B_{1z}\right] = -ik_x p_1, \tag{4.127}$$

$$i\omega\rho_0 U_{1y} + \frac{B_0}{4\pi}\left[ik_y B_{1z} - B_{1y}'\right] = -ik_y p_1, \tag{4.128}$$

$$i\omega\rho_0 U_{1z} = -p_1' - \rho_1 g, \tag{4.129}$$

$$i\omega\rho_1 + \rho_0' U_{1z} = 0, \tag{4.130}$$

$$\vec{\nabla}\cdot\vec{U}_1 = \vec{\nabla}\cdot\vec{B}_1 = 0, \tag{4.131}$$

$$i\omega\vec{B}_1 = B_0\vec{U}_1'. \tag{4.132}$$

The prime represents the derivative with respect to z. The elimination process consists of multiplying Equations (4.127) and (4.128) by $(-ik_x)$ and $(-ik_y)$ respectively, adding them and using Equations (4.131) and (4.132). We get:

$$\frac{\partial}{\partial z}\left[\rho_0\frac{\partial}{\partial z}U_{1z}\right] + \frac{B_0^2}{4\pi\omega^2}\left[\frac{\partial^2}{\partial z^2} - k^2\right]\frac{\partial^2 U_{1z}}{\partial z^2} = k^2\rho_0 U_{1z} + \frac{gk^2}{\omega^2}U_{1z}\frac{\partial\rho_0}{\partial z}, \tag{4.133}$$

where $k^2 = k_x^2 + k_y^2$.

We, now, need boundary conditions to solve this equation. Let the fluid be confined between two boundaries at $z = z_1$, and $z = z_2$ (Figure 4.10). If

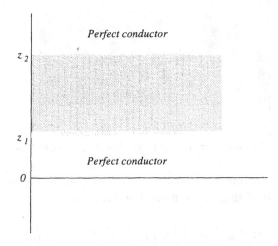

Figure 4.10.   Fluid Confined in the Region $z_1 \le z \le z_2$.

the boundaries are rigid, there can be no motion across them and $U_{1z}(z_1) = U_{1z}(z_2) = 0$. If the medium (at $z < z_1$ and $z > z_2$) adjacent to the fluid is a perfect conductor, then no magnetic field can cross the boundary (the fluid itself has been assumed to be a perfect conductor) and $B_{1z} = 0$ and $E_{1x} = E_{1y} = 0$ on the plane boundary. If the medium adjacent to the fluid is nonconducting, then no current can cross the boundary and $J_z = 0$; the magnetic field at $z < z_1$ and $z > z_2$ must correspond to a vacuum field. The continuity of the tangential stresses requires that B_{1x} and B_{1y}

are continuous which, due to Equations (4.131) and (4.132), implies the continuity of U'_{1z} and U''_{1z}.

By multiplying Equation (4.133) by U_{1z} and integrating from z_1 to z_2, it can be shown that $(i\omega)^2$ is real. Now, integrate Equation (4.133) across the boundary, say from $z_1 - \delta$ to $z_1 + \delta$, where $\delta \to 0$. We find:

$$\Delta \left[\rho_0 U'_{1z} + \frac{B_0^2}{4\pi\omega^2} \left(U'''_{1z} - k^2 U'_{1z} \right) \right] = \frac{gk^2}{\omega^2} U_{1z}(z = z_1)\Delta[\rho_0], \qquad (4.134)$$

where $\Delta[\ \ ]$ represents the jump in the value of the bracketed quantity across the boundary. Let us consider two fluids of uniform densities ρ_1 and

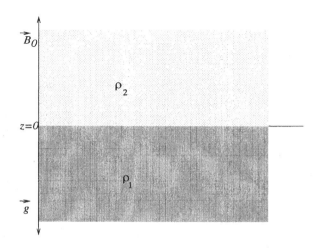

Figure 4.11. The Rayleigh-Taylor Instability in a Vertical Magnetic Field.

ρ_2 separated by an interface at $z = 0$. (Figure 4.11) Equation (4.133) is applicable to each of these regions and reduces to

$$U''''_{1z} - \left(k^2 - \frac{\omega^2}{V_A^2} \right) U''_{1z} - \frac{k^2 \omega^2}{V_A^2} U_{1z} = 0. \qquad (4.135)$$

The general solution of Equation (4.135) for $(i\omega) > 0$ can be written as:

$$\begin{aligned}
(U_{1z})_1 &= A_1 e^{kz} + B_1 e^{q_1 z} \quad \text{for } z < 0, \\
(U_{1z})_2 &= A_2 e^{-kz} + B_2 e^{-q_2 z} \quad \text{for } z > 0,
\end{aligned} \qquad (4.136)$$

where $q_1 = \dfrac{i\omega}{V_{A1}}$ and $q_2 = \dfrac{i\omega}{V_{A2}}$.

Recall that the V_A's are the Alfven speeds. The solutions (4.136) have been written to ensure that they vanish at $z = \pm\infty$. The constants A's and

B's are to be determined from the continuity of U_{1z}, U'_{1z} and U''_{1z} at the interface $z = 0$ along with the condition Equation (4.134), which can be written explicitly as:

$$\left[\rho_2 (U'_{1z})_2 - \rho_1 (U'_{1z})_1 - \left\{ \frac{1}{q_2^2} (U'''_{1z})_2 - k^2 (U'_{1z})_2 \right. \right.$$

$$\left. \left. - \frac{1}{q_1^2} (U'''_{1z})_1 + k^2 (U'_{1z})_1 \right\} \right]_{z=0} = \frac{gk^2}{\omega^2} (\rho_2 - \rho_1) U_{1z}(z = 0). \qquad (4.137)$$

The continuity of U_{1z}, for example, translates to

$$A_1 + B_1 = A_2 + B_2.$$

Similarly, substituting Equation (4.136) in the other continuity conditions as well as in Equation (4.137), we obtain four equations, whose solution requires that the determinant

$$D \equiv \begin{vmatrix} 1 & 1 & -1 & -1 \\ k & q_1 & k & q_2 \\ k^2 & q_1^2 & -k^2 & -q_2^2 \\ \dfrac{\bar{R}}{2} - \alpha_1 & \dfrac{\bar{R}}{2} - \dfrac{\alpha_1 k}{q_1} & \dfrac{\bar{R}}{2} - \alpha_2 & \dfrac{\bar{R}}{2} - \dfrac{\alpha_2 k}{q_2} \end{vmatrix} = 0, \qquad (4.138)$$

where $\alpha_1 = \dfrac{\rho_1}{\rho_1 + \rho_2}$, $\alpha_2 = \dfrac{\rho_2}{\rho_1 + \rho_2}$ and $\bar{R} = -\dfrac{gk}{\omega^2}(\alpha_2 - \alpha_1)$.

By using the standard properties of determinants, D can be simplified to get:

$$(q_1 - k)(q_2 - k) \left[(\bar{R} - 1)(q_1 + q_2 + 2k) - \right.$$

$$\left. 2k \left\{ \frac{\alpha_2}{q_2}(q_1 + k) + \frac{\alpha_1}{q_1}(q_2 + k) \right\} \right] = 0. \qquad (4.139)$$

Problem 4.18 Verify that the two roots $q_1 = k = q_2$ correspond to the solutions $(U_{1z})_1 = (U_{1z})_2 = 0$.

Therefore, we must have

$$(\bar{R} - 1)(q_1 + q_2 + 2k) - 2k \left\{ \frac{\alpha_2}{q_2}(q_1 + k) + \frac{\alpha_1}{q_1}(q_2 + k) \right\} = 0, \qquad (4.140)$$

or

$$-i\omega^3 - \omega^2 \left[2(\sqrt{\alpha_1} + \sqrt{\alpha_2})kV_{AT} \right] + i\omega \left[2k^2 V_{AT}^2 - gk(\alpha_2 - \alpha_1) \right]$$

$$+ 2gk^2 V_{AT} \left[\sqrt{\alpha_1} - \sqrt{\alpha_2} \right] = 0, \qquad (4.141)$$

where $V_{AT}^2 = \dfrac{B_0^2}{4\pi(\rho_1 + \rho_2)}$.

We can look at the asymptotic solutions of Equation (4.141):
for $k \to 0$

$$(i\omega)^2 \to gk(\alpha_2 - \alpha_1) \equiv \omega_I^2, \qquad (4.142)$$

which corresponds to the hydrodynamic R-T instability for $\alpha_2 > \alpha_1$. This shows that the large wavelength perturbations are unaffected by magnetic field.

For $k \to \infty$

$$(i\omega) \to \frac{g}{V_{AT}}(\sqrt{\alpha_2} - \sqrt{\alpha_1}) \equiv \omega_I, \qquad (4.143)$$

i.e., the growth rate tends to a fixed value independent of k.

Problem 4.19: Argue that the asymptotic solution (4.142) is correct.

Thus, the growth rate ω_I given by Equation (4.140) increases linearly with k for small values of k and becomes independent of k for large values of k.

Problem 4.20: Show that for $\alpha_2 < \alpha_1$, ω_I is imaginary.

The case $\alpha_2 < \alpha_1$ represents a stable configuration. The system responds to small perturbations by exciting Alfven oscillations. The propagation characteristics of these oscillations in an homogeneous medium are discussed in Chandrasekhar (1961).

Problem 4.21: Prove that, in the steady state, the perturbed velocity $\vec{U}_1$ cannot vary in the direction of the imposed uniform field $\vec{B}_0$. This shows that the magnetic field has a tendency to two-dimensionlize slow motions.

(ii) $\vec{g} = -g\hat{z}$; $\vec{B}_0 = B_0\hat{x}$ and $\rho_0 = \rho_0(z)$ (Figure 4.12)

The linearized MHD equations for this case are

$$i\omega\rho_0 U_{1x} = -ik_x p_1, \qquad (4.144)$$

$$i\omega\rho_0 U_{1y} = -ik_y p_1 + \frac{B_0}{4\pi}\left(ik_x B_{1y} - ik_y B_{1x}\right), \qquad (4.145)$$

$$i\omega\rho_0 U_{1z} = -p_1' - \frac{B_0}{4\pi}\left(B_{1x}' - ik_x B_{1z}\right) - \rho_1 g, \qquad (4.146)$$

$$i\omega\rho_0 + \rho_0' U_{1z} = 0, \qquad (4.147)$$

$$\vec{\nabla} \cdot \vec{U}_1 = \vec{\nabla} \cdot \vec{B}_1 = 0, \qquad (4.148)$$

$$i\omega\vec{B}_1 = ik_x B_0 \vec{U}_1. \qquad (4.149)$$

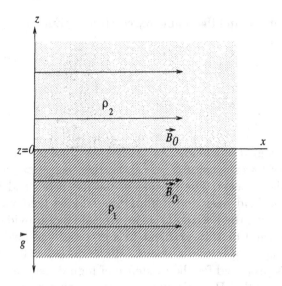

Figure 4.12.   The Rayleigh-Taylor Instability in a Horizontal Magnetic Field

Again, carrying out the elimination process, we get:

$$\frac{\partial}{\partial z}\left[\rho_0 \frac{\partial U_{1z}}{\partial z}\right] - \frac{B_0^2 k_x^2}{4\pi\omega^2}\left[\frac{\partial^2}{\partial z^2} - k^2\right]U_{1z} - k^2\rho_0 U_{1z} = \frac{gk^2}{\omega^2}\rho_0' U_{1z}. \quad (4.150)$$

The boundary conditions are the continuity of U_{1z} and B_{1z} at $z = 0$. Integrating Equation (4.150) across the boundary, as before, we find:

$$\Delta\left[\rho_0 U_{1z}'\right] - \frac{k_x^2 B_0^2}{4\pi\omega^2}\Delta\left[U_{1z}'\right] = \frac{gk^2}{\omega^2}U_{1z}(z=0)\Delta[\rho_0]. \quad (4.151)$$

For the case of two fluids of uniform densities ρ_1 (at $z < 0$) and ρ_2 (at $z > 0$) separated by an interface at $z = 0$, equation (4.150), for each fluid, becomes

$$\left(1 - \frac{k_x^2 V_A^2}{\omega^2}\right)\left[U_{1z}'' - k^2 U_{1z}\right] = 0. \quad (4.152)$$

The solutions of Equations (4.152) for each region can be written as:

$$(U_{1z})_1 = Ae^{kz} \quad \text{for} z < 0,$$

and

$$(U_{1z})_2 = Be^{-kz} \quad \text{for} z > 0. \quad (4.153)$$

Using these solutions and the boundary conditions, Equation (4.151) can be simplified to give:

$$(i\omega)^2 \equiv \omega_I^2 = \frac{1}{(\rho_1 + \rho_2)} \left[kg(\rho_2 - \rho_1) - \frac{2k_x^2 B_0^2}{4\pi} \right]. \tag{4.154}$$

We, first notice that the growth rate ω_I is reduced by the presence of the horizontal magnetic field B_0. As the heavier fluid of density ρ_2 tries to sink into fluid of lower density ρ_1, it carries the magnetic field also with it, bending it in the process. The tension in the magnetic field however tries to straighten the field lines and inhibits the sinking tendency of the fluid (Figure 4.13). Thus, the horizontal magnetic field can provide a support to a fluid with inverted density gradient in a gravitational field, against perturbations propagating in the horizontal direction ($k_x \neq 0$). This is one of the mechanisms proposed for the existence of high density and low temperature structures, called **Prominences**, in the solar corona. The density of prominence is about 100-1000 times and the temperature is about 0.01 that of the solar corona and they are embedded in a magnetic field of the order of 10 Gauss.

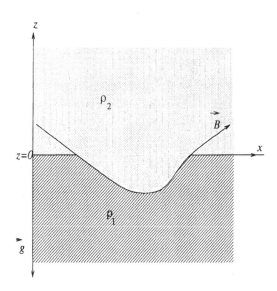

Figure 4.13. The Curved Field $\vec{B}$ Develops Tension Which Inhibits the Instability

Problem 4.22 Estimate the maximum wavelength of a perturbation for which a solar prominence is Rayleigh-Taylor stable.

4.18. The Kelvin Helmholtz Instability

So far, we have considered fluids of different densities lying over each other in the presence of gravitational and magnetic forces. We can give several examples where fluids of varying densities coexist in relative motion. Wind flowing over oceans, cometary tails whizzing against the solar wind, accreting flows around compact objects, propagating extragalactic jets and exploding supernovae ejecta are a few familiar sites. Such configurations of streaming fluids may have discontinuities in their flow speeds. This excess kinetic energy could drive the systems unstable. The resulting instability is known as the **Kelvin Helmholtz (K-H) Instability**. We shall study the development of K-H instability including the magnetic field. The direction of the magnetic field is specified in relation to the direction of streaming. We shall consider two cases (i) when the magnetic field is parallel to the flow velocity $\vec{U}_0$ and (ii) when it is perpendicular to the flow velocity.

(i) $\vec{B}_0 = B_0 \hat{x}; \quad \vec{U}_0 = U_0(z)\hat{x}; \quad \rho_0 = \rho_0(z); \quad \vec{g} = -g\hat{z}$ (Figure 4.14)

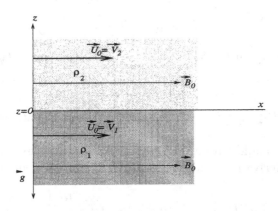

Figure 4.14.   The Kelvin-Helmholtz Instability for $\vec{U}_0 \parallel \vec{B}_0$

The linearized MHD equations for spatial and temporal dependence of the perturbed quantities given by Equation (4.126) are:

$$i\,\bar{\omega}\rho_0 U_{1x} + \rho_0 U_0' U_{1z} = -ik_x p_1, \tag{4.155}$$

$$i\bar{\omega}\rho_0 U_{1y} = -ik_y p_1 + \frac{B_0}{4\pi}\left[ik_x B_{1y} - ik_y B_{1x}\right], \tag{4.156}$$

$$i\bar{\omega}\rho_0 U_{1z} = -p_1' - \rho_1 g + \frac{B_0}{4\pi}\left[-B_{1x}' + ik_x B_{1z}\right], \tag{4.157}$$

$$i\bar{\omega}\vec{B}_1 = ik_x B_0 \vec{U}_1 + B_{1z}\vec{U}_0', \tag{4.158}$$

$$i\bar{\omega}\rho_1 + U_{1z}\rho_0' = 0, \tag{4.159}$$

$$ik_x U_{1x} + ik_y U_{1y} + U_{1z}' = 0, \tag{4.160}$$

$$ik_x B_{1x} + ik_y B_{1y} + B_{1z}' = 0, \tag{4.161}$$

where

$$\bar{\omega} \equiv \omega + k_x U_0. \tag{4.162}$$

The elimination procedure can be carried out by performing the following sequence of operations:

(1) Determine B_{1x}, B_{1y} and B_{1z} from Equations (4.158), substitute in Equations (4.156) and (4.157) and use Equation (4.159). This will give us equations we call as (4.156a) and (4.157a).

(2) Multiply equations (4.155) by (ik_y) and Equation (4.156a) by (ik_x) and add to get:

$$k_x U_{1y} - k_y U_{1x} = \frac{k_y U_{1z} U_0'}{i\bar{\omega}}. \tag{4.163}$$

Equation (4.156a) becomes

$$\bar{\omega}\rho_0 U_{1y} = -k_y p_1. \tag{4.164}$$

(3) Multiply Equation (4.155) by $(-k_x)$ and (4.164) by $(-k_y)$ and add to get:

$$i\bar{\omega}\rho_0 U_{1z}' - i\rho_0 k_x U_0' U_{1z} = -k^2 p_1. \tag{4.165}$$

(4) Determine U_{1x} and U_{1y} in terms of U_{1z} from Equations (4.163) and (4.160).

By now, we should have all the perturbed quantities expressed in terms of U_{1z} and its derivatives with respect to z. The master equation is:

$$
[i\rho_0 \bar{\omega} U_{1z}' - i\rho_0 k_x U_0' U_{1z}]' = i\rho_0 \bar{\omega} k^2 U_{1z} + i\frac{k_x^2 B_0^2}{4\pi}\left[\left(\frac{U_{1z}'}{\bar{\omega}}\right)' - \frac{k^2 U_{1z}}{\bar{\omega}}\right]
$$
$$
-\frac{k_x^3 B_0^2}{4\pi}\left[\frac{U_0' U_{1z}}{\bar{\omega}^2}\right]' + \frac{i\rho_0' g k^2 U_{1z}}{\bar{\omega}}. \tag{4.166}
$$

As for the case of R-T instability, let us consider a fluid of uniform density ρ_2 and uniform velocity $V_2\hat{x}$ in the region $z > 0$ streaming over a fluid of uniform density ρ_1 and uniform velocity $V_1\hat{x}$ in the region $z < 0$ with an interface at $z = 0$, so that in each region $U_0' = 0$ and $\rho_0' = 0$. Equation (4.166) valid for each fluid becomes:

$$
i\left[\rho_0\bar{\omega} - \frac{k_x^2 B_0^2}{4\pi\bar{\omega}}\right]\left[U_{1z}'' - k^2 U_{1z}\right] = 0. \tag{4.167}
$$

The solutions in the two regions can be written as:

$$(U_{1z})_1 = Ae^{kz} \text{ for } z < 0,$$
$$(U_{1z})_2 = Be^{-kz} \text{ for } z > 0. \tag{4.168}$$

The boundary conditions applicable at the interface $z = 0$ are (i) U_{1z} must be continuous and (ii) the displacement $S_z = U_{1z}/(i\omega + ik_x U_0)$ must be continuous at $z = 0$; the two boundary conditions become identical if U_0 is continuous. Thus, the solutions satisfying these conditions are:

$$(U_{1z})_1 = (i\omega + ik_x V_1)e^{kz},$$
$$(U_{1z})_2 = (i\omega + ik_x V_2)e^{-kz}. \tag{4.169}$$

The additional boundary condition to be satisfied is obtained by integrating Equation (4.166) across the interface. We find:

$$\Delta\left[\rho_0\left(i\omega + ik_x U_0\right)U'_{1z}\right] = \frac{-k_x^2 B_0^2}{4\pi}\Delta\left[\frac{U'_{1z}}{i\omega + ik_x U_0}\right]$$
$$-gk^2\left[\frac{U_{1z}}{i\omega + ik_x U_0}\right]_{z=0}\Delta[\rho_0]. \tag{4.170}$$

Substituting the solutions Equation (4.169) in Equation (4.170), we get:

$$\rho_2\left(\omega + k_x V_2\right)^2 + \rho_1\left(\omega + k_x V_1\right)^2 = \frac{k_x^2 B_0^2}{2\pi} - gk(\rho_2 - \rho_1). \tag{4.171}$$

The roots of Equation (4.171) are:

$$\omega = -k_x\left[\alpha_1 V_1 + \alpha_2 V_2\right] \pm \left[gk(\alpha_1 - \alpha_2) + 2k_x^2 V_A^2 - k_x^2\alpha_1\alpha_2(V_1 - V_2)^2\right]^{1/2}. \tag{4.172}$$

We notice that:
For $V_1 = V_2 = V_A = 0$, we recover the growth rate of the hydrodynamic R-T instability for $\alpha_2 > \alpha_1$;
For $V_1 = V_2 = 0$, we recover the growth rate for the R-T instability for $\vec{B}_0 \perp \vec{g}$. For the excitation of the **K-H Instability**, in the absence of a magnetic field, i.e., for $V_A = 0$, we must have

$$k_x^2\alpha_1\alpha_2(V_1 - V_2)^2 > gk(\alpha_1 - \alpha_2), \tag{4.173}$$

for $\alpha_1 > \alpha_2$. For $\alpha_1 < \alpha_2$, the expression inside the square root is negative and we have an imaginary part of the frequency ω. There is also a real part of ω. Thus this is an oscillatory instability - the amplitude of the perturbed quantity oscillates as well as grows with time. The real part ω_R is

$$\omega_R = -k_x\left[\alpha_1 V_1 + \alpha_2 V_2\right], \tag{4.174}$$

and the imaginary part ω_I is:

$$\omega_I = \pm\left[-gk(\alpha_1 - \alpha_2) + k_x^2\alpha_1\alpha_2(V_1 - V_2)^2\right]^{1/2}. \tag{4.175}$$

The instability corresponds to the negative sign in Equation (4.175). The **K-H Instability** exists even when $\alpha_1 = \alpha_2$. Thus, the energy for the excitation comes from the energy of the relative motion. We see from Equation (4.172) that the magnetic field has an inhibiting effect on the K-H instability. For $\alpha_1 = \alpha_2$ the system is stable if:

$$V_A^2 \geq \frac{\alpha_1^2}{2}(V_1 - V_2)^2. \tag{4.176}$$

It is again the tension of the magnetic field lines that provides the stabilizing force.

(ii) Magnetic Field $\vec{B}_0 \perp \vec{U}_0$ (Figure 4.15)

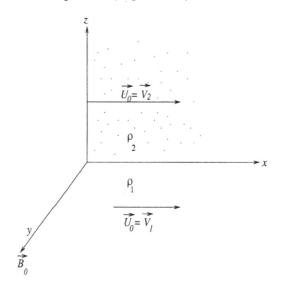

Figure 4.15. The Kelvin-Helmholtz Instability for $\vec{U}_0 \perp \vec{B}_0$

$$\vec{B}_0 = B_0\hat{y}; \quad \vec{U}_0 = U_0(z)\hat{x}; \quad \rho_0 = \rho_0(z); \quad \vec{g} = -g\hat{z}.$$

The linearized MHD equations are:

$$i\bar{\omega}\rho_0 U_{1x} + \rho_0 U_0' U_{1z} = -ik_x p_1 - \frac{B_0}{4\pi}\left[ik_x B_{1y} - ik_y B_{1x}\right], \tag{4.155'}$$

$$i\bar{\omega}\rho_0 U_{1y} = -ik_y p_1, \tag{4.156'}$$

$$i\bar{\omega}\rho_0 U_{1z} = -p'_1 - \rho_1 g + \frac{B_0}{4\pi}\left[ik_y B_{1z} - B'_{1y}\right],\qquad(4.157')$$

$$i\bar{\omega}\vec{B}_1 = ik_y B_0 \vec{U}_1 + B_{1z}\vec{U}'_0\qquad(4.158').$$

Equations (4.159), (4.160) and (4.161) remain valid for this case too. Following the procedure used in the previous case, we get the equivalent of Equation (4.166):

$$[i\rho_0\bar{\omega}U'_{1z} - i\rho_0 k_x U'_0 U_{1z}]' = i\rho_0\bar{\omega}k^2 U_{1z} + \frac{ik_y^2 B_0^2}{4\pi}\left[\left(\frac{U'_{1z}}{\bar{\omega}}\right)' - \frac{k^2 U_{1z}}{\bar{\omega}}\right]$$
$$- \frac{k_x k_y^2 B_0^2}{4\pi}\left[\frac{U'_0 U_{1z}}{\bar{\omega}^2}\right]' + \frac{i\rho'_0 g k^2 U_{1z}}{\bar{\omega}}\qquad(4.166').$$

We notice that for $k_y = 0$, the magnetic field has no effect.

Problem 4.23: By integrating Equation (4.166') across the interface $z = 0$ for the case of a fluid of uniform density ρ_2 and velocity $V_2\hat{x}$ occupying the region $z > 0$ and the fluid of uniform density ρ_1 and velocity $V_1\hat{x}$ occupying the region $z < 0$, show that Equation (4.170) with $(k_x^2 B_0^2)$ replaced by $(k_y^2 B_0^2)$ is obtained . Show that the frequency ω is given by:

$$\omega = -k_x\left[\alpha_1 V_1 + \alpha_2 V_2\right] \pm \left[gk(\alpha_1 - \alpha_2) + 2k_y^2 V_A^2 - k_x^2\alpha_1\alpha_2(V_1 - V_2)^2\right]^{1/2}.$$
$$(4.172')$$

We can again appreciate the stabilizing effect of the magnetic field on the K-H instability for $k_y \neq 0$, i.e., for obliquely propagating perturbations. The signatures of the K-H instability look like kinks and wavy shapes at the interface. The tendency is to reduce the shear at the interface and set up a smoother velocity gradient. The action of K-H instability has been seen at the surfaces of cometary tails which are in relative motion with the solar wind and at the edges of galactic and extragalactic jets (Figure 4.16).

Problem 4.24 Estimate the growth rate of the K-H instability for the solar wind- cometary system. The typical values of the density and velocity of the solar wind are ~ 10 protons/cm^3 and ~ 500 km/sec and of a cometary tail are ~ 100 protons/cm^3 and 1000 km/sec, respectively.

4.19. Current Driven Instabilities

Until now, we have considered the effect of a uniform magnetic field on the stability of a fluid possessing density, temperature, pressure or velocity gradients. There is another class of instabilities driven by an electric current flowing through a conducting fluid. The presence of the current is associated with an inhomogeneous magnetic field. The stability or otherwise of the fluid

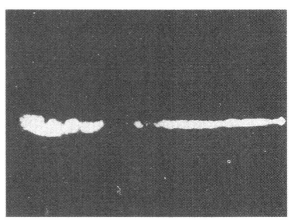

Figure 4.16.   (a) The Tail of the Comet Ikeya-Seki (b) Jet in the Source NGC 6251 Imaged at the Wavelength 22cm with VLA.

depends upon the magnitude and spatial variation of the current density. The instabilities deform the fluid surface which may develop 'waists' or the entire fluid may acquire serpentine characteristics and display DNA- like helices.

We shall study these instabilities in a cylindrical fluid column using the **Energy Principle**. All equilibrium quantities are functions of only the radial coordinates r of the cylindrical coordinates (r, θ, z). In the absence of gravitational forces and non-ideal effects, the equilibrium of a stationary conducting fluid is described by the balance of the pressure gradient and the Lorentz force as:

$$\frac{\partial p_0}{\partial r} = \frac{1}{c} \left[J_{0\theta} B_{0z} - J_{0z} B_{0\theta} \right], \tag{4.177}$$

where

$$J_{0\theta} = -\frac{c}{4\pi} \frac{\partial B_{0z}}{\partial r}, \tag{4.178}$$

and

$$J_{0z} = \frac{c}{4\pi} \frac{1}{r} \frac{\partial}{\partial r} (r B_{0\theta}).$$

Equation (4.177) can be written as:

$$\frac{\partial}{\partial r} \left(p_0 + \frac{B_{0z}^2}{8\pi} \right) + \frac{1}{4\pi r} B_{0\theta} \frac{\partial}{\partial r} (r B_{0\theta}) = 0. \tag{4.179}$$

It is instructive to define a quantity q related to the pitch of a magnetic helix of length L as:

$$q(r) = \frac{2\pi r B_{0z}(r)}{L B_{0\theta}(r)}. \tag{4.180}$$

The stability condition can then be expressed in terms of the value of q at any place in the fluid.

Problem 4.25: Show that the pitch of a magnetic field with components B_z and B_θ in a cylinder of radius R and length L is $(L B_\theta / 2\pi R B_z)$.

The Energy Principle of stability is best illustrated through the displacement vector $\vec{S}(\vec{r}, t)$ defined earlier. The distinction between the Lagrangian and the Eulerian coordinates becomes insignificant for the small displacements considered here when using the linearized MHD equations.

The linearized equations for the perturbed magnetic field

$$\frac{\partial \vec{B}_1}{\partial t} = \vec{\nabla} \times (\vec{U}_1 \times \vec{B}_0),$$

can be integrated once in order to express it in terms of the displacement vector $\vec{S}$ as:

$$\vec{B}_1(\vec{r}, t) = \vec{\nabla} \times (\vec{S} \times \vec{B}_0), \tag{4.181}$$

using the initial conditions that all perturbations vanish at $t = 0$. Similarly, the mass and energy conservation equations become:

$$p_1(\vec{r}, t) = -\vec{S} \cdot \vec{\nabla} p_0 - \gamma p_0 \vec{\nabla} \cdot \vec{S}. \tag{4.182}$$

The momentum conservation equation can be written as:

$$
\begin{aligned}
\rho_0 \frac{\partial^2 \vec{S}}{\partial t^2} =\ & \vec{\nabla}\left[\vec{S} \cdot \vec{\nabla} p_0 + \gamma p_0 \vec{\nabla} \cdot \vec{S}\right] + \frac{1}{4\pi}\left(\vec{\nabla} \times \vec{B}_0\right) \times \left[\vec{\nabla} \times \left(\vec{S} \times \vec{B}_0\right)\right] \\
& + \frac{1}{4\pi}\left[\vec{\nabla} \times \left\{\vec{\nabla} \times \left(\vec{S} \times \vec{B}_0\right)\right\}\right] \times \vec{B}_0.
\end{aligned}
\tag{4.183}
$$

Now, multiplying Equation(4.183) by the time derivative of the displacement vector $(\partial \vec{S}/\partial t)$ and integrating over the volume of the fluid, using the rigid boundary condition that the normal component of $\vec{S}$ vanishes at the boundary, we find:

$$\frac{\partial}{\partial t} \int \frac{1}{2} \rho_0 \left(\frac{\partial \vec{S}}{\partial t}\right)^2 d\vec{r} = \int \frac{\partial \vec{S}}{\partial t} \cdot \vec{F}(\vec{S}) d\vec{r}, \tag{4.184}$$

where $\vec{F}(\vec{S})$ is given by the right hand side of Equation (4.183). Since Equation (4.183) represents linear oscillations, the force density $\vec{F}(\vec{S})$ must be proportional to the displacement vector $\vec{S}$. Under these circumstances, Equation (4.184) can be written as:

$$\frac{\partial}{\partial t}\left[\int \frac{1}{2}\rho_0 \left(\frac{\partial \vec{S}}{\partial t}\right)^2 d\vec{r} - \int \frac{1}{2}\vec{S}\cdot\vec{F}(\vec{S})d\vec{r}\right] = 0. \qquad (4.185)$$

A more rigorous derivation of Equation (4.185) can be seen in Kadomtsev (1963). Equation (4.185) is a statement of the conservation of the total perturbed energy E_1 which is a sum of the perturbed kinetic energy T_1 and the perturbed potential energy W_1 defined as:

$$W_1 = -\frac{1}{2}\int \vec{S}\cdot\vec{F}(\vec{S})d\vec{r}. \qquad (4.186)$$

Thus, if $E_1 = T_1 + W_1 =$ constant, then any perturbation that causes a reduction in W_1 must enhance T_1 and the system is said to be unstable for this perturbation. The instability results whenever the system goes to a state of lower potential energy, with a net negative change in the potential energy, i.e., with $W_1 < 0$. Alternatively, if the perturbations lead to an increase of potential energy ($W_1 > 0$), then the system is stable. In reality, compared to stability, instability is easier to prove. The energy principle can also be used to determine the frequencies of oscillations. Assuming a time dependence for $\vec{S}$ of the form $\exp(-i\omega t)$, we find from Equation (4.183)

$$\omega^2 = -\frac{\int \vec{S}\cdot\vec{F}(\vec{S})d\vec{r}}{\int \rho_0(\vec{S}\cdot\vec{S})d\vec{r}}. \qquad (4.187)$$

Equation (4.183) can be solved for the eigenfunctions $\vec{S}$ and their corresponding eigenvalues ω. The components of the displacement vector $\vec{S}$ can be represented by trial functions having adjustable parameters and satisfying boundary conditions. We can write Equation (4.187) in a variational form:

$$\delta\left[\omega^2 \int \rho_0 \vec{S}\cdot\vec{S}d\vec{r} - 2W_1\right] = 0,$$

and try to maximize the growth rate ω_I or minimize the potential energy W_1. In practice, this procedure for determining eigenvalues and eigen-functions is useful only for a few simple cases. We shall not discuss this procedure any further. The energy principle is of great utility when we do not need to know the eigenvalues.

Problem 4.26: Argue that ω^2, given by Equation (4.187), is real.

We shall now determine stability criteria using the energy principle. We consider a cylindrical plasma column with $B_{0z} = 0; J_{0z} \neq 0$ and $B_{0\theta} \neq 0$

satisfying the equilibrium Equation (4.50):

$$p_0' = -\left(\frac{B_{0\theta}^2}{4\pi r} + \frac{B_{0\theta}B_{0\theta}'}{4\pi}\right), \qquad (4.188)$$

where the prime represents a derivative with respect to r. Let us perturb the plasma column with a displacement vector $\vec{S} = (S_r(r), 0, S_z(r))$ independent of the azimuthal angle θ. The perturbed potential energy W_1, determined by using Equation (4.186), can be written as:

$$W_1 = \frac{1}{2}\int\left[S_r p_0'\left\{(\vec{\nabla}\cdot\vec{S}) + \lambda_s\right\} + \gamma p_0(\vec{\nabla}\cdot\vec{S})^2 + \frac{B_{0\theta}^2}{4\pi}\lambda_s^2\right]d\vec{r}, \qquad (4.189)$$

where $\lambda_s \equiv \dfrac{\partial S_z}{\partial z} + \dfrac{1}{B_{0\theta}}\dfrac{\partial}{\partial r}(S_r B_{0\theta})$.

Using Equation (4.188), the quantity λ_s can be expressed in terms of $(\vec{\nabla}\cdot\vec{S})$ and S_r, and the potential energy W_1 can be cast in the form:

$$W_1 = \int\left[\left(\vec{\nabla}\cdot\vec{S}\sqrt{2+\gamma\beta} + \sqrt{2(4+\beta x)}\frac{S_r}{r}\right)^2 - \frac{2S_r}{r}\vec{\nabla}\cdot\vec{S}\times\right.$$
$$\left.\times\sqrt{2+\gamma\beta}\sqrt{2(4+\beta x)} - \frac{8S_r}{r}\vec{\nabla}\cdot\vec{S}\right]d\vec{r}, \qquad (4.190)$$

where $x = \dfrac{p_0' r}{p_0}$ and $\beta = \dfrac{8\pi p_0}{B_{0\theta}^2}$.

The stability condition $W_1 > 0$, therefore, gives

$$4\gamma + (2+\gamma\beta)x > 0. \qquad (4.191)$$

The condition (4.191) is a function of the position r and must be satisfied everywhere, for all r, for the stability of the column of a conducting fluid. The stability condition dictates the rate at which the fluid pressure p must fall with r for a stable column; stability requires a gentle fall.

Problem 4.27 Turning the inequality (4.191) into the equality, show that the limiting pressure variation is described by

$$p_0(r) = p_0(0)\left(\frac{\beta}{\beta+0.8}\right)^{5/2} \quad\text{and}\quad r = \frac{R}{\beta^{3/4}(\beta+0.8)^{1/4}}. \qquad (4.192)$$

Here, $p_0(0)$ is the pressure at $r = 0$, $\beta \to \infty$ as $r \to 0$ and R is the radius of the cylindrical fluid column. The stability condition (Equation (4.191)) implies that the pressure should not fall off faster than $r^{-2\gamma}$. Under larger pressure gradients, the fluid column becomes unstable against an axially

Figure 4.17. The Sausage Instability, the Arrows Represent the Flow of Fluid Into the Low Magnetic Pressure Regions.

symmetric perturbation $\vec{S} = (S_r(r), 0, S_z(r))$. A pinch with sharp boundaries develops 'waists' or 'necks' at the surface as shown in Figure(4.17). This is known as **The Sausage Instability**. For a uniform surface current density J_{0z} for a total current I, the azimuthal magnetic field $B_{0\theta} = 2I/cr$ for $r \geq R$. An axisymmetric perturbation that causes a reduction of the radius enhances the magnetic field $B_{0\theta}$, which further enhances the magnetic pressure and lowers the kinetic pressure p_0. As a result, the fluid is forced to move out of this region into a region of lower magnetic pressure. The whole column of fluid acquires bulges and waists. The presence of an axial magnetic field B_{0z} has a stabilizing effect. Thus, depending upon the magnitude of B_{0z}, the Sausage Instability can either saturate to finite values of the perturbed quantities like $\vec{S}$, $\vec{B}_1$, and p_1 or totally quench to vanishing values of $\vec{S}$, $\vec{B}_1$ and p_1. The radial pressure variation described by Equation (4.192) for $B_{0\theta} = 2I/cr$ is shown in Figure (4.18).

The action of the sausage instability has been seen in cometary tails, extragalactic jets and at other astrophysical sites showing filamentary structures.

Problem 4.28 Estimate the radial pressure gradient for the stability of a cometary tail against the Sausage Instability.

Let us now consider nonaxisymmetric perturbations such that:

$$
\begin{aligned}
S_r &= L_r(r) \sin m\theta e^{ikz}, \\
S_\theta &= L_\theta(r) \cos m\theta e^{ikz}, \\
S_z &= L_z(r) \sin m\theta e^{ikz},
\end{aligned}
$$

and $\qquad \vec{\nabla} \cdot \vec{S} = 0.$ \hfill (4.193)

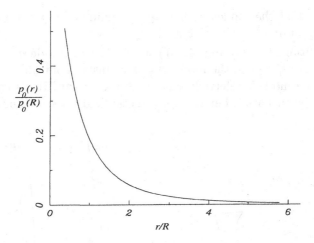

Figure 4.18.   The Limiting Radial Pressure Variation for Stability Against Sausage Instability (Equation 4.192).

We use the divergence free condition to determine S_θ and consider $m \neq 0$ so that

$$(\vec{\nabla} \cdot \vec{S})_\perp \equiv \frac{1}{r} \frac{\partial}{\partial r} (rL_r)) + ikL_z = \frac{m}{r} L_\theta. \qquad (4.194)$$

We find that the perturbed potential energy W_1 is now given as:

$$W_1 = \frac{1}{2} \int \left[L_r p_0' \left\{ (\vec{\nabla} \cdot \vec{S})_\perp + \lambda_s \right\} + \frac{m^2 B_{0\theta}^2}{4\pi r^2} L_r^2 + \frac{B_{0\theta}^2 \lambda_s^2}{4\pi} \right] d\vec{r}, \qquad (4.195)$$

in the limit $k \to \infty$ but (kL_z) remaining finite. Following the procedure outlined earlier, we arrive at the stability condition, $W_1 > 0$, in the form :

$$m^2 + \beta x > 0. \qquad (4.196)$$

Let us compare the two stability conditions given by Equations (4.191) and (4.196). For $x > 0$, i.e., for positive radial pressure gradient, the system is stable both for axi - and nonaxi - symmetric perturbations. For $x < 0$, i.e., for negative radial pressure gradient, Equation (4.191) gives us the maximum value of the magnitude $|x|_{max}$ of x as

$$|x|_{max} = \frac{4\gamma}{2 + \gamma\beta}. \qquad (4.197)$$

Substituting this value of $|x|$ in Equation (4.196) gives:

$$m^2 > 4 \left(\frac{\gamma\beta}{2 + \gamma\beta} \right). \qquad (4.198)$$

Thus, we see that the modes with $m \geq 2$ remain stable when the stability condition Equation (4.191) is satisfied.

The conditions (4.196) and (4.197) tell us that the mode $m = 1$ is stable if $\gamma\beta < 2/3$. For $\gamma = 5/3$, the mode $m = 1$ is unstable for all $\beta > 2/5$. This unstable $m = 1$ mode is referred to as the **Kink Instability** (Figure 4.19a). The $m = 1$ perturbation bends the magnetic field lines to produce convex

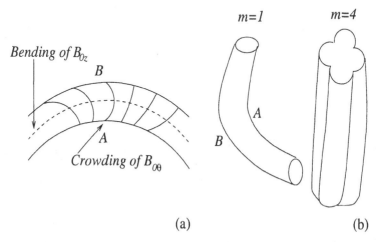

(a)   (b)

Figure 4.19. (a) The Kink Instability ($m = 1$ mode) (b) Multistranded Fluid Column for $m = 4$.

and concave curvatures. The lines of force experience a compression at the concave side (region A) and an expansion at the convex side (region B). The magnetic pressure is, therefore, stronger at A than at B. The resulting force from A to B pushes this region up to further enhance the bending and the perturbation grows. Again an axial magnetic field B_{0z} will try to straighten the bend, and the kink instability can either be entirely quenched or it can attain a final saturation level. The perturbation with high m values attribute a multistranded form to the cylindrical fluid (Figure 4.19b).

We can write the perturbed energy W_1 in other forms too. Here, we give two expressions which are the vector generalizations of the expressions given in Equation (4.189). These are:

$$W_1 = \frac{1}{2} \int \left[\frac{\vec{B}_1^2}{4\pi} + \frac{(\vec{\nabla} \times \vec{B}_0)}{4\pi} \cdot (\vec{S} \times \vec{B}_1) + \gamma p_0 (\vec{\nabla} \cdot \vec{S})^2 \right.$$

$$\left. + (\vec{S} \cdot \vec{\nabla}) p_0 (\vec{\nabla} \cdot \vec{S}) \right] d\vec{r}, \tag{4.199}$$

and

$$W_1 = \frac{1}{2} \int \left[\frac{\left| \vec{B}_{1\perp} \right|^2}{4\pi} + 4\pi \left| \frac{\vec{B}_{1\parallel}}{4\pi} - \frac{\vec{B}_0 \vec{S} \cdot \vec{\nabla} p_0}{B_0^2} \right|^2 + \gamma p_0 \left| \vec{\nabla} \cdot \vec{S} \right|^2 \right.$$

$$\left. + \frac{\vec{J}_0 \cdot \vec{B}_0}{\left| \vec{B}_0 \right|^2} \left(\vec{B}_0 \times \vec{S} \right) \cdot \vec{B}_1 - 2 \left(\vec{S} \cdot \vec{\nabla} p_0 \right) \frac{\left(\vec{S} \cdot \vec{R}c \right)}{\left| \vec{R}_c \right|} \right] d\vec{r}, \quad (4.200)$$

where $\vec{B}_{1\perp}$ and $\vec{B}_{1\parallel}$ are the components of the perturbed magnetic field $\vec{B}_1$ perpendicular and parallel to the zeroth order field $\vec{B}_0$, and $\vec{R}_c$ is the radius of curvature of the perturbed magnetic field lines. We can easily identify the first three terms, which are always positive, in Equation (4.200): these are the energies associated with the Alfven waves. The fourth term is responsible for driving kink modes since it depends upon the current $\vec{J}_0$. The fifth term can drive the system unstable if the curvature is in the the direction of the pressure gradient., i.e., for $\vec{R}_c \parallel \vec{\nabla} p_0$. The curvature vector $(\vec{R}_c)$ plays the role of the acceleration due to gravity $\vec{g}$. Thus, a fluid magnetic field configuration shown in Figure (4.20a) is unstable whereas the config-

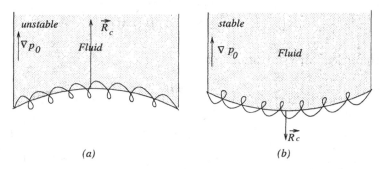

Figure 4.20. (a) The Rayleigh-Taylor Unstable Fluid-magnetic Field Configuration (b) The R-T Stable Fluid-magnetic Field Configuration

uration, shown in figure (4.20b) is stable. There is another way of looking at these situations. A charged particle traveling along the curved magnetic field line experiences a centripetal force which acts in a direction opposite to that of R_c. If we identify this force with the force due to gravity, we see that configuration in Figure (4.20a) is Rayleigh-Taylor unstable whereas the configuration in Figure (4.20b) is R-T stable.

4.20. Growth Rates of Current Driven Instabilities

We first consider the case where a cylindrical column of a conducting fluid carries a current density J_{0z} along its surface. The fluid lies in the region

$r = 0$ to $r = R$ and is surrounded by a vacuum region of an infinite extent. Ampere's law tells us that the constant current density J_{0z} produces a magnetic field $(0, B_{0\theta}(r), B_{0z}(r))$ and that at the surface $r = R$, the vacuum field $B_{0z}^V v$ and the field in the fluid B_{0z}^F are equal and

$$B_{0\theta}^F = 0, \quad B_{0\theta}^V = \frac{2I}{cr}, \tag{4.201}$$

with $I = \int_0^R 2\pi J_{0z} r dr$. The fluid equilibrium is governed by

$$-\vec{\nabla} p_0 + \frac{\vec{J}_0 \times \vec{B}_0}{c} = 0. \tag{4.202}$$

The equilibrium pressure balance at the vacuum fluid interface is described as:

$$p_0 = \frac{(B_{0\theta}^V(R))^2}{8\pi}. \tag{4.203}$$

In order to determine the instability condition and the growth rate of the instability of this system, we assume that all first order quantities vary as $e^{ikz+im\theta-i\omega t}$. The linearized equation of motion

$$-i\omega\rho_0 \vec{U}_1 = -\vec{\nabla} p_1^\star + \frac{1}{4\pi}(\vec{B}_0^F \cdot \vec{\nabla})\vec{B}_1^F. \tag{4.204}$$

The incompressibility condition

$$\vec{\nabla} \cdot \vec{U}_1 = 0, \tag{4.205}$$

gives

$$\nabla^2 p_1^\star = 0, \tag{4.206}$$

where

$$p_1^\star = p_1 + \frac{\vec{B}_0^F \cdot \vec{B}_1^F}{4\pi}.$$

The solution of equation (4.206) which is bounded at the origin is found to be:

$$p_1^\star = A_1 I_m(kr), \tag{4.207}$$

where I_m is the Bessel function of the first kind and A_1 is a constant. Substituting for $\vec{B}_1$ from the induction equation into Equation (4.204), we find:

$$\vec{U}_1 = \frac{-i\omega\vec{\nabla} p_1^\star}{\rho_0(\omega^2 - k^2 V_A^2)}, \tag{4.208}$$

where $V_A^2 = \frac{k^2 B_{0z}^2}{4\pi\rho_0}$.

The perturbed magnetic field $\vec{B}_1^V$ in the vacuum must satisfy the current-free condition.

$$\vec{\nabla} \cdot \times \vec{B}_1^V = 0, \tag{4.209}$$

so that

$$\vec{B}_1^V = \nabla \phi_B,$$

and from the divergence free condition, we have

$$\nabla^2 \phi_B = 0, \tag{4.210}$$

whose solution, which is bound at $r \to \infty$, is given by

$$\phi_B = A_2 K_m(kr), \tag{4.211}$$

where $K_m(kr)$ is the Bessel function of the second kind and A_2 is a constant.

In order to determine the two constants A_1 and A_2, we need two boundary conditions at the interface. The first condition can be derived by integrating the radial component of the total equation of motion across the fluid-vacuum boundary. This gives :

$$p + \frac{B^2}{8\pi}\bigg|_{R+S_r-\epsilon} = p + \frac{B^2}{8\pi}\bigg|_{R+S_r+\epsilon}, \tag{4.212}$$

where $\epsilon \to 0$ and $\vec{S} = (i\vec{U}_1/\omega)$ is the displacement vector. Since we are studying the linear instability, we must remember to evaluate the zeroth order quantities at the perturbed interface and the first order quantities at the unperturbed interface in Equation (4.212). We obtain, to the first order,:

$$\left[p_1 + \frac{1}{4\pi}\vec{B}_0^F \cdot \vec{B}_1^F\right]_R \equiv p_1^\star(R) \simeq \left[\frac{1}{4\pi}\vec{B}_0^V \cdot \vec{B}_1^V + \frac{1}{8\pi}S_r\frac{\partial}{\partial r}\left(B_{0\theta}^V\right)^2\right]_R. \tag{4.213}$$

This is the first boundary condition. The second boundary condition is obtained from the continuity equation of the tangential component of the electric field. Since, in a perfectly conducting fluid, the electric field E_1^F must satisfy the condition

$$E_1^F + \frac{\vec{U}_1 \times \vec{B}_0^F}{c} = 0, \tag{4.214}$$

the tangential component of the vacuum electric field $\vec{E}^V$, must also satisfy this condition. The corresponding vacuum magnetic field is then given by

$$\vec{B}_1^V = \vec{\nabla} \times (\vec{S} \times \vec{B}_0^V)\big|_{r=R}. \tag{4.215}$$

The second boundary condition is now expressed as the continuity of the normal component of the magnetic field given by:

$$B_{1r}^V = \left[\frac{im}{r} B_{0\theta}^V + k B_{0z}^V \right] S_r. \tag{4.216}$$

Using Equations (4.204), (4.207), (4.209) and (4.211) in (4.216), we find:

$$k K_m'(kR) A_2 - \frac{i k \vec{k} \cdot \vec{B}_0^V(R) I_m'(kR)}{\rho_0(\omega^2 - k^2 V_A^2)} A_1, \tag{4.217}$$

where

$$\vec{k} \cdot \vec{B}_0^V = k B_{0z}^V + \frac{m}{r} B_{0\theta}^V. \tag{4.218}$$

Equation (4.213) becomes, on using Equations (4.207), (4.208), and (4.209):

$$\frac{i(\vec{k} \cdot \vec{B}_0^V)}{4\pi} K_m(kR) A_2 + \left[\frac{k I_m'(kR)}{\rho_0(\omega^2 - k^2 V_A^2)} \frac{B_{0\theta}^V}{4\pi} \frac{\partial}{\partial r} \left(B_{0\theta}^V \right)_R - I_m'(kR) \right] A_1 = 0. \tag{4.219}$$

The nontrivial solution of Equations (4.217) and (4.219) gives us the dispersion relation:

$$\omega^2 - k^2 V_A^2 = -\frac{(\vec{k} \cdot \vec{B}_0^V)^2}{4\pi \rho_0} \frac{I_m'(kR)}{I_m(kR)} \frac{K_m(kR)}{K_m'(kR)} - \frac{k I_m'(kR)}{4\pi \rho_0 I_m(kR)} \frac{(B_{0\theta}^V)^2}{R}. \tag{4.220}$$

The fluid becomes unstable for $\omega^2 < 0$. Since

$$\frac{I_m'}{I_m} > 0 \quad \text{and} \quad \frac{K_m'}{K_m} < 0. \tag{4.221}$$

It is the second term on the right hand side of Equation (4.220) that can destabilize the fluid column. In the long wavelength limit, i.e., for $kR \ll 1$, using

$$\frac{I_m'}{I_m} \sim \frac{m}{kR} \quad \text{and} \quad \frac{K_m'}{K_m} = -\frac{m}{kR}, \tag{4.222}$$

and writing $\omega = i\omega_I$, we find the growth rate ω_I is given by:

$$\omega_I^2 = -k^2 V_A^2 - \frac{\left(k B_{0z}^V + \frac{m}{R} B_{0\theta}^V \right)^2}{4\pi \rho_0} + \frac{m(B_{0\theta}^V)^2}{4\pi \rho_0 R^2}. \tag{4.223}$$

We find that the wavelength of the perturbation with the largest growth rate is:

$$\lambda = \frac{2\pi}{k} = \frac{4\pi R}{m} \frac{B_{0z}^V}{B_{0\theta}^V}, \tag{4.224}$$

and the corresponding value of ω_I is given by

$$\omega_I^2 = \frac{m(2-m)}{2R^2} \frac{\left(B_{0\theta}^V\right)^2}{4\pi\rho_0}. \tag{4.225}$$

Problem 4.29 Derive Equations (4.224) and (4.225).

An inspection of equation (4.225) reveals that the perturbations with $m = 0$ and $m = 2$ are marginally stable $(\omega_I = 0)$ and the perturbation with $m = 1$ is unstable for $kR \ll 1$. We also find that the instability exists for $k < 0$.

Problem 4.30 Show that the range of wave vectors k, for $m = 1$ unstable perturbation, is given by $0 < |k| < \dfrac{B_{0\theta}^V(R)}{RB_{0z}^V}$.

The minimum value of k in a cylindrical column of length L is $\dfrac{2\pi}{L}$. Therefore, the criterion for $m = 1$ instability can be expressed as:

$$\frac{2\pi}{L} < \frac{B_{0\theta}^V(R)}{RB_{0z}^V},$$

or

$$\frac{2\pi RB_{0z}^V}{LB_{0\theta}^V(R)} < 1. \tag{4.226}$$

In terms of the quantity q defined in Equation (4.180), the instability condition becomes

$$q(R) < 1. \tag{4.227}$$

This is the famous **Kruskal-Shafranov Criterion**.

The loop like structures with high density, temperature and magnetic field in the solar corona are often seen as half tori, so that the length $L \simeq 2\pi R_T$, where R_T is the major radius of a torus and R is identified with the minor radius. The stability condition then becomes

$$\frac{B_{0z}^V}{B_{0\theta}^V(R)} > \frac{R_T}{R}. \tag{4.228}$$

This condition puts a limit on the current density that a stable loop can carry. Loops with larger current density will deconfine by developing radial flows over a time scale $(\omega_I)^{-1} \simeq$ Alfven crossing time over the major radius $= (R_T/V_A) \simeq 200$ seconds for typical coronal parameters.

We now consider the case where the fluid carries a **Uniform Current Density In The Entire Region** from $r = 0$ to $r = R$ (and not only on the surface) and is surrounded by a vacuum outside this region. The equilibrium is given by:

$$\frac{dp_0}{dr} + \frac{d}{dr}\frac{(B_{0\theta}^F)^2}{8\pi} + \frac{(B_{0\theta}^F)^2}{4\pi r} = 0, \tag{4.229}$$

where

$$B_{0\theta}^F(r) = \frac{4\pi}{cr}\int_0^r J_{0z} r dr = \frac{2I}{cR^2}r, \tag{4.230}$$

and the total current inside the cylinder

$$I = \int_0^R J_{0z} 2\pi r dr = \pi R^2 J_{0z}. \tag{4.231}$$

The equilibrium pressure profile is, then found to be:

$$p_0(r) = \frac{I^2}{\pi c^2 R^2}\left(1 - \frac{r^2}{R^2}\right), \tag{4.232}$$

which describes a parabolic fall of the pressure towards the surface $r = R$. The vacuum magnetic field due to the current I is given by

$$B_{0\theta}^V(r > R) = \frac{4\pi}{cr}\int_0^R J_{0z} r dr = \frac{2I}{cr}. \tag{4.233}$$

The system is also permeated by a uniform axial magnetic field $B_{0z}^F = B_{0z}^V = B_{0z}$. We also assume that $B_{0z} \gg B_{0\theta}$ so that the destabilizing perturbations are two dimensional and lie in the (r, θ) plane and the perturbed $U_{1z} \simeq 0$, $B_{1z}^F \simeq 0$. The perturbed field in the vacuum is, as before, given by Equations (4.209) and (4.211). In order to determine the perturbed magnetic field inside the fluid, we take curl of the equation of motion (4.204) and write its z component as:

$$-i\omega\rho_0\left[\frac{1}{r}\frac{\partial}{\partial r}(rU_{1\theta}) - \frac{im}{r}U_{1r}\right] = \frac{i}{c}(\vec{k} \cdot \vec{B}_0^F)J_{1z}. \tag{4.234}$$

From Ampere's law and the divergence free condition we find:

$$\frac{1}{r}\frac{\partial}{\partial r}\left[-\frac{r}{im}\frac{\partial}{\partial r}\left(rB_{1r}^F\right)\right] - \frac{im}{r}B_{1r}^F = \frac{4\pi}{c}J_{1z}. \tag{4.235}$$

From the incompressibility condition, we find:

$$U_{1\theta} = -\frac{1}{im}\frac{\partial}{\partial r}(rU_{1r}). \tag{4.236}$$

From the induction equation, we find:

$$U_{1r} = \frac{-\omega B_{1r}^F}{(\vec{k} \cdot \vec{B}_0^F)}. \tag{4.237}$$

Using equations (4.235 - 4.237) in Equation (4.234), we find:

$$\left[\frac{\partial}{\partial r} \left\{ r \frac{\partial}{\partial r} \left(r B_{1r}^F \right) \right\} - m^2 B_{1r}^F \right] \left[1 - \frac{4\pi \rho_0 \omega^2}{(\vec{k} \cdot \vec{B}_0^F)^2} \right] = 0. \tag{4.238}$$

The second factor in Equation (4.238) gives the dispersion relation of Alfven waves and cannot give rise to an instability. Therefore, for an instability, we must have

$$\frac{\partial}{\partial r} \left\{ r \frac{\partial}{\partial r} \left(r B_{1r}^F \right) \right\} - m^2 B_{1r}^F = 0, \tag{4.239}$$

from which we find that:

$$B_{1r}^F = A_3 r^{m-1}. \tag{4.240}$$

where the constant A_3 is related to the constant A_2 of Equation (4.211) through the requirement of the continuity of B_r across the interface, i.e.,

$$A_2 k K_m'(kR) = A_3 R^{m-1}. \tag{4.241}$$

Using Equation (4.240), we find

$$U_{1\theta} = i U_{1r} \quad \text{and} \quad B_{1\theta}^F = i B_{1r}^F. \tag{4.242}$$

The dispersion relation is finally obtained from matching pressures across the interface, i.e.,

$$\left. \frac{B_{0z}^V B_{1z}^V + B_{0\theta}^V B_{1\theta}^V}{4\pi} \right|_{r=R} = \left. \frac{p_1 + B_{0\theta}^F B_{1\theta}^F}{4\pi} \right|_{r=R}, \tag{4.243}$$

since $B_{1z}^F \simeq 0$. The perturbed pressure p_1 is determined from the θ component of the momentum Equation. We find:

$$p_1 = \frac{ir}{m} \left[\frac{-\omega^2 \rho_0}{(\vec{k} \cdot \vec{B}_0^F)} - \frac{J_{0z}}{c} + \frac{k B_{0z}}{4\pi} \right] B_{1r}^F. \tag{4.244}$$

Substituting for p_1, $B_{1\theta}^F$, $B_{1\theta}^V$ and B_{1z}^V in terms of B_{1r}^F in equation (4.243), we get the dispersion relation:

$$\omega^2 = -\frac{(\vec{k} \cdot \vec{B}_0^F) B_{0\theta}^F}{4\pi \rho_0 r} + \frac{(\vec{k} \cdot \vec{B}_0^F)^2}{4\pi \rho_0} \left[1 - \frac{m K_m(kR)}{kr K_m'(kR)} \right]. \tag{4.245}$$

Since, $(K_m/K'_m) < 0$, the first term in Equation (4.245) gives rise to an instability with growth rate

$$\omega_I = \left[\frac{(\vec{k} \cdot \vec{B}_0^F) B_{0\theta}^F}{4\pi\rho_0 r} - \frac{(\vec{k} \cdot \vec{B}_0^F)^2}{4\pi\rho_0} \left\{ 1 - \frac{mK_m(kR)}{krK'_m(kR)} \right\} \right]^{1/2}, \qquad (4.246)$$

provided that

$$0 < (\vec{k} \cdot \vec{B}_0^F) < \frac{2I}{cR^2(1 + R/r)}. \qquad (4.247)$$

A comparison of Equation (4.223), (4.246) and (4.247) shows that the conditions of instability are quite different in the two cases where the entire current flows only on the surface and where the current flows throughout the entire fluid. We can find the size of the marginally stable region for a given current I by putting $\omega_I = 0$ in Equation (4.246) along with the wavelength assumption $(kR \ll 1)$. We find

$$1 + \frac{R}{r} = \frac{2I}{cR^2(\vec{k} \cdot \vec{B}_0^F)}. \qquad (4.248)$$

Thus, a large value of I reduces the radius r of the marginally stable region or in other words, the fluid is said to be **Pinched**. As r becomes small, the vacuum field $B_{0\theta}^V$ becomes large and further pinches the fluid and the system goes unstable.

Problem 4.31 Determine the condition under which the perturbed motion is two dimensional $(U_{1z} \simeq 0; B_{1z}^F \simeq 0)$. Estimate the ratio of the axial magnetic field B_{0z}^F to the azimuthal magnetic field $B_{0\theta}^F$ for a stable coronal loop of radius $\sim 10^8$cm and length $\sim 10^9$ cm.

4.21. Resistive Instabilities

The presence of electrical resistivity allows a certain degree of freedom to the magnetic field to depart from the flow of the conducting fluid. The field and the fluid no longer remain frozen. The advantage is that the energy contained in complex fluid flow and fluid configurations can now be dissipated in the system, as a result of which the system becomes hot and may begin to radiate electromagnetic radiation. This is believed to be what happens, for example, during a solar flare. Enormous amounts of energy are released, in the form of mass motions and electromagnetic radiation over a wide spectral range, in an explosive manner. The magnetic field and the conducting fluid on the solar surface are continuously subjected to stresses caused by a variety of convective and wave-like motions. Beyond its endurance limit, the fluid-field configuration becomes unstable, and then relaxes to a lower state of energy, throwing out the excess energy in various forms. And it is the resistivity that brings about this relaxation. We will give one example of an instability in a resistive medium, called the **Tearing Mode Instability**.

Let us consider an incompressible flow given by:

$$
\begin{aligned}
U_x &= U_0 \cos ky && \text{for } |x| > d, \\
&= U_0 \frac{x}{d} \cos ky && \text{for } 0 < |x| < d, \\
\text{and} \quad U_y &= 0 \ \text{for } |x| > d, \\
&= -U_0 \frac{\sin ky}{kd} && \text{for } 0 < |x| < d.
\end{aligned}
\tag{4.249}
$$

Can you draw its streamlines? They are shown in Figure (4.21). The two

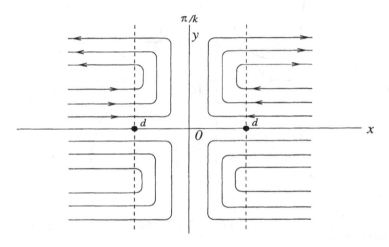

Figure 4.21. Streamlines of the Flow Described by Equation (4.249)

fluids approach the singular region $x \simeq 0$, stream towards each other for $y > 0$ and recoil back near $y < \pi/k$. The fluid initially at $y = 0$ has no velocity in the y direction and just stagnates at $x = y = 0$. Such a flow is known as the **Tearing Flow**. Where does it occur? At many astrophysical situations: on the solar surface where active regions containing fluid and field are pushed towards each other by, for example, the upheavals from below or in their neighbourhood; in planetary magnetospheres where the solar wind deflected by a planet acquires a tearing flow; and in accretion disks where small scale fluid and field regions may approach each other in a commonly turbulent environment.

What happens when such a flow carries a magnetic field given by:

$$
B_x = 0 \quad \text{and} \quad B_y = B_0 \frac{x}{D},
\tag{4.250}
$$

where D is a constant.

Since, in the ideal MHD, field is frozen to the fluid, it undergoes all the pushes and pulls of the flow. The magnetic field lines of the field (Equation (4.250)) are just straight lines parallel to the y-axis before they begin to be influenced by the flow. You may convince yourself that the flow described by Equation (4.249) reconfigures the field lines to what is shown in Figure (4.22). The field lines have developed sharp hair-pin bends thereby acquiring strong curvature; oppositely directed field lines pass close to each

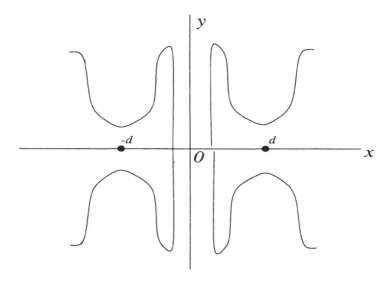

Figure 4.22.   The Magnetic Field Lines (Equation(4.250)) Reconfigured by the Flow (Equation (4.249))

other. The tension forces associated with the curved field lines tend to straighten them to their original form and thus act back on the flow to retard it.

If the fluid has a finite resistivity, the field does not remain frozen to the fluid. The field lines can break and reconnect to form a relaxed configuration with closed loops of field lines as shown in Figure (4.23). This new configuration does not posses strong gradients and therefore does not retard the flow as much as the configuration in the absence of the resistivity. Thus the magnetic field configurations which provide stability to the tearing type flows in the absence of resistivity can no longer do so in the presence of resistivity. The resistivity, therefore, can destabilize a system in the sense that flow velocity is enhanced.

We can investigate the linear stability of this resistive system using cartesian coordinates for convenience. We take the zeroth order magnetic field of

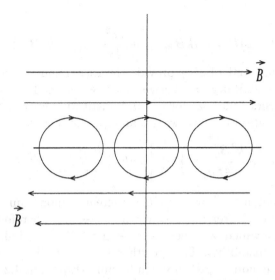

Figure 4.23. An Example of a Magnetic Field Configuration in the Resistive Region

the form:

$$\vec{B}_0 = B_{0y}(x)\hat{y} + B_{0z}\hat{z}, \tag{4.251}$$

where $B_{0y}(0) = 0$. The corresponding current density is

$$\vec{J}_0 = \frac{c}{4\pi}B_{0y}'\hat{z},$$

and the prime represents the derivative with respect to x. We assume that all first order quantities have a space and time dependence of the form $F_1(x)\exp[iky - i\omega t]$. The relation between the velocity and magnetic field can be found by taking the curl of the equation of motion which can be written as:

$$-i\omega\rho_0\vec{\nabla} \times \vec{U}_1 = \vec{\nabla} \times \left[\vec{J}_1 \times \vec{B}_0 + \vec{J}_0 \times \vec{B}_1\right]. \tag{4.252}$$

For two dimensional perturbations $\vec{U}_1 = U_{1x}\hat{x} + U_{1y}\hat{y}$ and $\vec{B}_1 = B_{1x}\hat{x} + B_{1y}\hat{y}$, we find, for incompressible motion, from the z component of Equation (4.252):

$$(-i\omega)\left[\frac{\rho_0}{ik}U_{1x}'' + ik\rho_0 U_{1x}\right] = \frac{1}{4\pi}\left[B_{0y}B_{1x}'' - B_{1x}B_{0y}'' - k^2 B_{0y}B_{1x}\right]. \tag{4.253}$$

From the induction equation including the resistivity η

$$\frac{\partial \vec{B}_1}{\partial t} = \vec{\nabla} \times (\vec{U}_1 \times \vec{B}_0) + \frac{\eta c^2}{4\pi}\nabla^2\vec{B}_1, \tag{4.254}$$

we find:

$$(-i\omega)B_{1x} = ikB_{0y}U_{1x} + \frac{\eta c^2}{4\pi}\left(B_{1x}'' - k^2 B_{1x}\right). \qquad (4.255)$$

Equations (4.253) and (4.255) provide the dispersion relation from which the conditions of stability or otherwise can be obtained.

We first consider the case for $\eta = 0$, Substituting for U_{1x} from Equation (4.255) into Equation (4.253), we get an equation for U_{1x}

$$\frac{d}{dx}\left[\left(-\omega^2 + \frac{k^2 B_{0y}^2(x)}{4\pi\rho_0}\right)U_{1x}'\right] = k^2\left(\frac{k^2 B_{0y}^2}{4\pi\rho_0} - \omega^2\right)U_{1x}, \qquad (4.256)$$

which shows that for $\omega^2 = -\omega_I^2$, U_{1x} increases monotonically with x as long as U_{1x} and U_{1x}' remain positive as $x \to \infty$. Therefore, there is no solution for U_{1x} which vanishes as $x \to \pm\infty$. We conclude that $\omega^2 > 0$ and there is no instability. The growth rate $\omega_I = 0$. This implies that the fluid and field remain in pressure equilibrium (Equation 4.252) in the zero resistivity region. Or if the equilibrium does get disturbed and magnetosonic waves or shocks are excited, the waves and shocks are damped out and the equilibrium is restored over a time scale much shorter than the growth time scale $(\omega_I)^{-1}$ of the tearing mode. In the limit $\omega_I = i\omega \to 0$, Equation (4.253) gives:

$$B_{1x}'' = \left(k^2 + \frac{B_{0y}''}{B_{0y}}\right)B_{1x}. \qquad (4.257)$$

For the ambient field chosen here, $B_{0y}'' = 0$ and we find:

$$\begin{aligned} B_{1x} &= Ae^{-kx} \text{ for } x > 0, \qquad (4.258) \\ &= Ae^{kx} \text{ for } x < 0, \end{aligned}$$

so that B_{1x} is continuous across the surface $x = 0$, but B_{1x}' is discontinuous. The same is true of the velocity U_{1x}. Thus only resistivity and inertia can connect the outer and inner $(x \simeq 0)$ solutions smoothly. It is in the inner regions with $\eta \neq 0$, that the tearing instability is excited. The free energy, however, lies in the fluid flow in the outer regions, which is finally dissipated in the resistive region $(x \simeq 0)$. We can estimate the rate of energy released $\dot{W}$ by the fluid as

$$\dot{W} = -\int \vec{F} \cdot \vec{U}^* d\vec{r}, \qquad (4.259)$$

where $\vec{F}$ is the force acting on the fluid and its curl is given by the right hand side of Equation (4.252). Equation (4.259) can be recast in terms of $curl\vec{F}$ by using the incompressibility condition on $\vec{U}$ which can, therefore, be expressed as

$$\vec{U} = \vec{\nabla} \times \vec{Q}, \quad \vec{Q} = Q\hat{z},$$

so that

$$\dot{W} = -\int Q(\vec{\nabla} \times \vec{F})_z \, d\vec{r}. \qquad (4.260)$$

Substituting for $\vec{U}$ and $(\vec{\nabla} \times \vec{F})_z$ and integrating from $x = 0 - \epsilon$ to $0 + \epsilon$ across the surface $x = 0$, we find for $\epsilon \to 0$

$$\dot{W} = \frac{\omega_I}{4\pi k^2} |B_{1x}(x = 0)|^2 \Delta, \qquad (4.261)$$

where $\Delta = \dfrac{1}{B_{1x}(x = 0)} \left[B'_{1x}|_{0+\epsilon} - B'_{1x}|_{0-\epsilon} \right]$ which shows that $\Delta > 0$ for exciting the tearing mode instability. An interesting point to appreciate here is that the free energy for the instability is contained in the motion of the fluid in the outer region which stresses the field, and due to resistivity the stressed fields release energy in the inner region.

The growth rate of the tearing instability can be estimated by equating the energy released by the fluid to that dissipated in the resistive region. The discontinuity in B'_{1x} gives rise to a current density

$$J_{1z} = \frac{c}{4\pi i k} B_{1x}(x = 0) \frac{\Delta}{L_c},$$

where L_c, a characteristic spatial scale of variation of B_{1y}, is defined as:

$$\frac{\partial B_{1y}}{\partial x} \simeq \frac{B_{1y}|_{0+\epsilon} - B_{1y}|_{0-\epsilon}}{L_c}. \qquad (4.262)$$

The power dissipated per unit area through ohmic dissipation is $\eta J_{1z}^2 L_c$, which, when equated to the power released gives for the growth rate:

$$\omega_I = \left[\frac{\eta^3 \Delta^4 c^6 k^2 B_0^2}{(4\pi)^4 \rho_0 D^2} \right]^{1/5}, \qquad (4.263)$$

and the characteristic spatial scale

$$L_c = \left[\frac{\eta^2 \Delta c^4 \rho_0 D^2}{4\pi k^2 B_0^2} \right]^{1/5}. \qquad (4.264)$$

We recall that the diffusion of the magnetic field takes over a time scale $t_d \propto \eta^{-1}$. Whereas, the relaxation time of the fluid $(\omega_I)^{-1}$ due to the tearing mode instability is proportional to $\eta^{-3/5}$. So, the relaxation through the tearing mode is much faster.

It seems possible to account for the energy released in a small flare through the mechanism of the tearing instability. A small flare releases, typically, about 10^{33} erg over a period of 100 seconds from a region of volume $\delta V \approx 10^{24}$ cm^3. The rate of energy release (from Equation (4.261)) can be estimated as

$$\dot{\epsilon} = (\dot{W}A) \simeq \frac{\omega_I B_0^2}{4\pi} \left(\frac{\Delta a}{k^2 a^2} \right) (Aa), \qquad (4.265)$$

where a is the spatial scale of the variation of the ambient magnetic field B_0; $|B_{1x}| \simeq B_0$, since it is in the stressed ambient field that the energy is stored, A is area such that $(Aa) \equiv \delta V$ is the volume. Thus, ignoring the numerical factors (Δa) and (ka), we rewrite the rate of energy release as:

$$\dot{\epsilon} \simeq \frac{\omega_I B_0^2 (\delta V)}{4\pi}. \qquad (4.266)$$

We see that ω_I can be recast in the form:

$$\omega_I = \left[\frac{1}{t_d^3 t_A^2} \right]^{1/5} \left[\frac{a^8 \Delta^4 k^2}{D^2} \right]^{1/5}, \qquad (4.267)$$

where $t_d = \dfrac{4\pi a^2}{\eta c^2}$ is the diffusion time of the magnetic field and $t_A = \dfrac{a}{(B_0/4\pi\rho)^{1/2}}$ is the Alfven travel time across the singular layer of size a. You may also see another approximate form for ω_I in the literature, given by

$$\omega_I \simeq \left[\frac{1}{t_d t_A} \right]^{1/2}. \qquad (4.268)$$

The actual estimates for the growth rate will depend on the profile of the ambient magnetic field. These approximate forms give us clues to the order of the time scales involved in the tearing mode process. Thus, again, for estimating ω_I from Equation (4.267), we will ignore the dimensionless quantity $(a^8 \Delta^4 k^2/D^2)^{1/5}$.

The typical values of the parameters for a flaring region on the sun are: $\rho = 10^{-12}$ gm/cm^3, $B_0 \simeq 500$ Gauss, $a \sim 10^4$ cm and the temperature $T \simeq 10^5$ K. The resistivity η due to binary electron-proton collisions is estimated to be $\sim 3 \times 10^{-15}$ sec^{-1} which gives an energy release rate $\dot{\epsilon} \sim 10^{28}$ erg/sec, a reasonably acceptable value for a solar flare! A complete account of the finite resistivity instabilities can be found in Furth, Killeen and Rosenbluth (1963) and their application to the solar phenomena can be seen in Spicer (1977, 1981).

We would like to emphasize here that the tearing mode instability can occur due to any non-ideal MHD circumstances, like electron inertia, charge separation, or displacement current in addition to the resistivity; in essence, any effect which violates the MHD constraint $\vec{E} + \vec{U} \times \vec{B}/c = 0$. Some of these cases have been discussed in Manheimer and Lashmore-Davis (1989).

4.22. MHD in Curved Space-Time

We begin by describing the electromagnetic field in a curved space-time or in a curvilinear coordinate system. We use the 'Four-notation'. The electromagnetic field tensor F_{ik} in special relativity is defined as:

$$F_{ik} = \frac{\partial A_k}{\partial x^i} - \frac{\partial A_i}{\partial x^k}, \tag{4.269}$$

where A_i is the four-vector potential whose time component is the electrostatic potential φ and the space components form the vector potential $\vec{A}$. The curvilinear coordinates (x^i) are denoted with a superscript. The general relativistic form of Equation (4.269) is obtained by changing the ordinary derivative to the covariant derivative usually denoted by ;. The covariant derivatives are defined as:

$$A^i_{;l} = \frac{\partial A^i}{\partial x^l} + \Gamma^i_{kl} A^k, \tag{4.270}$$

and

$$A_{i;l} = \frac{\partial A_i}{\partial x^l} - \Gamma^k_{il} A_k, \tag{4.271}$$

where Γ^i_{kl}, are known as the **Christoffel Symbols**. They are symmetric in the subscripts, i.e., $\Gamma^k_{il} = \Gamma^k_{li}$ and are given in terms of the space-time metric g_{ik} as:

$$\Gamma^i_{kl} = \frac{1}{2} g^{im} \left[\frac{\partial g_{mk}}{\partial x^l} + \frac{\partial g_{ml}}{\partial x^k} - \frac{\partial g_{kl}}{\partial x^m} \right]. \tag{4.272}$$

Problem 4.32: Show that Equation (4.269) remains valid even after replacing the ordinary derivative by the covariant derivative.

We can check that F_{ik} satisfies the equation:

$$\frac{\partial F_{ik}}{\partial x^l} + \frac{\partial F_{kl}}{\partial x^i} + \frac{\partial F_{li}}{\partial x^k} = 0, \tag{4.273}$$

which contains the following two Maxwell's equations

$$\vec{\nabla} \times \vec{E} = -\frac{1}{c} \frac{\partial \vec{B}}{\partial t}, \tag{4.274}$$

and

$$\vec{\nabla} \cdot \vec{B} = 0.$$

The four-current density J^i is defined as (Landau & Lifshitz 1962):

$$J^i = \frac{\rho_c c}{\sqrt{-g}} \frac{dx^i}{dx^0},$$

(4.275)

where g is the determinant of the metric g_{ik} and ρ_c is the spatial charge density. In special theory of relativity, the second pair of Maxwell's equations describing Poisson's law and Ampere's law are written as:

$$\frac{\partial F^{ik}}{\partial x^k} = \frac{4\pi}{c} J^i,$$

(4.276)

which in the **General Theory of Relativity**, becomes:

$$F^{ik}_{;k} = \frac{4\pi}{c} J^i,$$

(4.277)

or

$$\frac{1}{\sqrt{-g}} \frac{\partial}{\partial x^k} \left[\sqrt{-g} F^{ik} \right] = \frac{4\pi}{c} J^i.$$

(4.278)

The equation of continuity, correspondingly, can be written as:

$$J^i_{;i} = \frac{1}{\sqrt{-g}} \frac{\partial}{\partial x^i} \left(\sqrt{-g} J^i \right) = 0.$$

(4.279)

The energy-momentum tensor T^E_{ik} of the electromagnetic field defined as:

$$T^E_{ik} = \frac{1}{4\pi} \left[F^l_i F_{lk} + \frac{1}{4} g_{ik} F_{lm} F^{lm} \right],$$

(4.280)

satisfies its conservation law, which is written in the covariant form as:

$$(T^E)^{ik}_{;i} = -F^k_l J^l.$$

(4.281)

Finally, the four-dimensional wave equation of the electromagnetic field is found to be:

$$\left[\frac{\partial^2}{\partial x^{0^2}} - \nabla^2 \right] A^i - \left(A^k_{;k} \right)_{;i} = \frac{4\pi}{c} J^i.$$

(4.282)

This completes the description of the electromagnetic field in a curved space-time.

We will now write the mass, momentum and energy conservation laws of a conducting fluid in a magnetic field in a curved space-time.

If n_0 is the proper number of particles per unit proper volume, each of proper mass m_0, then the proper mass density ρ_0 in the comoving or the proper frame is given by:

$$\rho_0 = n_0 m_0.$$

(4.283)

In the laboratory frame, the mass density ρ and the number density n will be different. The number of particles N_0 in the comoving volume element δV_0 are

$$N_0 = n_0 \delta V_0. \tag{4.284}$$

If the number of particles is an invariant quantity, then

$$N_0 = n_0 \delta V_0 = n \delta V, \tag{4.285}$$

where δV is the volume element in the laboratory frame. But, due to the Lorentz contraction:

$$\delta V = \frac{\delta V_0}{\gamma_L}, \tag{4.286}$$

and therefore,

$$n_0 = \frac{n}{\gamma_L}, \tag{4.287}$$

where $\gamma_L = \left(1 - V^2/c^2\right)^{-1/2}$ is the Lorentz factor corresponding to the velocity $\vec{V}$ of the particles. The proper mass density in the laboratory frame is found to be:

$$\rho = n m_0 = \gamma_L \rho_0. \tag{4.288}$$

We could also define what is known as the relative mass density $\rho_r = nm = \gamma_L^2 \rho_0$. The relativistic form of the mass conservation law can be found by writing for mass density as $(\gamma_L \rho_0)$ and we get

$$\frac{\partial}{\partial t} (\gamma_L \rho_0) + \vec{\nabla} \cdot \left(\gamma_L \rho_0 \vec{U}\right) = 0, \tag{4.289}$$

which in terms of the four-derivative becomes:

$$(\rho_0 U^\alpha)_{,\alpha} = 0, \tag{4.290}$$

where the four-velocity $U^\alpha \equiv \left(\gamma_L c, \vec{V}\right)$. Replacing the ordinary derivative by the covariant derivative, we get the continuity equation in a curved space-time as:

$$(\rho_0 U^\alpha)_{;\alpha} = 0. \tag{4.291}$$

Thus, the particle flux density $(n_0 U^\alpha)$ is conserved in space-time. If there is generation or destruction of particles, as in nuclear reactions, a term representing source or sink must be added on the right hand side of Equation (4.291).

The momentum conservation law is expressed as:

$$\frac{\partial}{\partial t} (\rho V_i) = -\partial_j \Pi_{ij} + f_i, \tag{4.292}$$

where Π_{ij}, the momentum flux density tensor is given by:

$$\Pi_{ij} = \rho V_i V_j + p\delta_{ij},\qquad(4.293)$$

and f_i is the external force density. In order to convert Equation(4.292) into Four-notation, we must, first, define the four-dimensional generalizations of Π and $\vec{f}$. With the help of the four-dimensional momentum flux density tensor M_{ij}, Equation (4.292) can be cast in the form of Four-divergence of M equal to zero which will furnish four-conservation laws. The three space components of M_{ij} give the usual momentum conservation laws. The fourth relation corresponds to the energy conservation law. Thus, on the basis of Newtonian equations,we expect M to have the form

$$m_F = \begin{pmatrix} \rho_0 c^2 + e_T & c\rho\vec{V} \\ c\rho\vec{V} & \rho V_i V_j + p\delta_{ij} \end{pmatrix},\qquad(4.294)$$

which is not yet in the covariant form. Here, e_T is the specific internal energy of the fluid produced by the microscopic motions of the fluid particles and p is the mechanical pressure; ρ_0 and e_T are both defined in the covariant form. In trying to put m_F into a covariant form M, we must ensure that: M is a function of the world scalars (i.e., quantities that remain invariant under Lorentz transformation, for example, the interval ds between two space-time points); it must represent the correct fluid energy density and mechanical pressure in the comoving frame; and must reduce to the correct nonrelativistic form in the laboratory frame under the nonrelativistic limit $V/c \ll 1$. In the comoving frame, we see from Equation (4.294) that

$$M_0 = \begin{pmatrix} \rho_0 c^2 + e_T & 0 & 0 & 0 \\ 0 & p & 0 & 0 \\ 0 & 0 & p & 0 \\ 0 & 0 & 0 & p \end{pmatrix}.\qquad(4.295)$$

The matrix M_0 can be transformed to the laboratory frame by the Lorentz transformation. Another way is to realize that in the comoving frame, the four velocity $U^\alpha = (c,0,0,0)$ and the four-velocity product is found to be:

$$\left(U^\alpha U^\beta\right)_0 = \begin{pmatrix} c^2 & 0 & 0 & 0 \\ 0 & 0 & 0 & 0 \\ 0 & 0 & 0 & 0 \\ 0 & 0 & 0 & 0 \end{pmatrix}.\qquad(4.296)$$

We define a projection tensor as:

$$\left(P^{\alpha\beta}\right)_0 \equiv g_L^{\alpha\beta} + \frac{\left(U^\alpha U^\beta\right)_0}{c^2} = \begin{pmatrix} 0 & 0 & 0 & 0 \\ 0 & 1 & 0 & 0 \\ 0 & 0 & 1 & 0 \\ 0 & 0 & 0 & 1 \end{pmatrix},\qquad(4.297)$$

where $g_L^{\alpha\beta}$, the Lorentz metric is defined as:

$$g_L^{\alpha\beta} = \begin{pmatrix} -1 & 0 & 0 & 0 \\ 0 & 1 & 0 & 0 \\ 0 & 0 & 1 & 0 \\ 0 & 0 & 0 & 1 \end{pmatrix}. \qquad (4.298)$$

Substituting Equations (4.296) and (4.297) in Equation (4.295) we find:

$$\left(M^{\alpha\beta}\right)_0 = \frac{(\rho_0 c^2 + e_T + p)}{c^2} \left(U^\alpha U^\beta\right)_0 + p g_L^{\alpha\beta}. \qquad (4.299)$$

Now ρ_0 and e_T are world scalars and if p is too, then we find in the laboratory frame,

$$M^{\alpha\beta} = \left(\rho_0 c^2 + e_T + p\right) \frac{U^\alpha U^\beta}{c^2} + p g_L^{\alpha\beta}, \qquad (4.300)$$

which is in a fully covariant form. The form of $M^{\alpha\beta}$ in curved space-time is found by replacing the Lorentz metric by the space-time metric $g^{\alpha\beta}$.

The four-force density F^μ is defined as:

$$F^\mu = \frac{\gamma_L}{\delta V_0} \left(\frac{\vec{F} \cdot \vec{V}}{c}, \vec{F}\right), \qquad (4.301)$$

where $\vec{F}$ is the Newtonian force acting on a proper volume element δV_0, which is a world scalar. In terms of the three force density $\vec{f} = \vec{F}/\delta V$ in any frame, we find

$$F^\mu = \frac{\gamma_L \delta V}{\delta V_0} \left(\frac{\vec{f} \cdot \vec{V}}{c}, \vec{f}\right), \qquad (4.302)$$

or

$$F^\mu = \left(\frac{\vec{f} \cdot \vec{V}}{c}, \vec{f}\right), \qquad (4.303)$$

on using Equation (4.286). Thus, in the relativistic hydrodynamics, the four-force density in any frame has its space components equal to the Newtonian force density $\vec{f}$, and the fourth, the time component, is proportional to the rate of work done per unit volume by the Newtonian force density $\vec{f}$.

We can now write the relativistic MHD equations as:

1. Continuity equation

$$\left(\rho_0 U^\alpha\right)_{;\alpha} = 0; \qquad (4.304)$$

2. Energy - momentum conservation of the fluid and the electromagnetic field

$$\left(T_E^{\alpha\beta} + M^{\alpha\beta}\right)_{;\alpha} = 0; \qquad (4.305)$$

3. Condition for perfect electrical conductivity

$$U^{\alpha} F^{\alpha\beta} = 0; \qquad (4.306)$$

4. Equation of state for an ideal gas

$$S_0 = K_B (\Gamma - 1)^{-1} \ln(p\rho^{-\Gamma}), \qquad (4.307)$$

where S_0 is the proper entropy, and $\Gamma = 5/3$ for nonrelativistic motion in the comoving frame and $\Gamma = 4/3$ for relativistic motion.

Relativistic magnetohydrodynamics is used essentially to study the equilibria and stability of the region of accretion disks closest to the compact objects where general relativistic effects are most predominant. This field is still in an early stage of development. The intimidation caused by its inherent complexity can only be conquered by strong motivation or immediate necessity or perhaps by a love for aesthetics that the elders have seen in the General Theory of Relativity.

We have been particularly sketchy in this section since the idea is to let you know of the existence of this facet of plasmas and fluids and not to teach you all about it, not at this stage, any way. The motivated must go to the Classical Theory of Fields by L.D. Landau and E.M. Lifshitz; Foundations of Radiation Hydrodynamics by D. Mihalas and B. Mihalas; Relativity, The General Theory by J.L. Synge and of course the good old friend - Classical Electrodynamics by J.D. Jackson.

4.23. Virial Theorem

The magnetohydrodynamic equations are derived by taking the velocity moments of the Boltzmann equation for each species of particles and adding them appropriately. Now, we examine the consequences of taking spatial moments or averages of the momentum conservation law:

$$\rho \frac{d\vec{U}}{dt} = -\vec{\nabla} \cdot \Pi - \rho \vec{\nabla} \varphi_g + \rho_c \vec{E} + \frac{\vec{J} \times \vec{B}}{c}, \qquad (4.308)$$

where ρ_c the charge density is related to the electric field $\vec{E}$ through Poisson's equation.

Problem 4.33: By using Maxwell's equations, show that

$$\left(\rho_c \vec{E} + \frac{\vec{J} \times \vec{B}}{c} \right)_k = -\frac{\partial}{\partial x_i} T_{ik} - \frac{\partial G_k}{\partial t}, \qquad (4.309)$$

where the electromagnetic energy tensor

$$T_{ik} = \frac{1}{4\pi} \left[\frac{(E^2 + B^2)}{2} \delta_{ik} - E_i E_k - B_i B_k \right], \qquad (4.310)$$

and the momentum density of the electromagnetic field is:

$$G_k = \frac{(\vec{E} \times \vec{B})_k}{c}. \tag{4.311}$$

We shall now multiply the kth component of Equation (4.308) by x_k and integrate over volume. The left hand side becomes:

$$
\begin{aligned}
\int \rho x_k \frac{d^2 x_k}{dt^2} d\vec{r} &= \int x_k \frac{d^2 x_k}{dt^2} d^3 m \\
&= \frac{1}{2} \frac{d^2}{dt^2} \int x_k x_k d^3 m - \int \left(\frac{dx_k}{dt}\right)^2 d^3 m \\
&= \frac{1}{2} \frac{d^2 I}{dt^2} - 2 W_K \tag{4.312}
\end{aligned}
$$

where the mass element $d^3 m = \rho d\vec{r}$, I is the moment of inertia, and W_K is the kinetic energy of the fluid. Using Equation (4.309), the right hand side of Equation (4.308), on multiplying by x_k and integrating over $d\vec{r}$, becomes:

$$
\begin{aligned}
&-\int x_k \frac{\partial}{\partial x_i} \left(\Pi_{ik} + T_{ik}\right) d\vec{r} - \int \rho x_k \frac{\partial \varphi_g}{\partial x_k} d\vec{r} - \int x_k \frac{\partial G_k}{\partial t} d\vec{r} \\
&= -\int x_k \left(\Pi_{ik} + T_{ik}\right) dS_i + \int \frac{\partial x_k}{\partial x_i} \left(\Pi_{ik} + T_{ik}\right) d\vec{r} - \int \rho x_k \frac{\partial \varphi_g}{\partial x_k} d\vec{r} \\
&\quad - \int x_k \frac{\partial G_k}{\partial t} d\vec{r}, \tag{4.313}
\end{aligned}
$$

where we have used $d\vec{r} = dx_i dS_i$. The terms of the second and the third integrals in Equation (4.313) can be identified with total thermal energy W_T, electric energy W_E, magnetic energy W_B, and the gravitational energy W_G. Thus,

$$\sum_{k,i} \int \delta_{ik} \Pi_{ik} d\vec{r} = 2 W_T, \tag{4.314}$$

$$\sum_{k,i} \int \delta_{ik} T_{ik} d\vec{r} = W_E + W_B, \tag{4.315}$$

$$-\sum_k \int x_k \frac{\partial \varphi_g}{\partial x_k} \rho d\vec{r} = W_G. \tag{4.316}$$

Collecting all the terms from Equations (4.312), (4.313), (4.314), (4.315), and (4.316), we get:

$$\frac{1}{2} \frac{d^2 I}{dt^2} + \int x_k \frac{\partial G_k}{\partial t} d\vec{r} = 2\left(W_K + W_T\right) + W_E + W_B + W_G - \int x_k \left(\Pi_{ik} + T_{ik}\right) dS_i. \tag{4.317}$$

If we extend the volume to cover the entire plasma and field, the surface integral vanishes. In the steady state (no time derivatives), we find:

$$2\left(W_K + W_T\right) + W_E + W_B + W_G = 0. \tag{4.318}$$

We see that all the terms in Equation (4.318) except W_G are positive. Therefore, in the absence of gravitational forces, a plasma cannot remain in steady state and

$$\frac{d^2 I}{dt^2} > 0, \tag{4.319}$$

which implies that a plasma left to itself must expand. The integral including the Poynting vector is usually small. Radiation losses through this term certainly slow down the expansion.

The equilibrium condition (Equation 4.318) is known as the **Virial Theorem**. This theorem is extensively used assuming that most cosmic objects are in equilibrium in the absence of electromagnetic forces. Thus, with a knowledge of the gravitational potential energy W_G, we can determine the thermal energy or the root mean square thermal velocity of a gravitationally bound system with no ordered motion ($W_K = 0$) from the relationship

$$2W_T + W_G = 0. \tag{4.320}$$

For galaxies, for example, this root mean square velocity is identified with the velocity dispersion of the constituent stars. Now velocity dispersion is a measurable quantity. It is inferred from the Doppler widths of atomic lines, such as those of neutral hydrogen. The knowledge of the velocity dispersion is used to determine the total mass of the gravitationally bound system by using Equation (4.320). The thermal energy of a stellar system with mass M is

$$W_T = \frac{1}{2}M\left\langle V^2 \right\rangle, \tag{4.321}$$

where $\left\langle V^2 \right\rangle$ is the mean square speed of the system's stars. For a spherical system, the gravitational potential φ_g is determined from

$$\vec{\nabla}\varphi_g\Big|_r = \frac{d\varphi_g}{dr} = \frac{GM(r)}{r^2}, \tag{4.322}$$

where $M(r)$ is the total mass interior to r. The gravitational potential energy W_G can, therefore, be estimated as:

$$
\begin{aligned}
W_G &= -\int_0^R \vec{r} \cdot \vec{\nabla}\varphi_g \rho \, d\vec{r} \\
&= -\frac{(4\pi)^2}{3}G\rho^2 \int_0^R r^4 dr = -\frac{3}{5}\frac{GM^2}{R},
\end{aligned} \tag{4.323}
$$

where we have used $M(r) = \frac{4\pi}{3}\rho r^3$ for a uniform mass density ρ, and R is the radius of the spherical system of stars. The virial theorem (Equation 4.320) then gives:

$$\langle V^2 \rangle = \frac{3GM}{5R}. \tag{4.324}$$

This equation is used to determine M, known as the **Virial Mass**. This is the mass required by the system to remain gravitationally bound, in equilibrium, with a velocity dispersion $\langle V^2 \rangle$. For a galaxy consisting of, say, hundred billion stars, each of solar mass $M_\odot \simeq 2 \times 10^{33}$ gms, $[< V^2 >]^{1/2} \simeq 200$ km sec^{-1} at $R \simeq 10$ kpc. This agrees well with the observed properties of a typical galaxy. So far, so good! But the trouble began when velocity dispersion of $\simeq 200$ km sec^{-1} was observed to remain constant up to $R \simeq 25$ kpc, at a distance much beyond the observable galactic disk of stars. This result is referred to as the flat rotation curves of galaxies. (Figure 4.24). The virial theorem says that $\langle V^2 \rangle$ should fall as R^{-1}. Should we drop the virial theorem? Shall we conclude that galaxies are not gravitationally bound systems? No, instead, the astronomers proposed that the Mass M in Equation (4.324) is proportional to R, so that $\langle V^2 \rangle$ becomes independent of R, beyond a typical distance of 10 kpc. But this mass $M \propto R$ does not make any contribution to the luminosity of a galaxy, which is well accounted by summing up the stellar radiation. So, we call it **Dark Matter**, the matter that provides gravitational energy to sustain the velocity dispersion $\langle V^2 \rangle$ but does not radiate. The need for dark matter becomes more acute as we deal with bigger systems such as clusters of galaxies and superclusters of galaxies. Thus, a galaxy needs dark matter which is ten times as massive as the visible radiating matter. A supercluster of galaxies to remain in virial equilibrium needs dark matter which is nearly 100 times the mass of its visible matter. We are then forced to the conclusion that we live in a universe dominantly made up of dark matter. Can we find our way out of this darkness? It is our love of equilibrium that has led us into darkness. Shall we sacrifice equilibrium? What is our preference, an equilibrium dark universe, or a nonequilibrium bright universe, or a frank acceptance of our ignorance?

The velocity dispersion is only one of the problems that the dark matter helps us to solve. The formation of elements, nucleosynthesis, in the early stages of the universe, and the formation of galaxies are some of the other problems whose solutions are sought amidst dark matter. It appears that the more our need of this elusive dark matter, the less we know of its nature. Proposals, ranging from exotic high energy particles to Jupiter like objects as candidates for dark matter, abound. But, right now, dark matter holds us, we do not hold dark matter! Then, the ultimate question if the universe will continue to expand forever or turn around to contraction can be answered

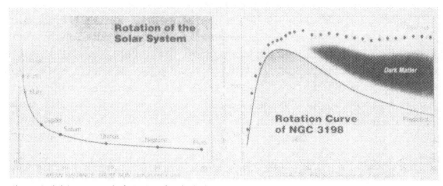

The case for dark matter in spiral galaxies. *Left.* The orbital velocities of the planets (dots) decrease with distance from the Sun exactly as predicted by Newtonian gravitation (line), assuming a system dominated by one solar mass at its center. *Right.* The cosmos is not as well behaved on galactic scales. Here a graph of orbital velocity versus radius has been computed for NGC 3198, a spiral galaxy in Ursa Major, assuming that the distribution of light serves as a good indicator of the distribution of mass. The failure of the observed velocities (dots) to match the predicted ones is striking and points to an unseen component of dark matter in the galaxy. Courtesy the author.

Figure 4.24.

only if we know, how much matter, of any colour, there is in the universe. All measurements and inferences of the total matter in the universe, to date, promise us an ever-expanding, an open universe. Claustrophobics can take a sigh of relief!

4.24. Magnetohydrodynamic Turbulence

A medium is said to be turbulent if it supports the interaction and exchange of energy over several spatial and temporal scales. We have seen that a magnetohydrodynamic system exhibits a variety of instabilities through which matter, motion and magnetic field get distributed over different spatio-temporal scales. Such a system can only be examined using statistical methods. Since the development of the magnetohydrodynamic turbulence followed that of the fluid turbulence, we postpone its discussion until the chapter on non-conducting fluids, where we shall briefly describe ways to characterize MHD turbulence and its role in the generation of large scale magnetic structures.

References

Bray, R.J. et al. , 1991, Plasma Loops in the Solar Corona, Cambridge University Press.

Chandrasekhar, S., 1961, Hydrodynamics and Hydromagnetic Stability, Oxford University Press.

Christensen - Dalsgaard, J. and Berthomieu, G., 1991, Solar Interior and Atmosphere, p401, Ed: A.N.Cox, W.C.Livingston & M.S. Mathews, The University of Arizona Press.

Deubner, F. -L. and Gough, D.O., 1984, Helioseismology: Oscillations as a diagnostic of the Solar interior, Ann.Rev.Astron. Astrophys. **22**, 593.

Furth, H.P., Killeen, J. and Rosenbluth, M.N., 1963, Phys. Fluids, **6**, 459.

Kadomtsev, B.B., 1966, Reviews of Plasma Physics, **2**, 153, Ed: M.A.Leontovich, Consultants Bureau.

Krishan, V., 1996, J.Plasma Physics, **56**, 427.

Lamb, H., 1932, Hydrodynamics, 6th ed., Cambridge University Press.

Landau, L.D. and Lifshitz, E.M., 1962, Classical Theory of Fields, Oxford: Pergamon.

Manheimer, W.M. and Lashmore - Davies, C.N., 1989, MHD And Microinstabilities in Confined Plasma, IOP Publishing Ltd.

Spicer, D.S., 1977, Solar Physics, **53**, 305.

Spicer, D.S., 1981, Solar Physics, **70**, 149.

CHAPTER 5

TWO-FLUID DESCRIPTION OF PLASMAS

5.1. Electron and Proton Plasmas

We have learnt in Chapter one that an electrically quasi-neutral system of negative and positive charges qualifies to be a plasma and that a plasma exhibits cooperative phenomena on certain spatial and temporal scales. Some of the consequences of the quasi-neutral nature of a plasma can be studied by treating each of its constituent components as a fluid. Thus, at this level of description, an electron - proton plasma consists of two fluids - the electron fluid and the proton fluid. Each fluid is allowed to have charge density fluctuations about the overall mean density. This is the most significant deviation from the MHD description. The charge density fluctuations produce current density fluctuations. The associated electric and magnetic fields can be determined from Maxwell's equations. The space and time dependences of these fields can manifest themselves in the forms of longitudinal and transverse waves (Chen 1974; Melrose 1986).

In the presence of free sources of energy, such as a relative streaming motion between the electron and the proton fluids or a temperature inequality and/or an anisotropy, the electric and magnetic fields may begin to grow exponentially with time or distance. Such circumstances produce instabilities. The energy contained in the growing fields could either leave the system as radiation, or be damped within the system. The periods of the waves and the growth rates of the instabilities carry information on the plasma parameters such as density, temperature and electric and magnetic fields. So, the observations of waves and instabilities help us to diagnose plasmas. We may wonder, if the quasineutrality, which exists over short space and time scales, has any role to play in the huge expanse of typically long-lived astrophysical plasmas. The very fact that most of the high energy astrophysical sources emit more radiation than their temperatures would permit is a pointer to the cooperative plasma phenomena. Further, the extremely short temporal variability of radiation (with or without polarization changes) can sometimes only be accounted for by plasma processes occurring over short time scales. So, if we wish to look for plasma phenomena in astrophysical sources, we must study the spectral, temporal and polarization characteristics of their radiation.

In this chapter, we will, first, study the static and dynamic equilibria of the electron and the proton fluids in the presence of electric and magnetic fields and then their stability under small departures from these equilibria. The mathematical tools needed for this investigation have already been developed in Chapter 2.

5.2. Static Equilibria of Electron and Proton Fluids

We begin with the two fluid equations derived in Chapter 2. The mass and momentum conservation laws of the electron fluid are:

$$\frac{\partial \rho_e}{\partial t} + \vec{\nabla} \cdot \left(\rho_e \vec{U}_e \right) = 0, \tag{5.1}$$

and

$$\rho_e \left[\frac{\partial \vec{U}_e}{\partial t} + \left(\vec{U}_e \cdot \vec{\nabla} \right) \vec{U}_e \right] = \frac{-e}{m_e} \rho_e \left[\vec{E} + \frac{\vec{U}_e \times \vec{B}}{c} \right] - \rho_e \vec{\nabla} \varphi_g - \vec{\nabla} \cdot \Pi_e + \vec{\Gamma}^{ei}. \tag{5.2}$$

The static equilibrium of the electron fluid ($\vec{U}_e = 0$) is found to be described by:

$$\frac{\partial \rho_e}{\partial t} = 0, \tag{5.3}$$

and

$$-\frac{e\rho_e}{m_e} \vec{E} - \rho_e \vec{\nabla} \varphi_g - \vec{\nabla} \cdot \vec{\Pi}_e + \vec{\Gamma}^{ei} = 0. \tag{5.4}$$

Similarly, the static equilibrium of the proton fluid ($\vec{U}_i = 0$) is described by

$$\frac{\partial \rho_i}{\partial t} = 0, \tag{5.5}$$

and

$$\frac{e\rho_i}{m_i} \vec{E} - \rho_i \vec{\nabla} \varphi_g - \vec{\nabla} \cdot \vec{\Pi}_i + \vec{\Gamma}^{ie} = 0. \tag{5.6}$$

In Chapter 2, we discussed one model of the the collision term $\vec{\Gamma}^{ie}$. According to this model $\vec{\Gamma}^{ie} = 0$ for a zero relative velocity of the two fluids. In static equilibrium the stress tensors Π_e and Π_i have only their diagonal parts nonzero representing the pressures. Now expressing the electric field $\vec{E}$ in terms of the electric potential φ, we find, by integrating Equations (5.4) and (5.6), that

$$n_e = \frac{\rho_e}{m_e} = n_0 \exp\left[(e\varphi - m_e \varphi_g) / K_B T_e \right], \tag{5.7}$$

and

$$n_i = \frac{\rho_i}{m_i} = n_0 \exp\left[-(e\varphi + m_i \varphi_g) / K_B T_i \right]. \tag{5.8}$$

Thus, in static equilibrium, the electron and proton densities n_e and n_i follow the **Maxwell - Boltzmann Distribution**, where n_0 is the particle density in the absence of forces. We find that the electron density n_e increases with an increase of electric potential φ whereas the proton density n_i decreases. Under the circumstances that the two fluids have equal temperatures ($T_e = T_i$) (though in the absence of collisions, unless there is some other plasma mechanism acting , the two temperatures can remain unequal) we see that the two fluids can sustain a net charge density given by:

$$\frac{e\,(n_i - n_e)}{n_0} = \frac{e\Delta n}{n_0} = \frac{2e^2\varphi}{K_B T}, \tag{5.9}$$

for weak potential $\varphi << K_B T/e$. Here, we have used the isothermal equation of state $p = nK_B T$ and the temperature T is assumed to be space and time independent. This is essentially the content of the energy conservation law of each fluid.

From Poisson's equation, we can estimate the electric field $\vec{E}$ produced by the net charge density $(e\Delta n)$ over a region of length x as

$$E = 4\pi(e\Delta n)x. \tag{5.10}$$

For a 1% change in electron density over a length $x = 1$ cm in a plasma of density n_0, we get

$$E = 4\pi e n_0 \times 10^{-2} \text{ C.G.S. units.} \tag{5.11}$$

In a solar coronal plasma with $n_0 \simeq 10^{10} \text{cm}^{-3}$, $E \simeq 0.6$ C.G.S. units or 1.8×10^4 volt/m; but in high density plasmas such as in the accretion disk of the X-ray binary source Cygnus X-1, where $n_0 \simeq 10^{20}$ cm^{-3}, the electric field could be as large as $\sim 9 \times 10^{13}$ volt/m. This exercise shows us that in dense plasmas, the charge separation must be extremely small and that plasmas are quasi-neutral.

Now, in the absence of electric potential ($\varphi = 0$), we see from Equations (5.7) and (5.8) that the gravitational potential can also create a charge imbalance due to the different scale heights of protons and electrons. However, while studying plasma phenomena, the relevant spatial scales are generally much less than both the scale heights, and therefore the difference between them may not be of much significance.

5.3. Equilibrium of Accreting and Radiating Fluids

Highly gravitating objects such as neutron stars, black holes or the nuclei of active galaxies accrete matter from their neighbourhood. A particle of mass m gains an energy (GMm/R) in falling from infinity to the surface of the object of mass M and radius R. Both electrons and protons feel the gravitational pull and are accelerated onto the object. Through interactions with the atmosphere of the object, the particles lose energy in the form of electromagnetic radiation. This radiation begins to exert a repulsive force on the accreted matter. The accretion stops when the radiation force balances the gravitational force. Now the magnitudes of these two forces are different for electrons and protons, as a result of which, a charge separation and therefore an electric field $\vec{E}$, arises. Soon, the system attains an equilibrium in which both electrons and protons experience equal forces and they move together. The radial force F_i on the proton fluid is:

$$F_i = n_0 \left[\frac{GM}{R^2} m_i - \sigma_T \left(\frac{m_e}{m_i} \right)^2 U_R + eE \right],$$

and F_e on the electron fluid is:

$$F_e = n_0 \left[\frac{GM}{R^2} m_e - \sigma_T U_R - eE \right],$$

where U_R is the radiation energy density and $\sigma_T = (8\pi/3) \left(e^2/m_e c^2 \right)^2$ is the **Thomson Cross Section**. Equating F_e and F_i fixes the electric field E to be:

$$eE \simeq -\frac{GM m_i}{2R^2} - \frac{\sigma_T U_R}{2}.$$

The limiting radiation energy density, therefore, for $E = 0$ is found to be:

$$U_R = \frac{GM m_i}{\sigma_T R^2},$$

and the corresponding isotropic luminosity L_E, known as the **Eddington Luminosity** is given by:

$$
\begin{aligned}
L_E &= 4\pi R^2 c U_R = \frac{4\pi G M m_i c}{\sigma_T} \\
&= 1.38 \times 10^{38} \left(\frac{M}{M_\odot} \right) \text{erg/sec},
\end{aligned}
$$

where $M_\odot = 2 \times 10^{33}$ gm is the mass of the Sun. Thus, L_E is the maximum luminosity that a spherical accreting object can produce. In this case of equilibrium of electron and proton fluids, we see an impressive interplay of gravitational and electromagnetic forces.

5.4. Equilibria of Flowing Fluids

When $\vec{U}_e \neq 0$ and $\vec{U}_i \neq 0$, the equilibria of the electron and the proton fluids are described by:

$$\rho_e \vec{\nabla} \cdot \vec{U}_e + \vec{U}_e \cdot \vec{\nabla} \rho_e = 0, \tag{5.12}$$

$$\rho_i \vec{\nabla} \cdot \vec{U}_i + \vec{U}_i \cdot \vec{\nabla} \rho_i = 0, \qquad (5.13)$$

and

$$\rho_e \left[\left(\vec{U}_e \cdot \vec{\nabla} \right) \vec{U}_e \right] = -\frac{e}{m_e} \rho_e \left[\vec{E} + \frac{\vec{U}_e \times \vec{B}}{c} \right] - \vec{\nabla} p_e + \vec{\Gamma}^{ei}, \qquad (5.14)$$

$$\rho_i \left[\left(\vec{U}_i \cdot \vec{\nabla} \right) \vec{U}_i \right] = \frac{e}{m_i} \rho_i \left[\vec{E} + \frac{\vec{U}_i \times \vec{B}}{c} \right] - \vec{\nabla} p_i + \vec{\Gamma}^{ie}. \qquad (5.15)$$

For uniform flows, mass densities do not vary in the directions of flow. From Equations (5.12) - (5.15), we can make inferences about equilibria under different conditions; such equilibria result from the balance of pressure gradient and Lorentz forces. Again, in the direction of flow and along the magnetic field, the particle densities vary according to Maxwell - Boltzmann distributions given by Equations (5.7) and (5.8).

5.5. Wave Motions of Electron and Proton Fluids

Analogous to the excitation of waves in a single conducting fluid, there are a variety of wave motions exhibited by two conducting fluids. We shall follow the standard procedure for the study of waves, i.e., we shall study the response of the fluids under small departures from their equilibria. There are essentially two major types of wave motions – high frequency waves governed by the response of the electron fluid and the low frequency waves governed by the response of the proton fluid. The presence of the magnetic field $\vec{B}_0$ introduces two more characteristic periods — the gyroperiods of the electrons and protons. The collisions between the two fluids result in the dissipation of these waves. We shall now consider the following cases.

1. Electron - Plasma Oscillations

In the absence of magnetic and gravitational fields, the static equilibria of the electron and the proton fluids are described by:

$$n_{e0} = n_{i0} = n_0 = \text{constant}; \quad \vec{E}_0 = 0; \quad p_{e0} = p_{i0} = p_0 = \text{constant};$$

$$T_{e0} = T_{i0} = T_e = \text{constant}; \quad \vec{U}_{e0} = \vec{U}_{i0} = 0; \quad \vec{B}_0 = 0.$$

The perturbations n_{e1} in the electron density and $\vec{U}_{e1}$ in the electron velocity satisfy the linearized mass and momentum conservation laws as

$$\frac{\partial n_{e1}}{\partial t} + \vec{\nabla} \cdot \left[n_0 \vec{U}_{e1} \right] = 0, \qquad (5.16)$$

and

$$\frac{\partial \vec{U}_{e1}}{\partial t} = -\frac{e}{m_e} \vec{E}_1 - \frac{K_B T}{n_0 m_e} \vec{\nabla} n_{e1} - \nu_{ei} \left(\vec{U}_{e1} - \vec{U}_{i1} \right), \qquad (5.17)$$

where we have substituted for Γ^{ei} (Section 2.17). The perturbation in electric field $\vec{E}_1$ is related to the perturbation in charge density through Poisson's Equation:

$$\vec{\nabla} \cdot \vec{E}_1 = -4\pi e(n_{e1} - n_{i1}).$$ (5.18)

The perturbations in the proton density n_{i1} and the proton velocity $\vec{U}_{i1}$ satisfy the linearized mass and momentum conservation laws as:

$$\frac{\partial n_{i1}}{\partial t} + \vec{\nabla} \cdot \left[n_0 \vec{U}_{i1} \right] = 0,$$ (5.19)

and

$$\frac{\partial \vec{U}_{i1}}{\partial t} = \frac{e}{m_i}\vec{E} - \frac{K_B T}{n_0 m_i}\vec{\nabla} n_{i1} + \nu_{ei}\left(\vec{U}_{e1} - \vec{U}_{i1} \right).$$ (5.20)

We now assume a plane wave type of variation for all the first order quantities and write:

$$n_{e1}(\vec{r}, t) = n'_{e1} \exp \left[i\vec{k} \cdot \vec{r} - i\omega t \right].$$ (5.21)

In order to determine the dispersion relation $\omega(k)$ of these waves, we substitute the solution, Equation(5.21), in Equations (5.16) - (5.20), subtract Equation (5.20) from Equation (5.17), take a dot product with $\vec{k}$, subtract Equation (5.19) from Equation (5.16) and use Equation (5.18) to get

$$\omega(\omega + 2i\nu_{ei})(n_{e1} - n_{i1}) = 4\pi n_0 e^2 \left(\frac{1}{m_e} - \frac{1}{m_i} \right)(n_{e1} - n_{i1})$$
$$+ K_B T \left(\frac{n_{e1}}{m_e} - \frac{n_{i1}}{m_i} \right) k^2.$$ (5.22)

If we assume $n'_{i1} = 0$ and use $m_i \gg m_e$, we find:

$$\omega(\omega + 2i\nu_{ei}) = \omega_{pe}^2 + \frac{K_B T}{m_e}k^2,$$ (5.23)

where

$$\omega_{pe} = \left[4\pi n_0 e^2 / m_e \right]^{1/2}$$

is known as the electron-plasma frequency. Equation (5.23) is the dispersion relation of the **Electron Plasma Waves** also called Langmuir waves. These waves represent oscillations of the net charge density n_{e1}.

The physics of these oscillations can be understood by referring to Figure (5.1). In a quasi-neutral plasma, local charge density fluctuations can arise. If there is an excess of, say, positive charge at some place, the negative charges would rush to that place and try to cancel it. However, in this attempt, the negative charges, due to their kinetic energy, may overshoot

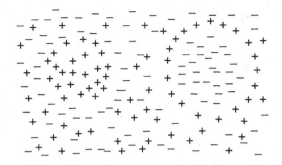

Figure 5.1.   Oscillations Set Up Due to Localized Regions of Excess Charge Density

the place of excess positive charge and create an excess of negative charge elsewhere, from where they will be pulled back by the positive charge. Thus, in an attempt to maintain quasi-neutrality, charge density oscillations set in. The protons form a static positively charged background. We see from the dispersion relation that the frequency ω of the electron plasma waves is a function of the ambient electron density n_0 and the temperature T. Further ω is a complex number due to the presence of collisions. In the limit $\nu_{ei} << \omega$, we find the real part ω_R of ω is given by:

$$\frac{\omega_R^2}{\omega_{pe}^2} = 1 + \frac{k^2}{k_D^2}, \tag{5.24}$$

where $k_D^2 = (4\pi n_0 e^2 / K_B T)$ and the imaginary part ω_I of ω is given by:

$$\omega_I = -2\nu_{ei}. \tag{5.25}$$

The collisions, therefore damp the wave amplitude. The variation of ω_R with wavevector k is shown in Figure (5.2). The damping rate ω_I (Equation 1.30) is nearly directly proportional to the electron density n_0 and inversely proportional to the cube-root of the temperature T, a behaviour expected of the Coulombic binary collisions. The electron plasma waves are longitudinal in their polarization – the displacement, and the electric field $\vec{E}_1$ are both parallel to the direction of the wave propagation vector $\vec{k}$.

Problem 5.1 (a) Find the phase and the group velocities of the electron plasma waves. (b) The Poisson equation for a dielectric medium is written as $\vec{\nabla} \cdot [\epsilon \vec{E}] = 0$. Show that the dielectric function ϵ for electron-plasma waves is given by $\epsilon = 1 - \omega_{pe}^2 / \omega^2$ for $T = 0$.

The cooperative behaviour of electrons is contained in the term ω_{pe}^2. The temperature dependent term is responsible for dispersion. We learnt in

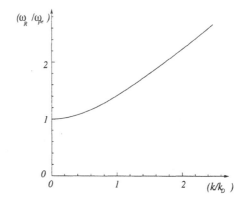

Figure 5.2. Dispersion Relation of the Electron-Plasma Waves (Equation 5.24)

Chapter 1 that the condition for the collective behaviour is that the characteristic spatial scale must be larger than the Debye wavelength. This implies
that in Equation (5.23)

$$\omega_{pe}^2 > \frac{K_B T}{m_e} k^2, \qquad (5.26)$$

from which we see see that the largest wavevector allowed for the electron
plasma waves is the Debye wavevector k_D.

Problem 5.2: Show that the dispersion relation of the electron- plasma
waves is given by $\omega^2 = \omega_{pe}^2 + (5K_B T k^2/3m_e)$ for the adiabatic equation of
state for electrons i.e., $p_e \propto \rho_e^{5/3}$.

The electron - plasma waves have been observed in laboratory plasmas.
It is impossible to see these waves in astrophysical plasmas since they are
localized oscillations and can only be picked up by in-situ probes. However,
their presence has been inferred in otherwise inaccessible regions by indirect
methods. One way, for example, is through the conversion of electron-plasma
waves into electromagnetic waves which can leave the heavenly plasma and
impinge on our telescopes. This is how some of the radio radiation from the
sun is believed to originate.

The first simultaneous detections of the electron-plasma waves as well
as the attendant radio emission were done by the solar orbiting Helios1 and
Helios2 spacecrafts (Gurnett and Anderson 1976).

2. Ion-Plasma Oscillations

We now study low frequency oscillations in which electrons and
protons both participate. In the absence of magnetic and gravitational fields

the static equilibria of the two fluids are described by:

$$n_{e0} = n_{i0} = n_0 = \text{constant}; \quad \vec{E}_0 = 0; \quad p_{e0} = n_0 K_B T_{e0};$$

$$p_{i0} = n_0 K_B T_{i0}; \quad \vec{U}_{e0} = \vec{U}_{i0} = 0.$$

We assume that the mass m_e of an electron is vanishingly small i.e., $m_e \to 0$. In this limit, Equation (5.17) for the linearized motion of the electron fluid becomes:

$$-e\vec{E}_1 - \frac{K_B T_{e0}}{n_0} \vec{\nabla} n_{e1} = 0, \tag{5.27}$$

where we have ignored the collisional forces. The solution of Equation (5.27) gives:

$$n_{e1} = \frac{n_0 e \varphi_1}{K_B T_{e0}}, \tag{5.28}$$

where we have expressed $\vec{E}_1 = -\vec{\nabla} \varphi_1$. For a plane-wave variation of the perturbed quantities, the mass conservation laws of the two fluids give:

$$n_{e1} = n_0 \frac{\vec{k} \cdot \vec{U}_{e1}}{\omega}, \tag{5.29}$$

and

$$n_{i1} = n_0 \frac{\vec{k} \cdot \vec{U}_{i1}}{\omega}. \tag{5.30}$$

The ion equation of motion dotted with the wave vector $\vec{k}$ gives:

$$\omega \, \vec{k} \cdot \vec{U}_{i1} = \frac{e}{m_i} k^2 \varphi_1 + \frac{K_B T_{i0}}{n_0 m_i} k^2 n_{i1}. \tag{5.31}$$

Finally, Poisson's equation relating the perturbations in charge density with the potential φ_1 is:

$$k^2 \varphi_1 = -4\pi e (n_{e1} - n_{i1}). \tag{5.32}$$

Substituting for $\vec{U}_{i1}$, $\vec{U}_{e1}$, φ_1 and n_{e1} in Equation (5.31), we find the dispersion relation for $n_{i1} \neq 0$ as:

$$\left(\frac{\omega^2}{\omega_{pi}^2} \right) = \left[1 - \left(1 + \frac{k^2}{k_D^2} \right)^{-1} \right] + \frac{k^2 T_{i0}}{k_D^2 T_{e0}}, \tag{5.33}$$

where $\omega_{pi} = \left(\dfrac{4\pi n_0 e^2}{m_i} \right)^{1/2}$ is the ion-plasma frequency and

$$k_D^2 = \frac{4\pi n_0 e^2}{K_B T_{e0}}. \tag{5.34}$$

is the Debye wavenumber. Equation (5.33) is the dispersion relation of the
Ion - Plasma Waves. In the short wavelength limit i.e., for $(k^2/k_D^2) \gg 1$,
the dispersion relation of the ion-plasma waves becomes:

$$\omega^2 = \omega_{pi}^2 + \frac{K_B T_{i0}}{m_i} k^2, \qquad (5.35)$$

which looks very much like the dispersion relation of the electron-plasma
waves. In the large wavelength limit i.e., for $(k^2/k_D^2) \ll 1$, the dispersion
relation of the ion-plasma waves becomes:

$$\omega^2 = \frac{K_B}{m_i} (T_{i0} + T_{e0}) k^2 = c_s^2 k^2, \qquad (5.36)$$

which looks like the dispersion relation of the sound waves. Here, c_s, is the
isothermal sound speed. For this reason, these waves are also known as the
Ion - Acoustic Waves.

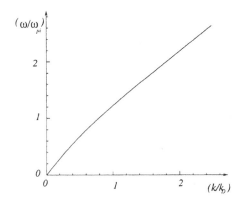

Figure 5.3. Dispersion Relation of the Ion-Plasma Waves (Equation 5.33) for $T_{i0} = T_{e0}$.

In contrast to the case of the electron-plasma waves where ions form
a static and uniform background, during ion-plasma wave excitations, the
electrons and ions both play a dynamic role. Electrons are pulled by a bunch
of ions and they screen the electric field produced by the bunching ions. As
for sound waves, here too, the ions form regions of high and low density.
The ion thermal motion produces a spreading of the condensation. Due to
the thermal motion of electrons, only a partial screening of the electric field
is achieved. These two effects are contained in the temperature dependence
of the dispersion relation (Equation 5.36). The full dispersion relation of the
ion-plasma waves is illustrated in Figure (5.3).

Problem 5.3: Derive the dispersion relation of ion-plasma waves by including the effect of electron-ion collisions. Discuss the polarization of these
waves

Problem 5.4: Show that the dispersion relation of the ion-plasma waves using the adiabatic equation of state for the the two fluids is given by

$$\left(\frac{\omega^2}{\omega_{pi}^2}\right) = \left[1 - \left(1 + \gamma_e \frac{k^2}{k_D^2}\right)^{-1}\right] + \frac{\gamma_i k^2 T_{i0}}{k_D^2 T_{e0}}.$$

where γ_e and γ_i are the adiabatic indices.

Problem 5.5: The equality $n_{e1} = n_{i1}$ along with nonzero $\vec{\nabla}\varphi_1$ is known as the **Plasma Approximation**. Show that the condition $(k/k_D) << 1$ corresponds to the plasma approximation.

3. Electron - Plasma Waves in Magnetized Fluids

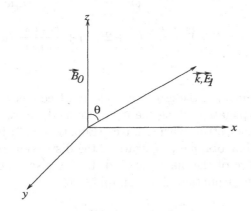

Figure 5.4. Electron-Plasma Waves Propagating at an Angle θ With $\vec{B}_0$.

We now investigate the effect of a uniform ambient magnetic field $\vec{B}_0$ on the characteristics of the electron plasma waves. The static equilibria of the electron and proton fluids are the same as before except that, now, $\vec{B}_0 = (0, 0, B_0)$. We, further, take for electron plasma oscillations, $n_{i1} = U_{i1} = 0$ and $\dfrac{m_e}{m_i} \to 0$ The wavevector $\vec{k} \parallel \vec{E}_1$ makes an angle θ with the magnetic field as shown in Figure (5.4). The mass conservation for electron fluid gives:

$$\frac{n_{e1}}{n_0} = \frac{\vec{k} \cdot \vec{U}_{e1}}{\omega}. \tag{5.37}$$

Poisson's equation gives:

$$\vec{k} \cdot \vec{E}_1 = i4\pi e n_{e1}. \tag{5.38}$$

The addition of x and z components of the momentum conservation laws for electron and ion fluids gives (Here U_e stands for U_{e1}):

$$(-i\omega + 2\nu_{ei})\vec{k} \cdot \vec{U}_e = -\frac{e}{m_e}\vec{k} \cdot \vec{E}_1 - \omega_{ce}U_{ey}k_x - \frac{iK_BT}{n_0 m_e}k^2 n_{e1}. \qquad (5.39)$$

The addition of the y component of the momentum conservation laws for the electron and ion fluids gives:

$$(-i\omega + 2\nu_{ei})U_{ey} = \omega_{ce}U_{ex}. \qquad (5.40)$$

From equations (5.37) − (5.40), by eliminating the various first order quantities, we get the dispersion relation:

$$\sin^2\theta\left[-(\omega + 2i\nu_{ei})^2 + \omega_{ce}^2 + \frac{\omega_{pe}^2}{\omega}(\omega + 2i\nu_{ei}) + \frac{K_BTk^2}{m_e\omega}(\omega + 2i\nu_{ei})\right]$$

$$+\cos^2\theta\left[-(\omega + 2i\nu_{ei})^2 + \frac{\omega_{pe}^2}{\omega}(\omega + 2i\nu_{ei}) + \frac{K_BTk^2}{m_e\omega}(\omega + 2i\nu_{ei})\right] = 0.$$
$$\qquad (5.41)$$

Here, $\omega_{ce} = (eB_0/m_e c)$ is the electron cyclotron frequency. We see that for $\theta = 0$, i.e., for propagation along the magnetic field, we recover the dispersion relation of the electron-plasma waves (Equation 5.23) in the absence of a magnetic field. For oblique propagation, the dispersion relation is modified by the presence of the magnetic field. In the absence of collisions and temperature effects, Equation (5.41) simplifies to:

$$\omega^2 = \omega_{pe}^2 + \omega_{ce}^2\sin^2\theta. \qquad (5.42)$$

This wave is known as the **Upper Hybrid Wave**, since its frequency ω is higher than the electron plasma frequency ω_{pe}. This is due to the additional restoring Lorentz force. The group velocity of these waves is zero in the absence of thermal effects.

4. Ion-Plasma Waves in Magnetized Fluids

In the presence of a uniform zeroth order magnetic field $\vec{B}_0 = (0, 0, B_0)$, we write the first order mass and momentum conservation laws for the hot electrons ($T_e \neq 0$) and cold ions ($T_i = 0$) assuming a plane wave for space and time dependence for the perturbations:

$$m_i(-i\omega)\vec{U}_{i1} = -e(i\vec{k})\varphi_1 + \frac{e}{c}(\vec{U}_{i1} \times \vec{B}_0), \qquad (5.43)$$

$$m_e(-i\omega)\vec{U}_{e1} = e(i\vec{k})\varphi_1 - \frac{K_BT_e}{n_0}(i\vec{k})n_{e1} - \frac{e}{c}(\vec{U}_{e1} \times \vec{B}_0), \qquad (5.44)$$

$$(-i\omega)n_{e1} + i\vec{k} \cdot \vec{U}_{e1}n_0 = 0, \tag{5.45}$$

and

$$(-i\omega)n_{i1} + i\vec{k} \cdot \vec{U}_{i1}n_0 = 0. \tag{5.46}$$

We take $\vec{k} = (k_x, 0, 0)$ (Figure 5.4) and use the plasma approximation $n_{e1} = n_{i1}$, but $\vec{E}_1 = -\vec{\nabla}\varphi_1 \neq 0$. Mass conservation then demands $\vec{U}_{e1} = \vec{U}_{i1}$. From the x and y components of Equation (5.43), we find

$$(\vec{U}_{i1})_x = \frac{ek_x}{m_i\omega}\left(1 - \frac{\omega_{ci}^2}{\omega^2}\right)^{-1}\varphi_1, \tag{5.47}$$

where $\omega_{ci} = \dfrac{eB_0}{m_i c}$ is the ion-cyclotron frequency. For $T_e = 0$, from the x and y components of Equation (5.44) we find:

$$(\vec{U}_{e1})_x = -\frac{ek_x}{m_e\omega}\left(1 - \frac{\omega_{ce}^2}{\omega^2}\right)^{-1}\varphi_1. \tag{5.48}$$

Using $(\vec{U}_{e1})_x = (\vec{U}_{i1})_x$, we find, in the limit $\dfrac{m_e}{m_i} \to 0$,

$$\omega = (\omega_{ce}\,\omega_{ci})^{1/2}. \tag{5.49}$$

This is the dispersion relation of the **Lower Hybrid Waves**. They have frequencies lower than the electron cyclotron frequency ω_{ce} but higher than the ion cyclotron frequency ω_{ci}. They propagate perpendicular to the magnetic field B_0. For a propagation vector $\vec{k}$ parallel to $\vec{B}_0$, we recover the dispersion relation of the ion acoustic wave.

We now investigate the case of oblique propagation i.e., for $\vec{k} = (k_x, 0, k_z)$ in the limit $m_e \to 0$.

The ion Equation (5.43) furnishes:

$$(U_{i1})_x = \frac{ek_x\varphi_1}{m_i\omega}\left(1 - \frac{\omega_{ci}^2}{\omega^2}\right)^{-1}, \quad U_{iz} = \frac{ek_z\varphi_1}{m_i\omega}. \tag{5.50}$$

The electron Equation (5.44) furnishes:

$$(U_{e1})_x = 0; \quad (U_{e1})_y = 0 \quad \text{and} \quad \frac{n_{e1}}{n_0} = \frac{e\varphi_1}{K_B T_e}. \tag{5.51}$$

The continuity equations for electrons and ions under the plasma approximation give:

$$\frac{n_{i1}}{n_0} = \frac{\vec{k} \cdot \vec{U}_{i1}}{\omega} = \frac{n_{e1}}{n_0}. \tag{5.52}$$

Eliminating $(\vec{U}_{i1})_x$ between Equations (5.50) and (5.52) using Equation (5.51), we find:

$$\omega^2 = \omega_{ci}^2 + \frac{k_x^2 c_s^2}{\left(1 - \dfrac{k_z^2 c_s^2}{\omega^2}\right)}. \tag{5.53}$$

This is the dispersion relation of the electrostatic **Ion-Cyclotron Waves**. In the limit

$$(\vec{k} \cdot \vec{U}_{i1}) \simeq k_x (\vec{U}_{i1})_x, \tag{5.54}$$

the dispersion relation for ion-cyclotron waves resembles the dispersion relation of the upper hybrid waves and predictably so, as the 'acoustic' motion of the ions is now modified by their cyclotron motion. We must, here, appreciate the need for $k_z \neq 0$. In order to preserve charge neutrality $n_{i1} = n_{e1}$, the electrons must move along the magnetic field, since their motion across the magnetic field is highly restricted. Thus, during the ion-cyclotron wave motion, the motion of the ions is predominantly perpendicular to the magnetic field while that of the electrons is essentially parallel to the magnetic field.

5. Electromagnetic Waves in Electron-Proton Fluids

So far, we have studied two examples of longitudinal waves which propagate only in a matter medium. We will now study the excitation of transverse electromagnetic waves which, though, they can propagate in vacuum, are modified in the presence of a medium. The static equilibria of the two fluids are described by:

$$n_{e0} = n_{i0} = n_0 = \text{constant}; \quad \vec{E}_0 = 0; \quad p_{e0} = n_0 K_B T_{e0},$$

$$p_{i0} = n_0 K_B T_{i0}; \quad \vec{U}_{e0} = \vec{U}_{i0} = 0.$$

Using Maxwell's equations, the wave equation for the electric field is found to be:

$$-\nabla^2 \vec{E} + \vec{\nabla}(\vec{\nabla} \cdot \vec{E}) = -\frac{4\pi}{c^2} \frac{\partial \vec{J}}{\partial t} - \frac{1}{c^2} \frac{\partial^2 \vec{E}}{\partial t^2}, \tag{5.55}$$

where $\vec{J}$ is the current density.

The linearized current density is given by

$$\vec{J}_1 = e n_0 \left[\vec{U}_{i1} - \vec{U}_{e1} \right]. \tag{5.56}$$

We wish to study transverse waves, for which

$$\vec{\nabla} \cdot \vec{E}_1 = i\vec{k} \cdot \vec{E}_1 = 0, \tag{5.57}$$

therefore we must put $n_{e1} = n_{i1}$. The mass conservation equations, then, could be satisfied with $\vec{U}_{e1} = \vec{U}_{i1}$. The wave equation (5.55) then describes propagation of electromagnetic waves in vacuum ($\vec{J}_1 = 0$) with dispersion relation

$$\omega^2 = k^2 c^2. \tag{5.58}$$

The linearized forms of the momentum conservation laws of the two fluids, describe a Boltzmann distribution of the density perturbations as

$$n_{e1} = n_{i1} = \frac{e n_0 \left(\dfrac{1}{m_e} + \dfrac{1}{m_i} \right) \varphi_1}{K_B \left(\dfrac{T_{e0}}{m_e} - \dfrac{T_{i0}}{m_i} \right)}. \tag{5.59}$$

The other way of satisfying Equation (5.57) is by putting $n_{e1} = n_{i1} = 0$. Mass conservation then gives:

$$\vec{\nabla} \cdot \vec{U}_{e1} = \vec{\nabla} \cdot \vec{U}_{i1} = 0, \tag{5.60}$$

i.e., the motion of the particles is transverse to the direction of the propagation vector $\vec{k}$. From Equations (5.17) and (5.20), we find for the current density

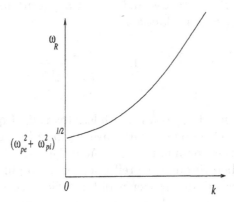

Figure 5.5. Dispersion Relation of the Electromagnetic Waves in an Electron-Proton Fluid.

$$\vec{J}_1 = \frac{n_0 e^2 \left(1/m_e + 1/m_i \right) \vec{E}_1}{(-i\omega + 2\nu_{ei})}. \tag{5.61}$$

On substituting for $\vec{J}_1$ in the wave equation (5.55), we get the dispersion relation for the transverse electromagnetic waves as:

$$\omega^2 = \frac{\omega_{pe}^2 + \omega_{pi}^2}{\left(1 + \dfrac{2i\nu_{ei}}{\omega} \right)} + k^2 c^2. \tag{5.62}$$

Again assuming that $\nu_{ei} << \omega$, we find the real part ω_R from:

$$\omega_R^2 \simeq \omega_{pe}^2 + \omega_{pi}^2 + k^2 c^2, \tag{5.63}$$

and the imaginary part

$$\omega_I \simeq -\frac{\nu_{ei}}{\omega_R^2} \left(\omega_{pe}^2 + \omega_{pi}^2\right). \tag{5.64}$$

We see that the phase and the group velocities of the electromagnetic waves become different in a plasma as there is a minimum value of $\omega_R = (\omega_{pe}^2 + \omega_{pi}^2)^{1/2}$ below which the waves cannot propagate in a plasma (the wave vector k becomes imaginary). The waves suffer damping due to collisions between electrons and ions. The dispersion relation of the electromagnetic waves is plotted in Figure (5.5).

Problem 5.6: Show that the phase velocity of the electromagnetic waves in a plasma exceeds the velocity of light c. Determine the group velocity. Does it also exceed c?

We can define the refractive index n of a plasma for electromagnetic waves from the dispersion relation as:

$$n^2 = \frac{k^2 c^2}{\omega^2} = 1 - \frac{\omega_{pe}^2}{\omega^2 \left(1 + \frac{2i\nu_{ei}}{\omega}\right)}, \tag{5.65}$$

where we have neglected ω_{pi}^2 as it is much less than ω_{pe}^2. Equation (5.65) provides the basis for reflection of short wavelength radio waves in the earth's ionosphere facilitating communication around the earth. The ionosphere, itself, has been studied through the reflection of radio pulses. The reflection occurs at a place, where the frequency of the radio pulse equals the electron plasma frequency. By this technique, the electron density in, as well as the the distance to, the reflection region can be estimated. Electron densities of $10^5 - 10^6$ cm^{-3} have been inferred at an altitude of 500 km in the ionosphere. These densities correspond to electron plasma frequencies of 17 - 54 MHz.

A note of caution is in order here. The reflection of electromagnetic waves with frequencies near the electron-plasma frequency is true only for low intensity radiation. High intensity radiation can change the properties of the plasma it propagates through. As a result of which, entirely novel processes or novel conditions for reflection or transmission may set in.

The dispersive properties of the interstellar medium have been put to good use for determining the distances of pulsars. Due to the difference in their group velocities, pulses of different frequencies take different durations of time to traverse the same distance. By measuring the time of delay, we can estimate the distance to the pulsar if the electron density of the interstellar medium is known. The time delay dt in the arrival of two pulses of frequencies ω_1 and ω_2 in traversing a distance ds is found to be

$$dt = \frac{ds}{c} \left[\omega_1 \left(\omega_1^2 - \omega_{pe}^2 \right)^{-1/2} - \omega_2 \left(\omega_2^2 - \omega_{pe}^2 \right)^{-1/2} \right]. \tag{5.66}$$

Under the assumption $\omega_1^2, \omega_2^2 >> \omega_{pe}^2$ and after integrating both sides of Equation (5.66), we find:

$$t_{\omega_1} - t_{\omega_2} = \frac{1}{2c} \left(\frac{1}{\omega_1^2} - \frac{1}{\omega_2^2} \right) \frac{4\pi e^2}{m_e} \int_0^s n_e ds, \tag{5.67}$$

where t_ω is the pulse travel time. The integral

$$\int_0^s n_e ds \equiv DM, \tag{5.68}$$

is known as the **Dispersion Measure**. Thus, for a given model of the interstellar medium which provides an estimate of the electron density, the distance s to a pulsar can be determined. For the pulsar CP1919, the observations made at $f_1 = \omega_1/2\pi = 73.8\text{MHz}$ and $f_2 = 111.5\text{MHz}$, furnished a time delay $\simeq 5.328$ sec for the lower frequency and a dispersion measure $\simeq 3.84 \times 10^{19}\text{cm}^{-2}$ was estimated. Taking $n_e \simeq 0.03\text{cm}^{-3}$ for the average electron density in the interstellar medium, we find that the distance s to the pulsar is $3.84 \times 10^{21}\text{cm}$ or about 10^3pc.

More often than not, astrophysical plasmas have particle densities varying over many spatial scales. These are known as density irregularities. Electromagnetic radiation traversing such a medium undergoes, what is known as the **Refractive Scintillation**. When two beams of radiation pass through two regions of typical size a with electron densities n_1 and n_2, they develop a phase difference $\delta\varphi = (\Delta k)a$, where Δk, the difference in the wave vectors k_1 and k_2 of the two beams at frequency ω is calculated as

$$\Delta k = (k_1 - k_2) = \frac{(\omega^2 - \omega_{p1}^2) - (\omega^2 - \omega_{p2}^2)}{c^2(k_1 + k_2)} \simeq \frac{e^2}{m_e c^2}\lambda\Delta N. \tag{5.69}$$

Here, we have used $(k_1 + k_2) \simeq 2k = 4\pi/\lambda$ and $\Delta N = n_1 - n_2$. If L is the size of the region containing irregularities of size a, then a random walk through this region will produce a root mean square phase difference $\Delta\varphi$ given by:

$$\Delta\varphi \simeq \frac{e^2 \Delta N \lambda a}{m_e c^2} \sqrt{\frac{L}{a}}. \tag{5.70}$$

It is assumed that in most situations, the scattering of radiation by the irregularities is equivalent to propagation through a thin screen causing phase variations $\Delta\varphi$ on a scale a. The thin screen is placed at somewhere in the middle of the depth L of the actual region. For $a >> \lambda$ and for large phase differences $\Delta\varphi > \pi$, we can use Snell's laws of refraction to determine the scattering angle. The relative refraction angle $(r_1 - r_2)$ or the scattering angle θ_s between the two beams for normal incidence is given by (Figure 5.6):

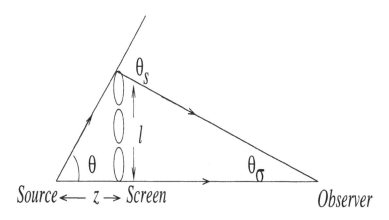

Source $\longleftarrow$ z $\longrightarrow$ Screen   Observer

Figure 5.6.   Refraction of Radiation in a Medium With Small Scale Density Inhomogeneities.

$$\theta_s = (r_1 - r_2) = \left(1 - \frac{\omega_{p1}^2}{\omega^2}\right)^{-1/2} - \left(1 - \frac{\omega_{p2}^2}{\omega^2}\right)^{-1/2}$$

$$\simeq \frac{1}{2\pi}\left(\frac{L}{a}\right)^{1/2}\frac{e^2 \Delta N \lambda^2}{m_e c^2}. \tag{5.71}$$

If the screen is placed at a distance z from the source where $z <<$ the distance of the source from the earth (Figure 5.6) scintillations will be observed if $\Delta\varphi > \pi$. Further, the signal will show variations of amplitude if several interfering beams reach the observer. This implies that the size of the screen l must be larger than the size of the irregularity. We find from Figure (5.6):

$$l \simeq z \tan\theta \simeq z\theta \simeq z(\theta_s - \theta_0) \simeq z\theta_s > a. \tag{5.72}$$

The scintillations will be similar at wavelengths λ and $\lambda + \Delta\lambda$, if the phase differences among the various interfering beams are same at λ and $(\lambda + \Delta\lambda)$ to within π radians. There are two contributions to the phase difference: (1) due to propagation through different parts of the screen with different values of ΔN, which is given by $\Delta \varphi$ and (2) due to different path lengths. The phase difference $\Delta \varphi_p$ due to the second contribution is estimated as:

$$\Delta \varphi_p = \frac{2\pi}{\lambda} \left[z - z \cos \theta_s \right] \simeq \frac{2\pi}{\lambda} \frac{z \theta_s^2}{2}, \qquad (5.73)$$

and is found to be larger than $\Delta \varphi$. Thus, the scintillations are correlated over a bandwidth $\Delta \lambda$ if

$$\frac{2\pi}{\lambda} \frac{z \theta_s^2}{2} + \pi \simeq \frac{2\pi}{(\lambda + \Delta \lambda)} \frac{z \theta_s^2}{2}, \qquad (5.74)$$

or

$$\left| \frac{\Delta \lambda}{\lambda} \right| \simeq \frac{\lambda}{z \theta_s^2}.$$

Equations (5.70), (5.72) and (5.74) describe the conditions under which large amplitude variations, coherent over a bandwidth $\Delta \lambda$, can be produced (Scheuer 1968).

Problem 5.7 : The apparent diameter of a source in Crab nebula has been measured to be $< 0.4"$ at 38MHz. Determine a and $\Delta \lambda$ if $z \simeq L \simeq 1.5 \times 10^{20}$ cm is the distance of the source and $\Delta N \simeq 10^{-3}$ cm^{-3} in the interstellar medium.

Since, a plasma has an index of refraction which is less than unity, electromagnetic waves diverge while passing through it. However, by tailoring the density, a part of the plasma can be made to work as a focusing device. The self-focusing of Laser beams results from such processes which fall in the category of nonlinear processes.

6. Electromagnetic Waves in Magnetized Fluids

For plane-wave-type space and time variations of all the first order quantities, the wave Equation (5.55) becomes:

$$\left(\frac{\omega^2}{c^2} - k^2 \right) \vec{E}_1 + \vec{k} \left(\vec{k} \cdot \vec{E}_1 \right) = \frac{4\pi i \omega n_0 e}{c^2} \vec{U}_{e1}, \qquad (5.75)$$

where we have assumed the ions to form a static positively charged background so that $\vec{U}_{i1} = n_{i1} = 0$ and the current density

$$\vec{J}_1 = -e n_0 \vec{U}_{e1}, \qquad (5.76)$$

is provided only by electrons.

All we have to do now is to determine the electron velocity $\vec{U}_{e1}$ in the presence of a uniform magnetic field, substitute it in the wave equation and

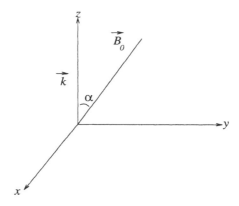

Figure 5.7. Oblique Propagation of Electromagnetic Waves.

we get the dispersion relation for electromagnetic waves in a magnetized plasma.

Let us take the propagation vector $\vec{k} = (0, 0, k)$ and the magnetic field $\vec{B}_0 = (0, B_0 \sin \alpha, B_0 \cos \alpha)$ where α is the angle between $\vec{k}$ and $\vec{B}_0$ (Figure 5.7). The electron velocity is then found to be:

$$U_{1x} = \frac{e}{m_e i \omega} \left[E_{1x} + \frac{U_{1y} B_0 \cos \alpha}{c} - \frac{U_{1z} B_0 \sin \alpha}{c} \right], \qquad (5.77)$$

$$U_{1y} = \frac{e}{m_e i \omega} \left[E_{1y} - \frac{U_{1x} B_0 \cos \alpha}{c} \right], \qquad (5.78)$$

and

$$U_{1z} = \frac{e}{m_e i \omega} \left[E_{1z} + \frac{U_{1x} B_0 \sin \alpha}{c} \right], \qquad (5.79)$$

where we have removed the subscript e from $\vec{U}_1$. We can, now, solve for $\vec{U}_1$ in terms of $\vec{E}_1$ and substitute in Equation (5.75). We get three homogeneous equations in E_{1x}, E_{1y} and E_{1z}:

$$\left[1 - n^2 - \frac{X}{1 - Y} \right] E_{1x} + \left[\frac{iX\sqrt{Y}}{1 - Y} \cos \alpha \right] E_{1y} - \left[\frac{iX\sqrt{Y}}{1 - Y} \sin \alpha \right] E_{1z} = 0, \quad (5.80)$$

$$\left[-\frac{iX\sqrt{Y}}{1 - Y} \cos \alpha \right] E_{1x} + \left[1 - n^2 - \frac{X(1 - Y\sin^2 \alpha)}{1 - Y} \right] E_{1y}$$

$$+ \left[\frac{XY\sin \alpha \cos \alpha}{1 - Y} \right] E_{1z} = 0, \qquad (5.81)$$

$$\left[\frac{iX\sqrt{Y}}{1 - Y} \sin \alpha \right] E_{1x} + \left[\frac{XY\sin \alpha \cos \alpha}{1 - Y} \right] E_{1y} + \left[1 - \frac{X(1 - Y\cos^2 \alpha)}{1 - Y} \right] E_{1z} = 0.$$

$$(5.82)$$

By putting the determinant of these equations to zero, we get the dispersion relation:

$$
\begin{vmatrix}
1 - n^2 - \dfrac{X}{1-Y} & \dfrac{iX\sqrt{Y}}{1-Y}\cos\alpha & \dfrac{-iX\sqrt{Y}}{1-Y}\sin\alpha \\[3mm]
\dfrac{-iX\sqrt{Y}}{1-Y}\cos\alpha & 1 - n^2 - \dfrac{X(1-Y\sin^2\alpha)}{1-Y} & \dfrac{XY}{1-Y}\sin\alpha\cos\alpha \\[3mm]
\dfrac{iX\sqrt{Y}}{1-Y}\sin\alpha & \dfrac{XY}{1-Y}\sin\alpha\cos\alpha & 1 - \dfrac{X(1-Y\cos^2\alpha)}{1-Y}
\end{vmatrix} = 0,
$$

(5.83)

where we have followed the notations usually used while studying wave propagation in the earth's ionosphere, i.e.,

$$
X = \frac{\omega_{pe}^2}{\omega^2}, \quad Y = \frac{\omega_{ce}^2}{\omega^2} \quad \text{and} \quad n^2 = \frac{k^2 c^2}{\omega^2}, \tag{5.84}
$$

We shall study a few special cases using Equation (5.83). First, notice that for $Y = 0$, the dispersion relation (Equation 5.63) for electromagnetic waves in the absence of magnetic fields and collisions is recovered.

For waves propagating perpendicular to the magnetic field i.e., for $\alpha = \pi/2$, we get:

$$
(1 - n^2 - x)\left[\left(1 - n^2 - \frac{X}{1-Y}\right)\left(1 - \frac{X}{1-Y}\right) - \frac{X^2 Y}{(1-Y)^2}\right] = 0. \tag{5.85}
$$

The two roots of n^2 given by Equation (5.84) describe two types of waves. The root

$$
n^2 = 1 - x = 1 - \frac{\omega_{pe}^2}{\omega^2}, \tag{5.86}
$$

describes what is known as the **Ordinary Wave**, since it remains unaffected by the presence of the magnetic field. By substituting for n^2 in Equation (5.80 - 5.82), we find $E_{1x} = E_{1z} = 0$ and $E_{1y} \neq 0$. Thus the ordinary wave is linearly polarized with its electric field parallel to the ambient magnetic field.

The other root of n^2 is:

$$
\begin{aligned}
n^2 &= \frac{(1 - x - Y) - x^2 Y(1 - x - Y)^{-1}}{1 - Y} \\[2mm]
&= 1 - \frac{\omega_{pe}^2(\omega^2 - \omega_{pe}^2)}{\omega^2(\omega^2 - \omega_{pe}^2 - \omega_{ce}^2)}
\end{aligned} \tag{5.87}
$$

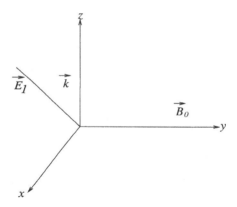

Figure 5.8.   Polarization of the Extraordinary Wave.

This is the dispersion relation of what is known as the **Extraordinary Wave**. From Equations (5.80) and (5.82), we find

$$\frac{E_{1x}}{E_{1z}} = \frac{iX\sqrt{Y}}{(1-Y)(1-n^2 - \frac{X}{1-Y})} = -\frac{\left(1 - \frac{X}{1-Y}\right)}{\frac{iX\sqrt{Y}}{1-Y}}, \qquad (5.88)$$

from which we, again recover the dispersion relation of the extraordinary wave, Equation (5.87). Thus, the extraordinary wave is elliptically polarized with its electric field $(E_{1x}, 0, E_{1z})$ perpendicular to the magnetic field $\vec{B}_0 = (0, B_0, 0)$. We must also acknowledge that this wave is not purely transverse as it has an electric field (E_{1z}) in the direction of propagation vector $\vec{k}$ (Figure 5.8).

From the dispersion relation of the extraordinary wave, we notice that the refractive index n becomes infinite for

$$\omega^2 = \omega_{pe}^2 + \omega_{ce}^2, \qquad (5.89)$$

which, we reckon, is the dispersion relation of the **Upper Hybrid Wave**. The frequency at which $n = \infty$ is known as the **Resonance Frequency**. At this frequency, the wavelength becomes zero. Had we included collisions, we would have found that the wavevector k is purely imaginary at the resonance. This implies that the wave is completely absorbed within the plasma, and its group and phase velocities are zero. We further see a transformation of the nature of the wave. The electromagnetic extraordinary wave has become an electrostatic upper hybrid wave. When we substitute Equation (5.89) in Equation (5.88), we find that $E_{1x} = 0$ and the extraordinary wave has become purely longitudinal with only $E_{1z} \neq 0$.

The extraordinary wave also has a **Cutoff Frequency**. This is the frequency at which the refractive index vanishes, so that the wavelength, the group and the phase velocities all become infinite. The wave at the cutoff frequency suffers a reflection. Although Equation (5.87) for $n^2 = 0$ gives four roots, we retain only the two positive frequency roots given by:

$$\omega_{RP} = \frac{1}{2}\left[\omega_{ce} + \left(\omega_{ce}^2 + 4\omega_{pe}^2\right)^{1/2}\right], \tag{5.90}$$

and

$$\omega_{LP} = \frac{1}{2}\left[-\omega_{ce} + \left(\omega_{ce}^2 + 4\omega_{pe}^2\right)^{1/2}\right]. \tag{5.91}$$

At the cutoff frequency ω_{RP}, the polarization of the extraordinary wave is found to be (Equation 5.88):

$$\frac{E_{1x}}{E_{1z}} = i, \tag{5.92}$$

and at the cutoff frequency ω_{LP}, the polarization of the extraordinary wave is found to be (Equation 5.88):

$$\frac{E_{1x}}{E_{1z}} = -i. \tag{5.93}$$

Obviously the subscripts R and L denote the right-handed and the left-handed circular polarizations. The pass band or the region of propagation of the extraordinary wave can be seen in a plot of n^2 vs ω (Figure 5.9). We see that as $\omega \to \infty$, $n^2 \to 1$. As ω decreases from ∞, n^2 decreases from 1 and becomes zero at $\omega = \omega_{RP}$, the higher cutoff frequency. For $\omega < \omega_{RP}$, $n^2 < 0$ until $\omega = \omega_h$, the upper hybrid frequency at which $n^2 = -\infty$. From $\omega = \omega_h$ to ω_{pe}, n^2 increases from $-\infty$ to 1. From $\omega = \omega_{pe}$ to ω_{LP}, n^2 decreases from 1 to zero. For $\omega < \omega_{LP}$, n^2 remains negative. Thus, the regions $\omega_{LP} < \omega < \omega_h$ and $\omega > \omega_{RP}$ for which $n^2 > 0$ are the pass bands of the extraordinary wave. It is circularly polarized at $\omega = \omega_{LP}$ and ω_{RP}; elliptically polarized at $\omega = \omega_{pe}$ and $\omega > \omega_h$ and longitudinal at $\omega = \omega_h$. So, we, now, know all about the extraordinary wave except its amplitude.

Let us now consider the case $\alpha = 0$ for the propagation of wave along the magnetic field, so that $\vec{k} = (0, 0, k)$ and $\vec{B}_0 = (0, 0, B_0)$. We find from Equation (5.83):

$$\left(1 - n^2 - \frac{X}{1-Y}\right)\left[\left(1 - n^2 - \frac{X}{1-Y}\right)(1-x)\right] - \frac{iX\sqrt{Y}}{1-Y}\left[\frac{-iX\sqrt{Y}}{1-Y}(1-x)\right] = 0,$$

or

$$n^2 = 1 - \frac{X(1 \pm \sqrt{Y})}{1-Y}. \tag{5.94}$$

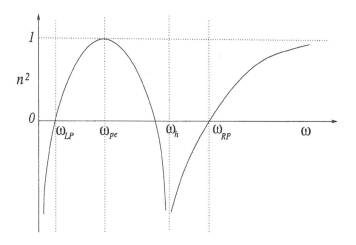

Figure 5.9. Pass-Band of the Extraordinary Wave.

We write

$$n_R^2 = 1 - \frac{X}{1 - \sqrt{Y}} = 1 - \frac{\omega_{pe}^2}{\omega^2 \left(1 - \frac{\omega_{ce}}{\omega}\right)} \tag{5.95}$$

and

$$n_L^2 = 1 - \frac{X}{1 + \sqrt{Y}} = 1 - \frac{\omega_{pe}^2}{\omega^2 \left(1 + \frac{\omega_{ce}}{\omega}\right)} \tag{5.96}$$

for the two roots of n^2 from Equation (5.94). These are the dispersion relations of the two waves propagating parallel to the magnetic field. The polarization of these waves found (from Equations 5.80- 5.82) is

$$\frac{E_{1x}}{E_{1y}} = \frac{-iX\sqrt{Y}}{(1 - n^2)(1 - Y) - X} \tag{5.97}$$

and $E_{1z} = 0$.

By substituting for $n^2 = n_R^2$ in Equation (5.97) we get

$$\frac{E_{1x}}{E_{1y}} = -i. \tag{5.98}$$

Referring to the coordinate system shown in Figure (5.7) we see that Equation (5.98) represents a right-handed or an anticlockwise circular polarization which is also the sense of polarization of the extraordinary wave at the cutoff frequency ω_{RP}.

By substituting for $n^2 = n_L^2$ in Equation (5.97) we get:

$$\frac{E_{1x}}{E_{1y}} = i, \qquad (5.99)$$

which represents a left-handed or a clockwise circular polarization which is also the sense of polarization of the extraordinary wave at the cutoff frequency ω_{LP}.

We, now investigate the pass-bands of the R-wave (Equation 5.95) and the L-wave (Equation 5.96), the way we did for the extraordinary wave.

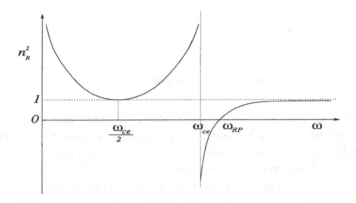

Figure 5.10. Pass-Band of R-Wave

The cutoff frequency of the R-wave is given by $n_R = 0$ and is found to be ω_{RP} defined in Equation (5.90). The resonance frequency of the R-wave is given by $n_R^2 = \infty$ and is found to be at $\omega = \omega_{ce}$. A plot of n_R^2 vs ω is shown in Figure (5.10). We find that n_R^2 has a minimum at $\omega = \omega_{ce}/2$. There is a low frequency pass band for $0 < \omega < \omega_{ce}/2$ in which n_R^2 decreases with an increase in ω and therefore the phase velocity is an increasing function of the frequency. It can be easily checked that the group velocity in this region is also an increasing function of ω. The waves in this pass-band have been named **Whistler Waves**. These waves propagate along the earth's magnetic field between the Northern and the Southern hemispheres and were detected in the ionosphere as radio waves in the audible range, producing a whistling sound. Due to the increase of the group and the phase velocities with ω, the low frequencies arrive later giving rise to descending tones. Thus, the pass band for the R-wave is $0 < \omega < \omega_{ce}$ and $\omega > \omega_{RP}$.

The cutoff frequency of the the L-wave is given by $n_L = 0$ and is found to be ω_{LP} defined in Equation (5.91). The resonance frequency of the L-wave is zero. A plot of n_L^2 vs ω (Figure 5.11) shows that L-wave propagates only for $\omega > \omega_{LP}$.

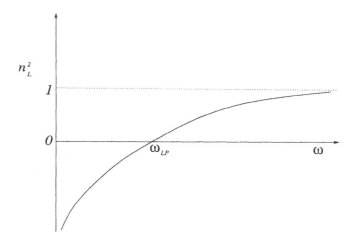

Figure 5.11.   Pass-Band of L-Wave

The propagation of electromagnetic waves at values of α other than zero and $\pi/2$ can be studied by using the general dispersion relation given by Equation (5.83). In real situations, such as in stellar atmospheres, the plasma density and the magnetic field continuously vary with distance. In such a medium, an electromagnetic wave undergoes continuous refraction and the angle α itself becomes position dependent. Under such circumstances, we could either divide the medium into small homogeneous regions and use the results obtained above in each region or go to full-throttle numerical computations.

Another important consequence of the propagation of electromagnetic waves in a magnetized plasma is the attendant variation of the polarization. We have seen that parallel to the magnetic field, waves with right and left-hand circular polarization propagate with different phase speeds since $n_R \neq n_L$. Due to this effect, a plane polarized wave propagating parallel to the magnetic field suffers rotation in its plane of polarization. Let us represent the electric field of a plane wave polarized in the $\hat{x}$ direction as:

$$\vec{E} = \hat{x} E_x e^{i(kz - \omega t)} = \frac{1}{2} \left[\vec{E}_R + \vec{E}_L \right], \tag{5.100}$$

where $\vec{E}_R = (E_x + iE_y) \exp i(kz - \omega t)$ and $\vec{E}_L = (E_x - iE_y) \exp i(kz - \omega t)$ as the superposition of a right E_R and a left E_L circularly polarized waves. The wave vector of the R-wave becomes k_R and that of the L-wave becomes k_L in the magnetized medium. After propagating a distance s in the medium, the electric field of the R-wave is given by

$$\vec{E}_R = (E_x + iE_y) \exp[i(k_R s - \omega t)], \tag{5.101}$$

and of the L-wave by

$$\vec{E}_L = (E_x - iE_y)\exp[i(k_L s - \omega t)].\tag{5.102}$$

If, we now superimpose the two waves, we find

$$\vec{E}_R + \vec{E}_L = \left[E_x \cos(k_L - k_R)\frac{s}{2} + E_y \sin(k_L - k_R)\frac{s}{2}\right] \times$$
$$\times \exp\left[i[(k_L + k_R)\frac{s}{2} - \omega t]\right],\tag{5.103}$$

which represents a plane-polarized wave with its electric field at an angle θ to the x-axis where

$$\theta = (k_L - k_R)\frac{s}{2} = \frac{s}{2c}\frac{\omega_{pe}^2\omega_{ce}}{\omega^2},\tag{5.104}$$

where k_L and k_R have been determined from Equations (5.96) and (5.95). Thus, the propagation through a magnetized medium of size s rotates the electric vector of the electromagnetic wave by the angle θ. This effect is known as the **Faraday Rotation** of the plane of polarization. This is an observable effect. We can estimate that $\theta \simeq 1$ radian when radiation at $\omega = 6 \times 10^8$ sec^{-1} passes through the interstellar medium of dimensions $\simeq 10^{19}$cm, the electron density $n \simeq 10^{-2}$cm^{-3} and the magnetic field $B_0 \simeq 3 \times 10^{-6}$ Gauss. The observations of polarization of radiation from a single source, for example, the Crab Nebula, at different frequencies can confirm the presence of the Faraday effect in addition to providing the parameters of the intervening medium.

With this we end our discussion of waves in a magnetized medium.

5.6. Instabilities of Electron and Proton Fluids

If the electron and or the proton fluids contain free energy in the form of density, temperature and pressure gradients or a relative streaming motion between them, the equilibrium of such a system could become unstable against small perturbations. The excess energy is then released through the growth of electric and magnetic fields, leading to macroscopic configurational changes or heating of plasma with or without emission of radiation. We illustrate the excitation of instabilities through a few simple examples.

1. Instabilities in Unmagnetized Fluids

Relative streaming between the electron and the proton fluids is the most common occurrence, especially in space and astrophysical environs, where the electrons and protons subjected to common acceleration mechanisms, end up with unequal velocities. The excess streaming energy is consumed by waves with their amplitudes growing at an exponential rate with time. Let us assume that in the equilibrium the proton fluid streams with a uniform velocity $\vec{V}_i$ and the electron fluid with a uniform velocity $\vec{V}_e$. We neglect the random component of motion and take $T_e = T_i = 0$. For a plane wave space-time variation of all the perturbed quantities, we obtain the linearized equations of mass and momentum conservation for the electron and the proton fluid as:

$$(-i\omega + i\vec{k} \cdot \vec{V}_e)n_{e1} + in_0\vec{k} \cdot \vec{U}_{e1} = 0, \tag{5.105}$$

$$n_0 m_e(-i\omega + i\vec{k} \cdot \vec{V}_e)\vec{U}_{e1} = -en_0\vec{E}_1, \tag{5.106}$$

$$(-i\omega + i\vec{k} \cdot \vec{V}_i)n_{i1} + in_0\vec{k} \cdot \vec{U}_{i1} = 0, \tag{5.107}$$

and

$$n_0 m_i(-i\omega + i\vec{k} \cdot \vec{V}_i)\vec{U}_{i1} = en_0\vec{E}_1. \tag{5.108}$$

Poisson's equation becomes

$$i\vec{k} \cdot \vec{E}_1 = 4\pi e(n_{i1} - n_{e1}). \tag{5.109}$$

Carrying out the usual elimination exercise, we find the dispersion relation:

$$1 - \frac{\omega_{pe}^2}{(\omega - \vec{k} \cdot \vec{V}_e)^2} - \frac{\omega_{pi}^2}{(\omega - \vec{k} \cdot \vec{V}_i)^2} = 0 \tag{5.110}$$

We can solve this polynomial, look for complex roots of ω; since they occur in pairs, one of them has a positive imaginary part. This root represents the instability as all the perturbed quantities grow exponentially with time in this case. We shall, here, illustrate an approximate way of solving Equation (5.110). We know that if one of the terms in Equation (5.110) becomes very large, the equation will have complex roots. Let us take $\vec{V}_i = 0$, so that $\vec{V}_e$ stands for the relative velocity between electrons and ions. Let us further assume that

$$(\omega - \vec{k} \cdot \vec{V}_e) \simeq \pm\omega_{pe}, \quad \omega << \vec{k} \cdot \vec{V}_e, \tag{5.111}$$

Equation (5.110) then gives:

$$\omega^3 = -\frac{m_e}{2m_i}(\vec{k} \cdot \vec{V}_e)^3, \tag{5.112}$$

from which, we find the real part

$$\omega_R = \left(\frac{m_e}{2^4 m_i}\right)^{1/3}\omega_{pe}, \tag{5.113}$$

and the imaginary part

$$\omega_{_I} = \sqrt{3} \left(\frac{m_e}{2^4 m_i} \right)^{1/3} \omega_{pe}. \tag{5.114}$$

The growth rate of the instability is $\omega_{_I}$. This is the **Two-Stream Instability** also called a **Buneman Type Instability**. We must remember that Equation (5.114) is valid only if $\vec{k} \cdot \vec{V_e} \simeq \omega_{pe}$. The source of energy for this instability is the kinetic energy density $(m_e n_0 V_e^2 / 2)$ of the electrons. Thus, the growth rate $\omega_{_I} = 0$ if $V_e = 0$.

There is another approximate way of solving Equation (5.110). We solve Equation (5.110) in the limit $(m_e/m_i) \to 0$ to find

$$(\omega - \vec{k} \cdot \vec{V_e}) = \pm \omega_{pe},$$

and substitute this in the term proportional to (m_e/m_i). We get

$$1 - \frac{\omega_{pe}^2}{(\omega - \vec{k} \cdot \vec{V_e})^2} - \frac{(m_e/m_i)\omega_{pe}^2}{(\vec{k} \cdot \vec{V_e} \pm \omega_{pe})^2} = 0. \tag{5.115}$$

The complex root with positive imaginary part is now given by

$$\omega = \vec{k} \cdot \vec{V_e} + \frac{i\omega_{pe}}{\left[\dfrac{\omega_{pi}^2}{(\vec{k} \cdot \vec{V_e} - \omega_{pe})^2} - 1 \right]^{1/2}}. \tag{5.116}$$

Thus, depending upon the approximations used, we get different values of the growth rate.

We see that in the approximate methods used above to determine the growth rate, we have used a matching of the Doppler shifted frequency $(\omega - \vec{k} \cdot \vec{V_e})$ with the frequency ω_{pe} of the normal mode — the electron plasma wave. Therefore, it appears that it is this resonance that drives the instability. There is the electron plasma wave associated with the motion of electrons and there is the ion-plasma wave associated with the motion of the ions. The Doppler shift of the proper sign brings these otherwise well separated frequencies to be nearly equal. It can be shown that in the presence of streaming, the electrons support what is known as a **Negative Energy Wave** i.e., the average energy density of the system in the presence of the wave is less than that in its absence or

$$\frac{1}{2} m_e n_0 V_e^2 > \frac{1}{2} \overline{m_e (n_0 + n_{e1})(\vec{V_e} + \vec{U}_{e1})^2}, \tag{5.117}$$

where the bar represents the average over space and time. This results due to the phase relation between the perturbed density n_{e1} and the perturbed velocity $\vec{U}_{e1}$ given by the mass conservation requirements. In the same way, the ions are associated with a positive energy wave. During the growth of the two stream instability, both the negative energy as well as the positive energy waves grow maintaining the constancy of the total energy.

Problem 5.8: Show that for $\omega_{pe}^2 > (\vec{k} \cdot \vec{V}_e)^2 >> \omega^2$, the two-stream instability is a purely growing instability.

The presence of finite amplitude low frequency waves plays an important role in modifying the electrical resistivity of the plasma. The usual Coulomb collisions among electrons and protons are replaced by the scattering of electrons by the low frequency waves which are manifestations of the collective behaviour of the ions. The resistivity in these circumstances could be larger by several orders of magnitude than for normal Coulombic interactions. An actual estimate of the resistivity would require a knowledge of the amplitudes of these low frequency waves. A large resistivity facilitates a fast release of magnetic energy through an ohmic dissipation type of mechanism. This kind of energy release, also known as flaring has been proposed to take place in situations as diverse as the Sun and accretion disks around compact objects.

The **Beam-Plasma Instability** is another instability of great importance for different astrophysical objects. This is excited when a beam of electrons propagates through a non-streaming two-fluid plasma of electrons and protons. The equilibrium of this system consists of a beam of electron density n_b beaming with a velocity $\vec{V}_b$ through a plasma of density n_0. We take the massive protons to only provide the positively charged uniform background. In order to determine the dispersion relation for this case, we can use Equations (5.105) and (5.106) with $\vec{V}_e = 0$ for the electron fluid. The linearized equations for the electron beam are:

$$(-i\omega + i\vec{k} \cdot \vec{V}_b)n_{b1} + in_b\vec{k} \cdot \vec{U}_{b1} = 0, \qquad (5.118)$$

and

$$n_b m_e(-i\omega + i\vec{k} \cdot \vec{V}_b)\vec{U}_{b1} = -en_b\vec{E}_1. \qquad (5.119)$$

The Poisson equation is

$$i\vec{k} \cdot \vec{E}_1 = -4\pi e(n_{b1} + n_{e1}). \qquad (5.120)$$

It is a simple task to find that the dispersion relation of the beam-plasma instability is given by:

$$1 - \frac{\omega_{pe}^2}{\omega^2} - \frac{\omega_b^2}{(\omega - \vec{k} \cdot \vec{V}_b)^2} = 0, \qquad (5.121)$$

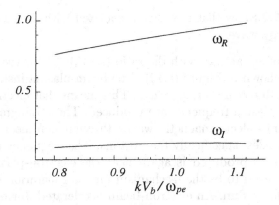

Figure 5.12. Variation of Oscillation Frequency ω_R and Growth Rate ω_I with (kV_b/ω_{pe}) for Beam Plasma Instability for $(\omega_b/\omega_{pe}) = 0.1$.

where $\omega_b^2 = \dfrac{4\pi n_b e^2}{m_e}$.

We can use the cues described during the discussion of the two-stream instability to approximately solve Equation (5.121) for complex roots. Thus, for

$$\omega << \vec{k} \cdot \vec{V_b} \quad \text{and} (\vec{k} \cdot \vec{V_b})^2 \simeq \left(\frac{n_b}{n_0}\right) \omega_{pe}^2,$$

we find:

$$\omega_R \simeq \frac{1}{2^{4/3}} \left(\frac{n_b}{n_0}\right)^{1/6} \omega_{pe}, \qquad (5.122)$$

and

$$\omega_I = \frac{\sqrt{3}}{2^{4/3}} \left(\frac{n_b}{n_0}\right)^{1/6} \omega_{pe}.$$

For $\omega = \vec{k} \cdot \vec{V_b} + i\omega_I < \omega_{pe}$ and $\left|\vec{k} \cdot \vec{V_b}\right| < \omega_{pe}$, $\omega_I << \omega_{pe}$, we find

$$\omega_R \simeq \vec{k} \cdot \vec{V_b}, \qquad (5.123)$$

and

$$\omega_I \simeq \left(\frac{n_b}{n_0}\right)^{1/2} \frac{\omega_{pe}}{\left(\frac{\omega_{pe}^2}{\omega^2} - 1\right)^{1/2}}.$$

The physical mechanism described for the excitation of the two-stream instability also holds for the beam-plasma instability except that, presently there is relative streaming between the two species of electrons instead of between electrons and protons.

Problem 5.9 Show that the wave associated with the electron beam is a negative energy wave.

Variations of ω_R and ω_I with the ratio $(\vec{k} \cdot \vec{V_b})\omega_{pe}^{-1}$ for the beam-plasma instability are shown in Figure (5.12). The beam-plasma instability has the maximum growth rate ω_I for $\omega_R \simeq \omega_{pe}$. This means that electrostatic waves at the electron plasma frequency are produced. These Langmuir waves can be converted into electromagnetic waves through nonlinear scattering on the plasma particles, specifically the protons. The frequency of the electromagnetic waves so produced is again near the electron-plasma frequency ω_{pe}. This is believed to be the mechanism for the generation of type III radio bursts from the Sun. An electron-beam accelerated during a solar flare propagates outwards in the solar corona (density n_0) with typical values of $(n_b/n_0) \sim 10^{-4}$ and $V_b \simeq 0.2c$. As the electron beam passes through the corona with continuously decreasing density n_0, electromagnetic waves of lower and lower frequency are excited. This gives rise to a drift rate of the frequency of radio emission. Drifting radio emission is taken as the signatures of the beam-plasma instability. The type III radio bursts have also been inferred to be emitted at twice the electron-plasma frequency. The emission at the second harmonic is believed to be generated by nonlinear interactions among the Langmuir waves.

Problem 5.10 Determine the growth rate of the instability excited in an electron-positron plasma in relative motion with a uniform velocity.

In high energy sources, such as pulsars and quasars, relativistic electrons are expected to exist along with an ambient non-relativistic plasma. Such a system gives rise to the **Relativistic Version** of the beam-plasma instability. We can determine the dispersion relation by using the relativistic equation of motion for the beam electrons. The linearized form of the relativistic equation of motion is found to be:

$$(-i\omega + i\vec{k}c \cdot \vec{\beta_0})\gamma_0(1 + \gamma_0^2\beta_0^2)\vec{U}_{b1} = -\frac{e}{m_e}\vec{E}_1. \qquad (5.124)$$

The other equations [(5.105) and (5.106) with $\vec{V_e} = 0$ and (5.119) and (5.120)] remain unaltered. The dispersion relation (Equation 5.121) is modified to:

$$1 - \frac{\omega_{pe}^2}{\omega^2} - \frac{\omega_b^2}{\gamma_0^3(\omega - c\vec{k} \cdot \vec{\beta_0})^2} = 0, \qquad (5.125)$$

where $\beta_0 = V_b/c$ and $\gamma_0 = (1 - \beta_0^2)^{-1/2}$.

We have already learnt how to solve Equation (5.125). The growth rates in the relativistic case can be obtained by replacing n_b by (n_b/γ_0^3) which shows that the relativistic beam-plasma instability grows at a reduced rate, since $\gamma_0 > 1$.

2. Instabilities in Magnetized Fluids

We have seen earlier that magnetic field behaves like a low density fluid and exerts pressure. It is due to this property that mechanical pressure can be balanced by magnetic pressure. We will now study the stability of the electron and proton fluids in the presence of a magnetic field $\vec{B}_0$ and acceleration due to gravity $\vec{g}$. The equilibrium of the proton fluid is described by:

$$m_i n_0 (\vec{U}_{i0} \cdot \vec{\nabla}) \vec{U}_{i0} = e n_0 \frac{\vec{U}_{i0} \times \vec{B}_0}{c} + m_i n_0 \vec{g}. \qquad (5.126)$$

For constant $\vec{g}$, $\vec{U}_{i0}$ is uniform and its component perpendicular to $\vec{B}_0$ is given by

$$(\vec{U}_{i0})_\perp = \frac{m_i}{e} \frac{\vec{g} \times \vec{B}_0}{B_0^2} \equiv \vec{U}_0. \qquad (5.127)$$

The electrons have a drift velocity opposite to and smaller than that of the protons and we assume it to be zero in the limit $(m_e/m_i) \to 0$.

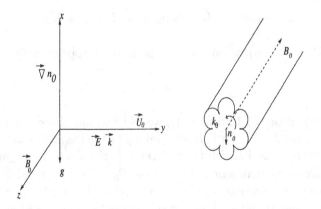

Figure 5.13.   Circumstances of the Flute Instability

The equation of motion of ions is:

$$(n_0 + n_1) m_i \left[\frac{\partial \vec{U}_{i1}}{\partial t} + (\vec{U}_{i0} + \vec{U}_{i1}) \cdot \vec{\nabla} (\vec{U}_{i0} + \vec{U}_{i1}) \right] = m_i (n_0 + n_1) \vec{g} +$$

$$(n_0 + n_1) e \left[\vec{E}_1 + (\vec{U}_{i0} + \vec{U}_{i1}) \times \frac{\vec{B}_0}{c} \right] \qquad (5.128)$$

A simplified form of this equation can be obtained by subtracting Equation (5.126), after multiplying it with $(1 + n_{i1}/n_0)$, from Equation (5.128) as:

$$m_i n_0 \left[-i\omega + i\vec{k} \cdot \vec{U}_0 \right] \vec{U}_{i1} = e n_0 \left[\vec{E}_1 + \frac{\vec{U}_{i1} \times \vec{B}_0}{c} \right], \qquad (5.129)$$

where we have not introduced any perturbation in the magnetic field. The mass conservation for ions is described as:

$$(-i\omega + i\vec{k} \cdot \vec{U}_0) n_{i1} = -i n_0 (\vec{k} \cdot \vec{U}_{i1}) - \vec{U}_{i1} \cdot \vec{\nabla} n_0, \qquad (5.130)$$

where now, the fluids have an equilibrium density gradient $\vec{\nabla} n_0$. This is essential in the presence of gravitational forces. The linearized form of the momentum conservation for the electron fluid, in the limit $(m_e/m_i) \to 0$ is given by:

$$-e \left[\vec{E}_1 + \frac{\vec{U}_{e1} \times \vec{B}_0}{c} \right] = 0. \qquad (5.131)$$

The continuity equation for the electron fluid is:

$$-i\omega n_{e1} + \vec{U}_{e1} \cdot \vec{\nabla} n_0 + i\vec{k} \cdot \vec{U}_{e1} n_0 = 0. \qquad (5.132)$$

Instead of Poisson's equation, we use the plasma approximation

$$n_{i1} = n_{e1}. \qquad (5.133)$$

By solving Equations (5.129)-(5.133), we can find the dispersion relation and determine the conditions for complex values of the frequency ω. We wish to study an electrostatic instability, we take $\vec{k} \parallel \vec{E}_1$. Further, for propagation perpendicular to the magnetic field, we take $\vec{B}_0 = (0, 0, B_0)$ and $\vec{k} = (0, k, 0)$. Assuming the acceleration due to gravity $\vec{g}$ to be in the $(-x)$ direction (Figure 5.13) and therefore $\vec{\nabla} n_0 (\equiv n_0')$ is also in the x direction, we find the dispersion relation:

$$\omega(\omega - kU_0) = \left(\frac{n_0'}{n_0 k} \right) \left[\frac{(\omega - kU_0)}{\omega_{ci}} \left\{ (\omega - kU_0)^2 - \omega_{ci}^2 \right\} + \omega\omega_{ci} \right] = 0. \qquad (5.134)$$

This is a cubic in ω and can be solved analytically to determine the three roots. The condition for complex roots is given by:

$$\frac{a^3}{27} + \frac{b^2}{4} > 0, \qquad (5.135)$$

where

$$a = \frac{1}{3}\alpha\, \omega_{ci}^2\, (3\beta - \alpha),$$

$$b = \frac{1}{27}\omega_{ci}^3\left(-36\beta^2\alpha - 9\beta\alpha^2 + 27\beta + 2\alpha^3\right),$$

$$\alpha = \frac{n_0 k}{n_0'},$$

$$\beta = \frac{kU_0}{\omega_{ci}}.$$

Here, however, we solve Equation (5.134) under the assumption

$$(\omega - kU_0)^2 \ll \omega_{ci}^2. \tag{5.136}$$

Equation (5.134) reduces to obtain a quadratic:

$$\omega^2 - (kU_0)\omega + \frac{n_0' g}{n_0} = 0, \tag{5.137}$$

which has the roots

$$\omega = \frac{kU_0}{2} \pm \frac{1}{2}\left[(kU_0)^2 - \frac{4n_0' g}{n_0}\right]^{1/2}. \tag{5.138}$$

Thus, there is an instability for

$$\frac{4n_0' g}{n_0} > (kU_0)^2 \tag{5.139}$$

with growth rate

$$\omega_I \simeq \left(\frac{n_0' g}{n_0}\right)^{1/2}. \tag{5.140}$$

Doesn't this look familiar? This is the growth rate of the R-T instability in a medium with density gradient opposite to the direction of $\vec{g}$. We notice that all the motions of the two fluids are perpendicular to the magnetic field. These cyclotron and streaming motions are of unequal magnitudes for electrons and protons, as a result of which a charge separation takes place giving rise to the electric field $\vec{E}_1$. The $\vec{E}_1 \times \vec{B}_0$ drift further enhances the charge separation and the instability is excited. This instability is also known as the **Flute Instability** when studied in a cylindrical fluid column, in which the wavevector k is in the θ direction for an axial magnetic field $\vec{B}_0$. The undulations produced in the originally circular cross section of the cylindrical column give it a fluted appearance, and hence the name.

Problem 5.11 Discuss the Flute Instability for $\vec{k} \parallel \vec{B}_0$.

Problem 5.12 Estimate the length scale of the inverted density gradient required for the flute instability in the atmospheres of pulsars, which have, both, large $\vec{g}$ and large $\vec{B}_0$.

We, now, give an example of an **Electromagnetic Instability**. In some astrophysical objects such as pulsars and quasars, a pair plasma consisting of electrons and positrons is generated through the interactions of γ-ray photons. The two fluids of electrons and positrons may stream with speed V_0 in opposite directions along the ambient magnetic field. The dispersion relation for such a system can easily be written down, now that we have had some experience with the procedures. The dispersion relation for the right-hand polarization including two species of particles, electrons and positrons counterstreaming with speed V_0 is given by [cf Equation 5.95]:

$$\frac{k^2 c^2}{\omega^2} = 1 - \frac{\omega_{pe}^2(\omega + kV_0)}{\omega^2(\omega + kV_0 - \omega_{ce})} - \frac{\omega_{pe}^2(\omega - kV_0)}{\omega^2(\omega - kV_0 + \omega_{ce})}. \tag{5.141}$$

The complex roots of this equation can be determined as before. We find that an electromagnetic instability exists for $\omega_R \ll \omega_{pe}$, $\omega_{ce} > kV_0$ and

$$\frac{2kV_0}{\omega_{ce} - kV_0} > \frac{k^2 c^2}{\omega_{pe}^2}. \tag{5.142}$$

We shall discuss more cases of high frequency electromagnetic instabilities in the next chapter based on the kinetic treatment of plasmas since these microscopic instabilities leading to emission of radiation are manifestations of non-Maxwellian particle distributions.

5.7. Ambipolar Diffusion

The charged particles in a plasma are never free. A movement of, say, electrons immediately leads to a movement of positively charged ions in order to maintain the overall charge neutrality of the plasma. Therefore, diffusion of particles is also a collective phenomena in which both electrons and massive ions participate. It may come as a surprise that ions are the dominating partner. The continuity equation for electrons and ions can be written as:

$$\frac{\partial n_j}{\partial t} + \vec{\nabla} \cdot \vec{P}_j = 0, \tag{5.143}$$

where n_j is the particle density, $\vec{P}_j = n_j \vec{U}_j$ is the particle flux and j stands for the species of particles. We take electron density n_e equal to the ion density $n_i = n$. The steady state velocity $\vec{U}_j$ of each species is determined from their respective fluid equations as:

$$0 = -en\vec{E} - \vec{\nabla}p_e - m_e n\nu_{ei}\vec{U}_e, \tag{5.144}$$

or

$$\vec{U}_e = -\mu_e\vec{E} - D_e\frac{\vec{\nabla}n}{n}, \tag{5.145}$$

where $\mu_e = (e/m_e\nu_{ei})$ is called the electron mobility coefficient and

$$D_e = \frac{K_B T_e}{m_e \nu_{ei}}, \tag{5.146}$$

is the electron diffusion coefficient. The electron flux $\vec{P}_e$ is then found to be:

$$\vec{P}_e = n\vec{U}_e = -n\mu_e\vec{E} - D_e\vec{\nabla}n. \tag{5.147}$$

The steady state ion (i.e., proton) fluid equation is

$$0 = en\vec{E} - \vec{\nabla}p_i - m_i n\nu_{ie}\vec{U}_i, \tag{5.148}$$

which gives:

$$\vec{U}_i = \mu_i\vec{E} - D_i\vec{\nabla}n, \tag{5.149}$$

where

$$\mu_i = e/m_i\nu_{ie},$$

$$D_i = \frac{K_B T_i}{m_i \nu_{ie}}, \tag{5.150}$$

and the proton flux

$$\vec{P}_i = n\mu_i\vec{E} - D_i\vec{\nabla}n. \tag{5.151}$$

As we have discussed in Chapter 1, electrons being faster than ions, tend to diffuse out of the plasma first, leaving behind an excess of positive charge. This sets up an electric field $\vec{E}$ which retards the electrons but accelerates the ions. We can determine $\vec{E}$ by equating the electron and ion fluxes. This gives

$$\vec{E} = \frac{(D_i - D_e)}{(\mu_e + \mu_i)}\frac{\vec{\nabla}n}{n}, \tag{5.152}$$

and the flux

$$\vec{P} = \vec{P}_e = \vec{P}_i = -\frac{(\mu_i D_e + \mu_e D_i)}{(\mu_e + \mu_i)}\vec{\nabla}n. \tag{5.153}$$

The effective diffusion coefficient D_A is now defined using Fick's law of diffusion for the case of zero electric field, for then,

$$\vec{P} = -D\vec{\nabla}n, \qquad (5.154)$$

so that

$$D_A = \frac{\mu_i D_e + \mu_e D_i}{\mu_e + \mu_i}. \qquad (5.155)$$

This is the coefficient of the **Ambipolar Diffusion**. We see that $\mu_e >> \mu_i$ and the ambipolar diffusion coefficient

$$D_A \simeq D_i + \frac{\mu_i}{\mu_e}D_e \qquad (5.156)$$

$$= D_i \left[1 + \frac{T_e}{T_i}\right],$$

is nearly twice the ion diffusion coefficient D_i for equal electron and ion temperatures. Thus, the ion diffusion is enhanced from D_i to D_A, since $(D_A/D_i) \simeq 2$. The diffusion of electrons is slowed down from D_e to D_A, since $(D_A/D_e) \simeq 2\,(m_e/m_i) << 1$. So ions call the shots!

In the presence of a magnetic field, the ion diffusion rate perpendicular to the magnetic field is much larger than that for electrons. The ambipolar diffusion coefficient $D_{A\perp}$ perpendicular to the magnetic field is therefore nearly twice the electron diffusion rate $D_{e\perp}$.

Problem 5.13: Show that $D_{A\perp} \simeq 2D_{e\perp}$.

References

Chen, F.F.: 1974, Introduction to Plasma Physics, Plenum Press.
Gurnett, D.A. & Anderson, R.R.: 1976, Science, **194**, 1159.
Melrose, D.B.: 1986, Instabilities in Space and Laboratory Plasmas, Cambridge University Press.
Scheuer, P.A.G.: 1968, Nature, **218**, 920.

CHAPTER 6

KINETIC DESCRIPTION OF PLASMAS

6.1. Back to the Vlasov-Maxwell Way

We have come back full circle! In Chapter 2, we started with the phase - space description of N discrete particles and then transformed it into a continuum two-fluid and finally one fluid descriptions using several averaging processes. After investigating some characteristics of the one-fluid and two-fluid descriptions, we now deal head-on with N discrete particles, electrons and protons, using the Vlasov equation. In this description, we work with particle distribution functions in the phase space of velocities and positions. The time evolution of the distribution function defines the stability or otherwise of the system. Plasmas are particularly interesting because they often submit to, or support, or generate, nonthermal (non-Maxwellian) and non-equilibrium distributions for finite durations of time. In other words, different species of particles can have unequal temperatures. Even a single species of particles can have different temperatures corresponding to different degrees of freedom. The free energy contained in these non-equilibrium distribution functions is then released in the form of heat and radiation. Plasmas are valued for their intrinsic cooperative nature due to which the transport, dissipative and radiative processes proceed at anomalously large rates as compared to single particle processes. Several astrophysical sources with extremely high luminosities with spectral energy distribution far from that of the blackbody, often showing variability on extremely short time scales, warrant the operation of coherent plasma processes. In this chapter, we shall study what is known as the kinetic or microscopic equilibrium and stability of an electron - proton plasma

6.2. Kinetic-Equilibrium of an Electron-Proton Plasma

The equilibrium is now determined from the Vlasov equation, one each for electron and proton species, and the Maxwell's equations. Neglecting collisional processes, we write the Vlasov equation for electrons as (Chapter 2):

$$\frac{\partial f_e}{\partial t}(\vec{r},\vec{V},t) + \vec{V}\cdot\frac{\partial f_e(\vec{r},\vec{V},t)}{\partial \vec{r}} - \left[\frac{e}{m_e}\vec{E} + e\frac{\vec{V}\times\vec{B}}{m_e c}\right]\cdot\frac{\partial f_e(\vec{r},\vec{V},t)}{\partial \vec{V}} = 0. \quad (6.1)$$

In the absence of electric and magnetic fields, in equilibrium, Equation (6.1) reduces to:

$$\vec{V} \cdot \frac{\partial f_e}{\partial \vec{r}} = 0, \tag{6.2}$$

which implies that the equilibrium single particle electron distribution function, f_e, must be independent of the space and the time coordinates and is a function only of velocity. For example, the **Maxwell-Boltzmann Distribution** of velocities expressed as:

$$f_e(\vec{V}) = n_0 \left(\frac{m_e}{2\pi K_B T_e} \right)^{3/2} \exp\left(-\frac{m_e V^2}{2 K_B T_e} \right), \tag{6.3}$$

is a solution of Equation (6.2). As a matter of fact, we can choose for f_e, any function which depends only on the constants a_i of motion of a particle, since

$$\frac{d}{dt} f_e(a_1, a_2, \ldots) = \sum_i \frac{\partial f_e}{\partial a_i} \frac{da_i}{dt} = 0. \tag{6.4}$$

We must remember that constants of motion are functions of $(\vec{r}, \vec{V})$ and are independent of time only for each single particle's motion. In general, a_i are functions of $(\vec{r}, \vec{V}, t)$. When a_i are independent of time, so is f_e - the equilibrium distribution function.

The first example, perhaps, of a constant of motion independent of time is the total energy. For a free particle, the total energy is $(mv^2/2)$ and the Maxwell-Boltzmann distribution function (Equation 6.3) is realized. For electrons executing circular motion in a uniform magnetic field, the total energy, the energy associated with motion parallel to the magnetic field, the energy associated with motion perpendicular to the magnetic field and the angular momentum are all constants of motion. Thus, the electron distribution function in the presence of magnetic field could be represented as:

$$f_e(\vec{V}) = \frac{n_0 m_e^{3/2}}{(2\pi K_B T_\parallel)^{1/2}(2\pi K_B T_\perp)} \exp\left(-\frac{m_e V_\parallel^2}{2 K_B T_\parallel} - \frac{m_e V_\perp^2}{2 K_B T_\perp} \right), \tag{6.5}$$

where $\parallel$ and $\perp$ are with respect to the direction of the magnetic field.

In slowly varying fields, the adiabatic invariants play the role of the constants of motion. We have studied in Chapter 3 that in a magnetic mirror, charged particles with velocities inclined at small angles to the magnetic field escape from the system; the resulting phase space distribution of the particles is known as the **Loss-Cone Distribution**.

The condition for the escape of particles from a magnetic mirror has been derived in Chapter 3. It says that for a given value of V_z, particles with $V_\perp < pV_z$ are absent from the system, and the loss cone angle θ_M is given by

$$\theta_M = \tan^{-1} p.$$

One representation of the Loss-cone distribution function is (Figure 6.1):

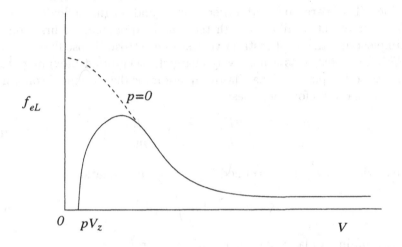

Figure 6.1.   Representation of the Loss-Cone Distribution Function f_{eL}.

$$f_{eL} = \left(\frac{m_e}{2\pi K_B T_e}\right)^{3/2} \left(p^2 + 1\right)^{1/2} \exp\left[-\frac{m_e V^2}{2K_B T_e}\right] \Theta\left(V_\perp^2 - V_z^2 p^2\right), \quad (6.6)$$

which reduces to a Maxwellian for $p = 0$. Here Θ is the unit step function which is unity if its argument is greater than zero and zero otherwise.

6.3.   Kinetic Description of Electron-Plasma Waves and Instabilities

After determining the kinetic equilibrium, we would like to find out if this equilibrium is stable or not. For this purpose, we perturb the plasma so that its two distribution functions, one for electrons (f_e) and the other for protons (f_i), are given by

$$f_s(\vec{r}, \vec{V}, t) = f_{s0}(\vec{r}, \vec{V}) + f_{s1}(\vec{r}, \vec{V}, t), \quad (6.7)$$

where the species index s stands for e (electrons) and i (protons). The linearized Vlasov equation

$$\frac{\partial f_{s1}(\vec{r}, \vec{V}, t)}{\partial t} + \vec{V} \cdot \frac{\partial f_{s1}(\vec{r}, \vec{V}, t)}{\partial \vec{r}} + \frac{e_s}{m_s} \left[\vec{E}_1 + \frac{\vec{V} \times \vec{B}_1}{c} \right] \cdot \frac{\partial f_{s0}(\vec{r}, \vec{V})}{\partial \vec{V}} = 0, \quad (6.8)$$

is then obtained by assuming $\mid f_{s1} \mid << \mid f_{s0} \mid$. Here, $\vec{E}_1$ and $\vec{B}_1$ are the first order electric and magnetic fields to be determined from Maxwell's equations. There are no zeroth order electric and magnetic fields.

In order to get familiarity with the kinetic approach, we first consider the simplest and the most instructive case of electrostatic oscillations ($\vec{B}_1 = 0$). We treat protons as a positively charged background providing charge neutrality to the plasma. The Vlasov equation for the electronic component of the plasma, therefore becomes:

$$\frac{\partial f_{e1}}{\partial t} + \vec{V} \cdot \frac{\partial f_{e1}}{\partial \vec{r}} - \frac{e}{m_e} \vec{E}_1 \cdot \frac{\partial f_{e0}}{\partial \vec{V}} = 0. \quad (6.9)$$

The electric field $\vec{E}_1$ is determined from Poisson's equation:

$$\vec{\nabla} \cdot \vec{E}_1 = -4\pi e \int f_{e1} d\vec{V}. \quad (6.10)$$

Again assuming a plane-wave type variation for f_{e1} as

$$f_{e1}(\vec{r}, \vec{V}, t) = f_{e1}(\vec{V}) \exp\left[i(k \cdot r - \omega t)\right], \quad (6.11)$$

we find from Equations (6.9) and (6.10):

$$\frac{4\pi e^2}{m_e k} \int \frac{\hat{k} \cdot \frac{\partial f_{e0}}{\partial \vec{V}}}{(\vec{k} \cdot \vec{V} - \omega)} d\vec{V} = 1, \quad (6.12)$$

where for electrostatic perturbations, $\vec{k} \parallel \vec{E}_1$ and $\hat{k} = \frac{\vec{k}}{\mid k \mid}$ have been used.

The evaluation of the integral in Equation (6.12) is a trifle tricky since the integrand diverges for $\vec{k} \cdot \vec{V} = \omega$. It was the Russian physicist Lev Landau (Landau 1946) who realized the importance of the singularity at $\vec{k} \cdot \vec{V} = \omega$ and showed a way to handle it. He stressed that this problem must be treated as an initial value problem, which means that the perturbations can be Fourier decomposed in space but we must use the Laplace transform for the time coordinate (Schmidt 1966, Ichimaru 1973, Harris 1975).

For

$$f_{e1}(\vec{r}, \vec{V}, t) = f_{e1}(\vec{k}, \vec{V}, t)e^{i\vec{k} \cdot \vec{r}},$$

and

$$\vec{E}_1(\vec{r}, t) = \vec{E}_{1k}(t)e^{i\vec{k}\cdot\vec{r}}, \tag{6.13}$$

Equations (6.9) and (6.10) become

$$\left[\frac{\partial}{\partial t} + i\vec{k}\cdot\vec{V}\right] f_{e1}(\vec{k}, \vec{V}, t) = \frac{e}{m_e}\vec{E}_{ik}\cdot\frac{\partial f_{e0}}{\partial\vec{V}}, \tag{6.14}$$

and

$$\vec{k}\cdot\vec{E}_{1k} = 4\pi ei \int f_{e1}(\vec{k}, \vec{V}, t)d\vec{V}. \tag{6.15}$$

On substituting Equation (6.15) into Equation (6.14), we get, for $\vec{k} \parallel \vec{E}_1$:

$$\left[\frac{\partial}{\partial t} + i\vec{k}\cdot\vec{V}\right] f_{e1}(\vec{k}, \vec{V}, t) = \frac{4\pi e^2 i}{m_e k^2}\left(\vec{k}\cdot\frac{\partial f_{e0}}{\partial\vec{V}}\right)\int f_{e1}(\vec{k}, \vec{V}, t)d\vec{V}. \tag{6.16}$$

We now write the Laplace transform $\overline{f}_{e1}(\vec{k}, \vec{V}, \omega)$ of $f_{e1}(\vec{k}, \vec{V}, t)$ as:

$$\overline{f}_{e1}(\vec{k}, \vec{V}, \omega) = \int_0^\infty e^{i\omega t} f_{e1}(\vec{k}, \vec{V}, t)dt, \tag{6.17}$$

where ω is a parameter with positive imaginary part. The inverse Laplace transform is defined as:

$$f_{e1}(\vec{k}, \vec{V}, t) = \frac{1}{2\pi}\int_C e^{-i\omega t}\overline{f}_{e1}(\vec{k}, \vec{V}, \omega)d\omega, \tag{6.18}$$

where C is the path of integration in the complex ω - plane going from $\omega = -\infty$ to $\omega = \infty$ and lying in the upper half-plane including all singularities of $\overline{f}_{e1}(\vec{k}, \vec{V}, \omega)$ (Figure 6.2). In order to take the Laplace transform of Equation (6.16), we multiply both sides by $e^{i\omega t}$ and integrate over t from $t = 0$ to $t = \infty$ by parts. We find:

$$\overline{f}_{e1} = -\frac{4\pi e^2}{m_e k^2}\frac{\vec{k}\cdot\dfrac{\partial f_{e0}}{\partial\vec{V}}}{(\omega - \vec{k}\cdot\vec{V})}\int \overline{f}_{e1}d\vec{V} + i\frac{f_{e1}(\vec{k}, \vec{V}, 0)}{(\omega - \vec{k}\cdot\vec{V})}.$$

Now, integrating both sides over velocity, we get

$$\int \overline{f}_{e1}d\vec{V} = \frac{i}{\epsilon(\vec{k}, \omega)}\int \frac{f_{e1}(\vec{k}, \vec{V}, 0)}{(\omega - \vec{k}\cdot\vec{V})}d\vec{V}, \tag{6.19}$$

where

$$\epsilon(\vec{k}, \omega) = 1 + \frac{4\pi e^2}{m_e k^2}\int \frac{\vec{k}\cdot\dfrac{\partial f_{e0}}{\partial\vec{V}}}{(\omega - \vec{k}\cdot\vec{V})}d\vec{V}, \tag{6.20}$$

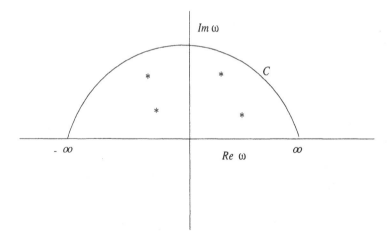

Figure 6.2. Path of Integration C for Taking the Inverse Laplace Transform, the Stars Represent the Singularities of $\overline{f_{e1}}(\vec{k}, \vec{V}, \omega)$.

is known as the **Plasma Dielectric Function.**

Using Equations (6.15),(6.19) and (6.20) we find the Laplace transform $\overline{E}_{1k}$ of $E_{1k}(t)$ to be:

$$\overline{E}_{1k}(\omega) = -\frac{4\pi e}{k\epsilon(\vec{k},\omega)} \int \frac{f_{e1}(\vec{k},\vec{V},0)}{(\omega - \vec{k}\cdot\vec{V})} d\vec{V}, \qquad (6.21)$$

and the Laplace transform $\overline{f}_{e1}(\vec{k},\vec{V},\omega)$ of $f_{e1}(\vec{k},\vec{V},t)$ to be:

$$\overline{f}_{e1}(\vec{k},\vec{V},\omega) = \frac{if_{e1}(\vec{k},\vec{V},0)}{\omega - \vec{k}\cdot\vec{V}} - \frac{i4\pi e^2}{m_e k^2 \epsilon(\vec{k},\omega)} \frac{\vec{k}\cdot\dfrac{\partial f_{e0}}{\partial \vec{V}}}{(\omega - \vec{k}\cdot\vec{V})} \int \frac{f_{e1}(\vec{k},\vec{V},0)}{(\omega - \vec{k}\cdot\vec{V})} d\vec{V}. \tag{6.22}$$

The time dependence of the field $\vec{E}_1$ and the distribution function f_{e1}, are recovered by taking the inverse Laplace transforms (defined in Equation 6.18 for f_{e1}). We then have

$$\vec{E}_{1k}(t) = \frac{1}{2\pi} \int_C e^{-i\omega t} \vec{E}_{1k}(\omega) d\omega. \tag{6.23}$$

If $E_{1k}(\omega)$ has a number of simple poles at $\omega = \omega_l$ then Cauchy's formula of contour integration gives

$$\vec{E}_{1k}(t) = \frac{1}{2\pi} \cdot 2\pi i \sum_l e^{-i\omega_l t} \vec{E}_{1k}(\omega_l). \tag{6.24}$$

The electric field $\vec{E}_{1k}(t)$ represents a superposition of waves of different frequencies ω_l but with the same wave vector $\vec{k}$. The poles of $\vec{E}_{1k}(\omega)$ may be either due to the initial perturbation f_{e1} $(\vec{k}, \vec{V}, 0)$ or due to the dielectric function $\epsilon(\vec{k}, \omega)$. The latter contains information on the properties of the medium. These poles are given by

$$\epsilon(\vec{k}, \omega) = 0. \tag{6.25}$$

We see from Equation (6.20) that the integral blows up at $\omega = \vec{k} \cdot \vec{V}$ if ω is real. This does not happen in the Laplace transform method used above, since, here, ω is complex with a positive imaginary part and the denominator $\omega - \vec{k} \cdot \vec{V}$ can never vanish. In order to determine $\vec{E}_{1k}(t)$, however, we must analytically continue $\vec{E}_{1k}(\omega)$ in the lower half - plane of ω, i.e., for negative imaginary values of ω, which correspond to the stable behaviour of a plasma.

We now evaluate an approximate value of $\epsilon(\vec{k}, \omega)$. In order to perform the integration in Equation (6.20), we must remember that ω lies in the upper half plane. Therefore, we write:

$$\frac{1}{\omega - \vec{k} \cdot \vec{V}} \equiv \frac{1}{\omega - \vec{k} \cdot \vec{V} + i\alpha}$$

$$= \lim_{\alpha \to 0} P \frac{1}{\omega - \vec{k} \cdot \vec{V}} - i\pi\delta(\omega - \vec{k} \cdot \vec{V}), \tag{6.26}$$

where P stands for the principal value of the integral.

Then the real part, $\epsilon_R(\vec{k}, \omega)$ and the imaginary part, $\epsilon_I(\vec{k}, \omega)$, for real ω are found to be:

$$\epsilon_R(\vec{k}, \omega) = 1 + \frac{4\pi e^2}{m_e k^2} P \int \frac{\vec{k} \cdot \frac{\partial f_{e0}}{\partial \vec{V}}}{\left(\omega - \vec{k} \cdot \vec{V}\right)} d\vec{V}, \tag{6.27}$$

$$\epsilon_I(\vec{k}, \omega) = -\frac{4\pi^2 e^2}{m_e k^2} \int \vec{k} \cdot \frac{\partial f_{e0}}{\partial \vec{V}} \delta(\omega - \vec{k} \cdot \vec{V}) d\vec{V}. \tag{6.28}$$

We note that if Equation (6.25) is to be satisfied for real ω then ϵ_I must vanish, i.e., there should be no particles in the distribution f_{e0} with velocities equal to the phase velocity (ω/k) of the wave. In general, therefore $\epsilon_I \neq 0$. In order to determine the roots of Equation (6.25), we write

$$\omega = \omega_R + i\omega_I, \tag{6.29}$$

and expand $\epsilon(\vec{k}, \omega_R + i\omega_I)$, assuming $\mid \omega_I \mid << \mid \omega_R \mid$ as

$$\epsilon(\vec{k}, \omega_R + i\omega_I) = \epsilon_R(\vec{k}, \omega_R + i\omega_I) + i\epsilon_I(\vec{k}, \omega_R + i\omega_I),$$

$$\simeq \left[\epsilon_R(\vec{k}, \omega_R) + i\omega_I \frac{\partial \epsilon_R(\vec{k}, \omega_R)}{\partial \omega_R} \right]$$

$$+ i \left[\epsilon_I(\vec{k}, \omega_R) + i\omega_I \frac{\partial \epsilon_I(\vec{k}, \omega_R)}{\partial \omega_R} \right],$$

or

$$\epsilon(\vec{k}, \omega) \simeq \epsilon_R(\vec{k}, \omega_R) + i \left[\omega_I \frac{\partial \epsilon_R(\vec{k}, \omega_R)}{\partial \omega_R} + \epsilon_I(\vec{k}, \omega_R) \right]. \tag{6.30}$$

Thus, the approximate roots of Equation (6.25) are given by

$$\epsilon_R(\vec{k}, \omega_R) = 0, \tag{6.31}$$

and

$$\omega_I = -\frac{\epsilon_I(\vec{k}, \omega_R)}{\left[\dfrac{\partial \epsilon_R(\vec{k}, \omega_R)}{\partial \omega_R} \right]}. \tag{6.32}$$

Equation (6.31) determines the real part ω_R. [Caution: In Equations (6.27) and (6.28) ω stands for ω_R] The principal part of the integral

$$P \int \frac{\vec{k} \cdot \dfrac{\partial f_{e0}}{\partial \vec{V}}}{(\omega_R - \vec{k} \cdot \vec{V})} d\vec{V},$$

should be evaluated from $V_x = -\infty$ to $V_x = \omega_R/k$ where we have chosen $\vec{k}$ to be in the x direction. The integration over the other two components is carried out over their entire range $(-\infty, \infty)$. In the case when the phase velocity (ω_R/k) is much larger than the thermal velocity of the electrons, the integration over V_x can be extended to the entire range $(-\infty, \infty)$ without introducing too large an error. Since in an equilibrium distribution, there are not many particles with velocities much larger than the phase velocity (ω_R/k), their contribution to the principal value integral is, therefore, small. Let us write the equilibrium distribution function as

$$f_{e0}(\vec{V}) = n_0 f_{e0}(V_x)\varphi_{e0}(V_y, V_z), \tag{6.33}$$

and substitute in Equation (6.27). After integrating over V_y and V_z and partially integrating over V_x, we get:

$$\epsilon_R(k, \omega_R) = 1 + \omega_{pe}^2 \int_{-\infty}^{\infty} \frac{f_{e0}(V_x)}{(\omega_R - kV_x)^2} dV_x, \tag{6.34}$$

where we have used

$$\int_{-\infty}^{\infty} \varphi_{e0}(V_y, V_z)dV_ydV_z = 1.$$

For a Maxwellian f_{e0} (V_x) and using the assumption $\mid \omega_R \mid >> \mid kV_x \mid$ we obtain:

$$\epsilon_R = 1 - \frac{\omega_{pe}^2}{\omega_R^2} - \frac{3k^2 V_{Te}^2 \omega_{pe}^2}{2\omega_R^4}, \tag{6.35}$$

and

$$\epsilon_I = -\frac{\omega_{pe}^2 \pi}{k^2}\left[\frac{\partial f_{e0}(V_x)}{\partial V_x}\right]_{V_x = \omega_R/k}, \tag{6.36}$$

where

$$f_{e0}(V_x) = \left(\frac{m_e}{2\pi K_B T_e}\right)^{1/2} \exp\left[-\frac{V_x^2}{V_{Te}^2}\right], \tag{6.37}$$

and

$$V_{Te}^2 = \frac{2K_B T_e}{m_e},$$

is the thermal velocity of the Maxwellian electrons. From Equations (6.31) and (6.32), we find the real part ω_R of the frequency ω to be:

$$\omega_R^2 = \omega_{pe}^2 + \frac{3}{2}k^2 V_{Te}^2, \tag{6.38}$$

under the approximation that the thermal term $(3k^2 V_{Te}^2 /2) << \omega_{pe}^2$ and the imaginary part ω_I to be:

$$\omega_I = -\sqrt{\pi}\frac{\omega_R^4}{k^3 V_{Te}^3} \exp\left[-\frac{\omega_R^2}{k^2 V_{Te}^2}\right]. \tag{6.39}$$

The value of ω_R is identical to that obtained with the two-fluid description of plasmas. But we now have an imaginary part ω_I with a negative value even in the absence of collisions. This is a major outcome of the kinetic approach. The electron-plasma waves suffer damping, known as **Collisionless** or **Landau Damping**. It has originated from the presence of electrons with velocity equal to the phase velocity of the wave. Such electrons are called **Resonant Electrons**. They move with the phase velocity of the wave and therefore see an almost static electric field $\vec{E}_1$. Under such conditions, electrons and the wave can exchange energy between themselves: the electrons can gain energy from the wave, resulting in the damping of the wave, or the electrons can lose energy to the wave, resulting in the amplification of the wave. Which of the two processes occurs is decided by the electron distribution function $f_{e0}(\vec{V})$. We have found that the wave damps for the Maxwellian distribution function because it has

$$\frac{\partial f_{e0}(V_x)}{\partial V_x} < 0. \tag{6.40}$$

A velocity distribution with

$$\frac{\partial f_{e0}(\vec{V})}{\partial V} > 0, \tag{6.41}$$

gives a positive value of ω_I and the wave amplitude $\vec{E}_1$ grows as $\exp{(\omega_I t)}$. This produces the circumstances of an **Instability**. We have seen earlier that an electron beam passing through an electron-proton plasma gives rise to an instability. The one dimensional velocity distribution function of an electron beam of velocity V_0 in the x direction and temperature T_b can be represented by the drifted Maxwellian as:

$$f_{b0}(V_x) = n_b \left(\frac{m_e}{2\pi K_B T_b}\right)^{1/2} \exp\left[-\frac{m_e(V_x - V_0)^2}{2K_B T_b}\right]. \tag{6.42}$$

The total equilibrium electron distribution function f_{e0} is, therefore, given by (Figure 6.3):

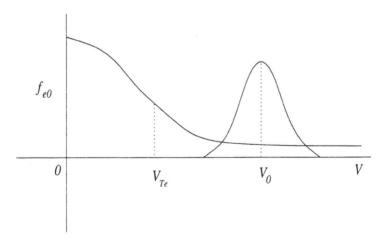

Figure 6.3. The Electron Distribution Function Consisting of a Superposition of Two Maxwellian Velocity Distributions has a Region of Positive Velocity Gradient (e.g. Equation 6.43)

$$f_{e0}(\vec{V}) = \frac{n_0}{\pi^{3/2} V_{Te}^3} \exp\left[-\frac{V^2}{V_{Te}^2}\right] + \frac{n_b}{\sqrt{\pi} V_{T_b}} \exp\left[-\frac{(V_x - V_0)^2}{V_{T_b}^2}\right], \tag{6.43}$$

where V_{T_b} is the thermal speed of the beam electrons. We can evaluate $\epsilon_R(k, \omega_R)$ and $\epsilon_I(k, \omega_R)$ by substituting the distribution (6.43) in Equations (6.27) and (6.28) respectively, to find:

$$\epsilon_R = 1 - \frac{\omega_{pe}^2}{\omega_R^2} - \frac{\omega_b^2}{(\omega_R - kV_0)^2}, \qquad (6.44)$$

and

$$\epsilon_I = 2\sqrt{\pi} \frac{\omega_{pe}^2}{k^3} \frac{\omega_R}{V_{Te}^3} \exp\left(-\frac{\omega_R^2}{k^2 V_{Te}^2}\right) + \frac{2\sqrt{\pi}\omega_b^2(\omega_R - kV_0)}{k^3 V_{T_b}^3} \times$$

$$\times \exp\left[-\frac{(\omega_R - kV_0)^2}{k^2 V_{T_b}^2}\right], \qquad (6.45)$$

where we have neglected the thermal terms in Equation (6.44). Again ω_R is determined from $\epsilon_R = 0$ and ω_I from Equation (6.32). We observe from Equation (6.45) that its second term gives rise to a positive contribution to ω_I for $\omega_R < kV_0$, which is the condition for positive velocity gradient of the beam distribution function. We find from Equation (6.45) that

$$\omega_I = -\frac{\sqrt{\pi}\omega_R^4}{k^3 V_{Te}^3} \exp\left[-\frac{\omega_R^2}{k^2 V_{Te}^2}\right] + \frac{\sqrt{\pi}\omega_R^3 n_b \delta_R}{k^3 V_{T_b}^3 n_0} \exp\left[-\frac{\delta_R^2}{k^2 V_{T_b}^2}\right], \qquad (6.46)$$

where

$$\delta_R = kV_0 - \omega_R, |\delta_R| <<| kV_0 |, |\delta_R| <<| \omega_R |,$$

$$1 >>\left|\left(\frac{\delta_R}{\omega_R}\right)\right| >> \left(\frac{n_b}{n_0}\right)^{1/3},$$

and $\omega_R \simeq \omega_{pe}$. Thus, if the second term in Equation (6.46) exceeds the first, we get an instability in which waves with frequency near the electron plasma frequency grow with a growth rate ω_I. In the fluid treatment of the beam - plasma instability we did not have the Landau damping term and therefore the growth rate did not depend on the plasma temperature T_e.

Problem 6.1: The sign of ω_I depends both on the sign of the velocity derivative of the distribution function as well as the frequency derivative of ϵ_R. Discuss the beam - plasma instability when $(\partial\epsilon_R/\partial\omega_R) < 0$ and $(\partial f_{e0}/\partial V) < 0$.

The maximum growth rate of the **Beam - Plasma Instability** is found for $\delta_R = kV_{T_b}$ so that (neglecting Landau damping)

$$\omega_I = \frac{\sqrt{\pi}}{e}\left(\frac{n_b}{n_0}\right)\left(\frac{V_0}{V_{T_b}}\right)^2 \omega_{pe}, \qquad (6.47)$$

and

$$\left(\frac{V_{T_b}}{V_0}\right) >> \left(\frac{n_b}{n_0}\right)^{1/3}.$$

This also says that $|\omega_I| <<| \delta_R |$ — the bandwidth of the growing waves. The growth rate given by Equation (6.47) is quite different from what we found in the fluid limit for a cold electron beam ($V_{T_b} = 0$). We can recover the zero temperature result by evaluating ω_I for $\delta_R^2 >> k^2 V_{T_b}^2$.

Problem 6.2: Show that in the limit $\delta_R^2 \simeq (n_b/n_0)^{2/3} \omega_R^2 >> k^2 V_{T_b}^2$, the growth rate $\omega_I \propto \omega_{pe}(n_b/n_0)^{1/3}$.

We discover that for this case $\left(V_{T_b}/V_0\right) << (n_b/n_0)^{1/3}$. Thus $\delta_R/\omega_R \simeq (n_b/n_0)^{1/3} \simeq \left(V_{T_b}/V_0\right)$ defines the transition region between the fluid and the kinetic treatments.

Problem 6.3: Determine the growth rate of the beam - plasma instability for a monoenergetic electron beam.

Problem 6.4: If we replace the electrostatic force by the gravitational force, we can investigate Jean's instability using Vlasov equation. Show that the dispersion relation for Jean's instability is given by:

$$k^2 = \left(4\pi G\rho_0/\sigma^2\right) \left\{1 - \sqrt{\pi}s\exp(s^2)\left[1 - \operatorname{erf}(s)\right]\right\}$$

where the zeroth order distribution function

$$f_0 = \left\{\rho_0/(2\pi\sigma^2)^{3/2}\right\}\exp\left(-V^2/2\sigma^2\right),$$

$s = \left(\omega_I/\sqrt{2}\sigma k\right)$ and ω_I is the growth rate (Binney 1995).

In the linear analysis given so far, the electric field $\vec{E}_1$ grows at an exponential rate. How long will this continue? At some later time, $\vec{E}_1$ may become so large that it starts influencing the orbits of the beam and ambient electrons. This is when the linear behaviour breaks down. Our neglect of the nonlinear term $\left(\vec{E}_1 \cdot \frac{\partial f_{e1}}{\partial \vec{V}}\right)$ is no longer justified and the system has moved into the nonlinear regime. In this regime, there is no dispersion relation — the frequency ω begins to depend on $\vec{E}_1$.

We may imagine that the nonlinearity can also be dealt with the perturbation methods. We may write the distribution function as:

$$f = f_0 + af_1 + a^2 f^2 + \dots, \tag{6.48}$$

the electric field as:

$$\vec{E} = a\vec{E}_1 + a^2\vec{E}_2 + \ldots, \tag{6.49}$$

substitute in the Vlasov and the Poisson equations and equate terms of the same order in the expansion parameter a, supposed to be $\ll 1$. The hierarchy of equations so obtained enable us to determine the distribution function and the electric field at a given order in terms of their values at one order lower. Each of these equations is linear and can be solved by the methods outlined above. We find that the dispersion relation is again quite similar to that obtained from Equation (6.34), since there is no change in the denominators and hence in their poles. Therefore, to all orders the dispersion relation remains unchanged Nonlinearity, if handled this way, does not give rise to any novelty. The stability or the instability found in the linearized treatment continues to exist in all orders. This is true only for the case $a \ll 1$, where the higher order terms are small and Equations (6.48) and (6.49) represent converging series. For all other cases of arbitrary magnitudes of perturbations, what the nonlinear system would exhibit cannot be predicted with any generality; it is system specific. There is no unique way of dealing with a nonlinear problem. Here, we shall illustrate one method of estimating the effects of nonlinearities, through what is known as the **Quasi-Linear Relaxation**.

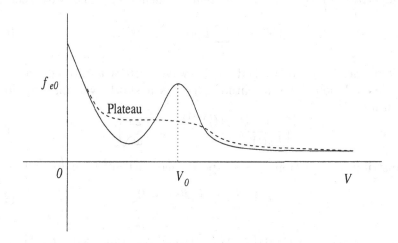

Figure 6.4. The Distribution Function Shown in Figure (6.3) Develops a Plateau Due to the Quasi-Linear Relaxation.

From Equation (6.46), we see that the Landau damping term begins to reduce the growth rate ω_I for wave vector $k \simeq \omega_R/V_{T_e}$. This is where **Quasi-Linear Relaxation** takes place. Plasma waves are excited in the region of positive velocity gradient of the distribution function (Figure 6.3)

The particles in this region of the phase space lose their energy and move to the lower velocity region. Since high energy particles also lose energy at a higher rate, the distribution function of the beam particles changes from a Gaussian to more like a table-shaped gaining in width and losing in height, to conserve the total number of particles. The portion of the beam distribution with negative slope remains unchanged as it contains particles which are not in resonance with the plasma wave. Eventually the beam particles thermalize with the ambient plasma. At this stage, a plateau is formed in the total distribution and there is no longer a region with positive velocity gradient (Figure 6.4).

As a result, now the generation of plasma waves comes to a halt. These circumstances define the **Quasi-Linear Regime**, which can be studied using the Vlasov equation. We now wish to study the time evolution of the zeroth order velocity distribution function due to the presence of the finite electric field of the electron plasma waves. We write the Vlasov equation for the total distribution function as:

$$\frac{\partial f_0}{\partial t} + \frac{\partial f_1}{\partial t} + \vec{V} \cdot \left(\frac{\partial f_0}{\partial \vec{r}} + \frac{\partial f_1}{\partial \vec{r}} \right) - \frac{e}{m_e} \vec{E} \cdot \left(\frac{\partial f_0}{\partial \vec{V}} + \frac{\partial f_1}{\partial \vec{V}} \right) = 0. \qquad (6.50)$$

Here, E is the electric field of the electron plasma waves and is expressed as a superposition of waves of different frequencies ω_k and wave vectors $\vec{k}$ as:

$$\vec{E}(\vec{r}, t) = \sum_{\vec{k}} E_{\vec{k}}(t) e^{i(\vec{k} \cdot \vec{r} - \omega_k t)}. \qquad (6.51)$$

The frequencies ω_k are related to the wave vectors k through the linear dispersion relation. The amplitude $\vec{E}_k(t)$ is a slowly varying function of time such that:

$$\left| \left[\frac{1}{E_k(t)} \frac{\partial E_k(t)}{\partial t} \right]^{-1} \right| >> \left(\frac{2\pi}{\omega_k} \right). \qquad (6.52)$$

The distribution function f_1 is the perturbed part and is expressed as:

$$f_1 = \sum_k f_k(t) e^{i(k \cdot r - \omega_k t)}. \qquad (6.53)$$

Its average over short spatial scale (k^{-1}) and fast time scales (ω_k^{-1}) is:

$$\langle f_1 \rangle = 0 \qquad (6.54)$$

The function f_0 is the zeroth order velocity distribution function; for example, for a beam - plasma system, it is represented by a drifting Maxwellian. It becomes a function of time as the slowly varying electric field $E_k(t)$ begins to act on the resonant particles. Taking the average over a fast time scale (ω_k^{-1}) and short spatial scale (k^{-1}) of Equation (6.50) we get:

$$\frac{\partial f_0}{\partial t} + \vec{V} \cdot \frac{\partial f_0}{\partial \vec{r}} - \frac{e}{m_e} \left\langle \vec{E} \cdot \frac{\partial f_1}{\partial \vec{V}} \right\rangle = 0. \tag{6.55}$$

Subtracting Equation (6.55) from (6.50), we find the equation which determines the time evolution of the distribution function f_1 as:

$$\frac{\partial f_1}{\partial t} + \vec{V} \cdot \frac{\partial f_1}{\partial \vec{r}} - \frac{e}{m_e} \vec{E} \cdot \frac{\partial f_0}{\partial \vec{V}} = 0, \tag{6.56}$$

where we have neglected the nonlinear terms responsible for wave-wave interactions, since these processes are not included in the quasi-linear treatment. Assuming f_0 to be independent of space, we find

$$\frac{\partial f_0}{\partial t} = \frac{e}{m_e} \left\langle \vec{E} \cdot \frac{\partial f_1}{\partial \vec{V}} \right\rangle. \tag{6.57}$$

Equation (6.56) can be solved to find f_1 in terms of $\left(\partial f_0 / \partial \vec{V} \right)$ neglecting $(\partial f_1 / \partial \vec{r})$, and on substituting this in Equation (6.57), begets us:

$$\frac{\partial f_0}{\partial t} = \frac{e}{m_e} \left\langle \vec{E} \cdot \frac{\partial}{\partial \vec{V}} \left\{ \frac{e}{m_e} \vec{E} \cdot \int e^{\omega_I t} \frac{\partial f_0}{\partial \vec{V}} dt \right\} \right\rangle, \tag{6.58}$$

where the slow $(\omega_I \ll \omega_k)$ dependence of $E(t)$ on time has been expressed as

$$E(t) = E e^{\omega_I t}, \tag{6.59}$$

with ω_I the linear growth rate of the beam - plasma stability. Equation (6.58) is a diffusion equation describing the diffusion of particles in the velocity space. It can be cast in the form:

$$\frac{\partial f_0}{\partial t} = \frac{\partial}{\partial V} \left(D \frac{\partial f_0}{\partial V} \right), \tag{6.60}$$

where D is the diffusion coefficient. We find that D is approximately given by:

$$D \simeq \frac{e^2 E^2}{m_e^2 \omega_I}. \tag{6.61}$$

The characteristic time scale t_D of diffusion of electrons of a typical velocity V is then given by

$$t_D \simeq \frac{V^2}{D}. \tag{6.62}$$

Thus, in the quasi-linear regime, the principle of superposition of oscillations (a linear feature) is combined with the effect of waves on the distribution function of the particles (a nonlinear feature).

We still have to estimate E^2. One way of estimating E^2 is to consider the motion of the beam electrons in the wave electric field $E(\vec{x}, t)$. The equation of motion is (Davidson 1972):

$$m_e \frac{d^2 x}{dt^2} = -eE \sin(kx - \omega t). \tag{6.63}$$

In the limit $E \to 0$, the electron motion is that of a free particle. For finite E, it is convenient to define its position as:

$$\theta \equiv kx'(t) = kx(t) - \omega t, \tag{6.64}$$

so that Equation (6.63) becomes

$$\frac{d^2 \theta(t)}{dt^2} + \frac{eEk}{m_e} \sin \theta(t) = 0, \tag{6.65}$$

which reduces to the equation of a simple harmonic oscillator for small displacements $\mid \theta \mid << 1$. The electrons in the field of the wave execute oscillations at the bottom of the corresponding potential φ (Figure 6.5) with a frequency ω_B given by

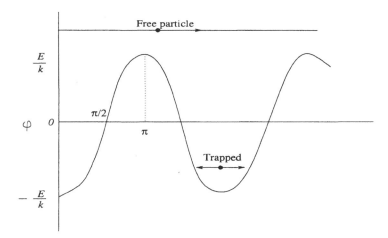

Figure 6.5.   An Electron in the Energy Range $-\dfrac{eE}{k} < W < \dfrac{eE}{k}$ is Trapped in the Potential Well and Executes Oscillations with Frequency ω_B.

$$\omega_B = \left(\frac{eEk}{m_e} \right)^{1/2}. \tag{6.66}$$

Equation (6.65) can be integrated once to find that the total energy

$$W = \frac{m_e}{2} \left(\frac{dx'(t)}{dt} \right)^2 - e\varphi(x'), \tag{6.67}$$

is a constant, where $\vec{E} = -\vec{\nabla}\varphi$ is the electrostatic potential in the wave frame. We see that an electron in the energy range $-\frac{eE}{k} < W < \frac{eE}{k}$ is trapped by the finite amplitude of the wave. Thus, the linear theory is valid only until the trapping effects become important. We can have a very rough estimate of E by assuming that the growth rate ω_I (Equation 6.47) is comparable to the oscillation frequency also called the bounce frequency ω_B since once the particles are trapped, they just oscillate without giving any energy to the wave. The diffusion time t_D is then given by:

$$t_D \simeq \frac{\omega_{pe}^2}{\omega_I^3} \simeq \omega_{pe}^{-1} \left(\frac{n_0}{n_b}\right)^3 \left(\frac{V_{T_b}}{V_0}\right)^6. \tag{6.68}$$

As discussed earlier, sporadic radio emission from the Sun, called type III radio bursts (Figure 1.11) are believed to be excited by an electron beam propagating through the solar corona. For typical parameters on the sun, i.e. $n_0 \simeq 10^9 \mathrm{cm}^{-3}$, $(n_b/n_0) \simeq 10^{-5}$, $(V_{T_b}/V_0) \simeq 10^{-1}$, we find $t_D \simeq 1$ sec. What this means is that the electron beam undergoes quasilinear relaxation in a very short time of the order of one second (Kaplan and Tsytovich 1973). This poses a problem, since high energy electrons as well as radio emission far from the solar atmosphere have been observed through satellite measurements. If the beam thermalizes in one second, then how does it maintain itself during its travel to the earth?

Well, this is a subject of active research. Several different scenarios for the recycling of the beam i.e. in situ reacceleration of beam electrons have been proposed. They are based on the inhomogeneities in the beam and ambient coronal plasma as well as on nonlinear plasma processes. Without going into these aspects, we conclude our discussion of the beam - plasma system and the excitation of electron - plasma waves.

6.4. Kinetic Description of Ion-Acoustic Waves and Instabilities

In order to find what happens to the low frequency ion-acoustic waves in the kinetic description, we will have to write Vlasov equations both for electrons and protons. The linearized Vlasov equations and Poisson's equations are:

$$\left[\frac{\partial}{\partial t} + i\vec{k} \cdot \vec{V}\right] f_{e1}(\vec{k}, \vec{V}, t) = \frac{e}{m_e} \vec{E}_{1k} \cdot \frac{\partial f_{e0}}{\partial \vec{V}}, \tag{6.69}$$

$$\left[\frac{\partial}{\partial t} + i\vec{k} \cdot \vec{V}\right] f_{i1}(\vec{k}, \vec{V}, t) = -\frac{e}{m_i} \vec{E}_{1k} \cdot \frac{\partial f_{i0}}{\partial \vec{V}}, \tag{6.70}$$

$$\vec{k} \cdot \vec{E}_{1k} = 4\pi e i \int (f_{e1} - f_{i1}) d\vec{V}. \tag{6.71}$$

After carrying out all the steps illustrated for the electron - plasma waves, we find

$$\epsilon_R(\vec{k}, \omega_R) = 1 \; + \; \frac{4\pi e^2}{m_e k^2} P \int \frac{\vec{k} \cdot \dfrac{\partial f_{e0}}{\partial \vec{V}}}{(\omega_R - \vec{k} \cdot \vec{V})} d\vec{V}$$

$$- \frac{4\pi e^2}{m_i k^2} P \int \frac{\vec{k} \cdot \dfrac{\partial f_{i0}}{\partial \vec{V}}}{(\omega_R - \vec{k} \cdot \vec{V})} d\vec{V}. \qquad (6.72)$$

For Maxwellian velocity distributions for electrons and protons and in the low frequency limit $kV_{T_i} < |\omega_R| < kV_{T_e}$ we find:

$$\epsilon_R(\vec{k}, \omega_R) = 1 - \frac{\omega_{pe}^2}{k^2 V_{T_e}^2} + \frac{\omega_{pi}^2}{\omega_R^2}, \qquad (6.73)$$

with roots

$$\omega_R^2 = \frac{\omega_{pi}^2 k^2 V_{T_e}^2}{\omega_{pe}^2 (1 - k^2 \lambda_{De}^2)} = \frac{k^2 c_s^2}{(1 - k^2 \lambda_{De}^2)}, \qquad (6.74)$$

where V_{T_i} is the proton thermal speed. Thus, in the kinetic treatment, the frequency ω_R of the ion-acoustic wave is the same as in the fluid treatment, as expected. The imaginary part of the dielectric function ϵ_I has contribution only from the protons as there are almost no electrons which are in resonance with this low-frequency wave. The resonant protons give rise to the Landau damping ω_I of the ion - acoustic wave where

$$\epsilon_I = \frac{4\pi^2 e^2}{m_i k^2} \int \vec{k} \cdot \frac{\partial f_{i0}}{\partial \vec{V}} \delta(\omega_R - \vec{k} \cdot \vec{V}) d\vec{V}, \qquad (6.75)$$

and

$$\omega_I = -\frac{\sqrt{\pi} \omega_R^4}{k^3 V_{T_i}^3} \exp\left[-\frac{\omega_R^2}{k^2 V_{T_i}^2}\right]. \qquad (6.76)$$

For the ion-acoustic waves to be weakly Landau-damped, we must have

$$\omega_R^2 = k^2 c_s^2 >> k^2 V_{T_i}^2, \qquad (6.77)$$

or

$$T_e >> 2T_i.$$

where T_i is the proton temperature. Whatever, we said about resonant electrons with respect to electron - plasma waves is also true for resonant protons with respect to the ion-acoustic waves.

We can also consider the excitation of the **Ion-Acoustic Instability** through the relative streaming of electrons and protons. Let us take a drifted Maxwellian distribution function for electrons and a Maxwellian for protons. We find the real part of the dielectric function

$$\epsilon_R(\omega_R, k) = 1 - \frac{\omega_{pe}^2}{(\omega_R - kV_0)^2} + \frac{\omega_{pi}^2}{\omega_R^2}. \tag{6.78}$$

Again the roots are

$$\omega_R^2 = \frac{\omega_{pi}^2(\omega_R - kV_0)^2}{\omega_{pe}^2\left(1 - \frac{(\omega_R - kV_0)^2}{\omega_{pe}^2}\right)}, \tag{6.79}$$

and the growth rate

$$\omega_I = \frac{\sqrt{\pi}\omega_{pe}^2}{k^3 V_{Te}^3} \frac{(\omega_R - kV_0)}{\left[-\frac{\omega_{pi}^2}{\omega_R^3} + \frac{\omega_{pe}^2}{(\omega_R - kV_0)^3}\right]} \exp\left[-\frac{(\omega_R - kV_0)^2}{k^2 V_{Te}^2}\right], \tag{6.80}$$

if $\omega_R < kV_0$ or $V_0 > c_s$.

Thus the ion-acoustic instability is excited if the relative drift speed V_0 between electrons and protons exceeds the ion sound speed c_s. This instability can also be excited through the propagation of an ion beam in analogy to the excitation of the beam - plasma instability.

Problem 6.5: Determine the growth rate of the ion-acoustic instability when an ion beam of density n_b and velocity V_0 propagates through a plasma.

As mentioned earlier, the ion-acoustic instability plays a very important role in effectively increasing the electron–ion collision frequency which provides the plasma with an anomalously large resistivity.

We can also investigate the excitation of electrostatic waves in the presence of a magnetic field: the upper and the lower hybrid waves using the kinetic approach. Of course, this needs to be done only if we are interested in finding the Landau damping or growth of these waves. We shall come back to this topic after we discuss the excitation of electromagnetic waves and instabilities in magnetized plasmas.

6.5. Electromagnetic Waves and Instabilities in a Magnetized Plasma

Conventionally, one would first study electromagnetic waves in an unmagnetized plasma. However, with the experience gained by this time, we can aspire to jump two steps at a time and take up this more complicated problem. We will limit ourselves to the study of the response of electrons to electromagnetic perturbations in the presence of a uniform zero order magnetic field $\vec{B}_0 = (0, 0, B_0)$ pointing in the z direction. The protons will be treated as a stationary uniform positively charged background. The excitation of electromagnetic waves and instabilities can be investigated by using the Vlasov equation along with the electromagnetic wave equation. The zeroth order electron velocity distribution function f_{eo} must satisfy the zeroth order Vlasov equation:

$$-\frac{e}{m_e}\left((\vec{V} \times \vec{B}_0) \cdot \frac{\partial f_{e0}}{\partial \vec{V}}\right) = \frac{eB_0}{m_ec}\frac{\partial f_{e0}}{\partial \phi} = 0, \tag{6.81}$$

where ϕ is the polar angle in the cylindrical coordinates (Akhiezer et al. 1975).

As discussed in the beginning of this Chapter, f_{eo} is a function only of the constants of motion, which, in the presence of a magnetic field, are V_z and $V_\perp$ — the components of velocity parallel and perpendicular to the magnetic field, respectively. Therefore, the solution of Equation (6.81) is

$$f_{e0}(\vec{V}) = f_{e0}(V_\parallel, V_\perp). \tag{6.82}$$

The linearized Vlasov equation with $(\vec{E}_1, \vec{B}_1)$ as the first order electric and magnetic fields is:

$$\left(\frac{\partial}{\partial t} + i\vec{k} \cdot \vec{V}\right) f_{e1} - \frac{e}{m_e}\left[\vec{E}_1 + \frac{\vec{V} \times \vec{B}_1}{c}\right] \cdot \frac{\partial f_{e0}}{\partial V} - \frac{e}{m_ec}\left(\vec{V} \times \vec{B}_0\right) \cdot \frac{\partial f_{e1}}{\partial \vec{V}} = 0. \tag{6.83}$$

The wave equation is:

$$\left(k^2 + \frac{1}{c^2}\frac{\partial^2}{\partial t^2}\right)\vec{E}_1 = \frac{4\pi e}{c^2}\int \vec{V}\frac{\partial f_{e1}}{\partial t}d\vec{V}. \tag{6.84}$$

We did not use the Fourier transform for the time coordinate. If we did, we would encounter the problem of dealing with resonant particles, as in the case of the electron plasma waves. In order to be prepared for the eventuality that the electromagnetic waves could interact resonantly with electrons, we would follow the same procedure of treating it as an initial value problem and taking the Laplace transform for the time coordinate.

We shall consider propagation of the electromagnetic waves with the wave vector $\vec{k} = (k_x, 0, k_z)$ in a magnetized plasma with the ambient magnetic field $\vec{B}_0 = (0, 0, B_0)$ in the z direction (Figure 6.6).

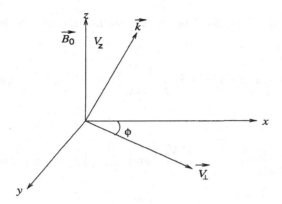

Figure 6.6. Orientations of Velocity, Magnetic Field and Propagation Vector.

Using the relation

$$\vec{\nabla} \times \vec{E}_1 = i\vec{k} \times \vec{E}_1 = -\frac{1}{c}\frac{\partial \vec{B}_1}{\partial t}, \tag{6.85}$$

and defining the Laplace transform of $\vec{E}_1, \vec{B}_1$ and f_{e1} as

$$\overline{E_1}(\vec{k}, \omega) = \int_0^\infty e^{i\omega t} \vec{E}_1(\vec{k}, t) dt, \tag{6.86}$$

we find

$$\frac{e}{m_e}\left[\vec{E}_1 + \frac{\vec{V} \times \vec{B}_1}{c}\right] \cdot \frac{\partial f_{e0}}{\partial \vec{V}} = \frac{e}{m_e}\left[\frac{(\omega - \vec{k} \cdot \vec{V})\vec{E}_1}{\omega} \cdot \frac{\partial f_{e0}}{\partial \vec{V}}\right.$$
$$\left. + \frac{\vec{V} \cdot \vec{E}_1}{\omega}\vec{k} \cdot \frac{\partial f_{e0}}{\partial \vec{V}}\right] \equiv F(\vec{V}, \phi), \tag{6.87}$$

and

$$\frac{e}{m_e c}\left(\vec{V} \times \vec{B}_0\right) \cdot \frac{\partial f_{e1}}{\partial \vec{V}} = -\omega_{ce}\frac{\partial f_{e1}}{\partial \phi}, \tag{6.88}$$

where $\omega_{ce} = \dfrac{eB_0}{m_e c}$ is the electron cyclotron frequency and we have used the cylindrical coordinate system

$$V_x = V_\perp \cos\phi, \quad V_y = V_\perp \sin\phi. \tag{6.89}$$

The Vlasov equation (6.83) takes the form

$$-i(\omega - \vec{k} \cdot \vec{V})f_{e1} + \omega_{ce}\frac{\partial f_{e1}}{\partial \phi} = F(\vec{V}, \phi). \tag{6.90}$$

Integration of the Vlasov equation determines the first order distribution function as:

$$f_{e1} = \exp\left\{i\omega'\phi - i\lambda\sin\phi\right\} \int \frac{F(\vec{V}, \phi)}{\omega_{ce}} \exp\left\{-i\omega'\phi + i\lambda\sin\phi\right\} d\phi,$$

where

$$\omega' = \frac{\omega - \vec{k}\cdot\vec{V}}{\omega_{ce}}, \quad \lambda = \frac{k_x V_\perp}{\omega_{ce}} \quad \text{and} \quad f_{e1}(\phi) = f_{e1}(\phi + 2\pi). \tag{6.91}$$

Writing

$$e^{-i\lambda\sin\phi} = \sum_{l=-\infty}^{\infty} J_l(\lambda)e^{-il\phi}, \tag{6.92}$$

we can perform the integration over ϕ to get

$$f_{e1} = \frac{ie}{m_e} \sum_{l'} \sum_{l} \frac{J_{l'}(\lambda)\vec{F}_l \cdot \vec{E}_1}{(\omega - k_z V_z - l\omega_{ce})} e^{i(l-l')\phi}, \tag{6.93}$$

where the components of the vector $\vec{F}_l$ are:

$$\begin{aligned}
F_{lx} &= \left[\left(1 - \frac{k_z V_z}{\omega}\right)\frac{\partial f_{e0}}{\partial V_\perp} + \frac{k_z V_\perp}{\omega}\frac{\partial f_{e0}}{\partial V_z}\right]\frac{l}{\lambda}J_l(\lambda), \\
F_{ly} &= \left[\left(1 - \frac{k_z V_z}{\omega}\right)\frac{\partial f_{e0}}{\partial V_\perp} + \frac{k_z V_\perp}{\omega}\frac{\partial f_{e0}}{\partial V_z}\right]iJ_l'(\lambda), \\
F_{lz} &= \left[\frac{k_x V_z}{\omega}\frac{\partial f_{e0}}{\partial V_\perp} - \frac{k_x V_\perp}{\omega}\frac{\partial f_{e0}}{\partial V_z}\right]\frac{l}{\lambda}J_l(\lambda) + J_l(\lambda)\frac{\partial f_{e0}}{\partial V_z},
\end{aligned} \tag{6.94}$$

and $J_l'(\lambda)$ is the derivative of the Bessel function $J_l(\lambda)$ with respect to its argument λ. We can now substitute for f_{e1} in the wave Equation (6.84) to determine the dielectric function for electromagnetic waves in a magnetized plasma. While determining the first order current density, we can easily perform the integration over the angle ϕ. This gives a relation between l and l', using which one summation can be carried out. We can then decide how many terms (l values) we should retain in the expression for the dielectric function. For waves near harmonics of the electron cyclotron frequency, we have to include terms with $l \geq 2$. For $l \leq 2$, we can make the hydrodynamic approximation i.e., $\lambda \ll 1$ and use series representations for the Bessel functions. With these remarks and keeping in mind that ω has a small imaginary part (Laplace transform), we now deal with the case of circularly polarized waves ($E_x = \pm iE_y$) propagating parallel to the magnetic field

$(k_x = 0, k_z = k)$ for $l = 1$ in more detail. As discussed for the case of electrostatic waves, we can express the wave equation as:

$$\int \bar{f}_{e1}(\vec{k}, \vec{V}, \omega) V_\perp d\vec{V} = \frac{i}{G(k, \omega)} \int \frac{\bar{f}_{e1}(k, V, 0) V_\perp d\vec{V}}{(\omega - kV_z + \omega_{ce})}, \qquad (6.95)$$

where

$$G_\pm(k, \omega) = 1 + \frac{4\pi^2 e^2}{m_e c^2 \left(k^2 - \dfrac{\omega^2}{c^2}\right)} \int \frac{(\omega - kV_z)\dfrac{\partial f_{e0}}{\partial V_\perp} + kV_\perp \dfrac{\partial f_{e0}}{\partial V_z}}{(\omega - kV_z \pm \omega_{ce})} V_\perp^2 dV_\perp dV_z, $$

$$(6.96)$$

now carries the response of the plasma medium to electromagnetic perturbations. As before, the zeros of $G_\pm(k, \omega)$ determine the dispersion relation of the electromagnetic waves (+ for left and – for right circular polarization) which is:

$$\frac{k^2 c^2}{\omega^2} = 1 + \frac{4\pi^2 e^2}{m_e} \int \frac{(\omega - kV_z)\dfrac{\partial f_{e0}}{\partial V_\perp} + kV_\perp \dfrac{\partial f_{e0}}{\partial V_z}}{(\omega - kV_z \pm \omega_{ce})} V_\perp^2 dV_\perp dV_z \equiv \varepsilon_\pm(k, \omega).$$

$$(6.97)$$

Here, $\varepsilon_\pm$ is the dielectric function of the medium in the presence of magnetic field. Following the procedure already discussed for evaluating the integral in Equation (6.97), we find for the real part $\varepsilon_{\pm R}$ of $\varepsilon_\pm$

$$\varepsilon_{\pm R} = 1 + \frac{4\pi^2 e^2}{m_e} P \int \frac{(\omega_R - kV_z)\dfrac{\partial f_{e0}}{\partial V_\perp} + kV_\perp \dfrac{\partial f_{e0}}{\partial V_z}}{(\omega_R - kV_z \pm \omega_{ce})} V_\perp^2 dV_\perp dV_z, \qquad (6.98)$$

and the imaginary part $\epsilon_{\pm I}$

$$\varepsilon_{\pm I}(k, \omega_R) = -\frac{4\pi^3 e^2}{m_e} \int \left[(\omega_R - kV_z)\frac{\partial f_{e0}}{\partial V_\perp} + kV_\perp \frac{\partial f_{e0}}{\partial V_z}\right] \times$$
$$\times \delta(\omega_R - kV_z \pm \omega_{ce}) V_\perp^2 dV_\perp dV_z, \qquad (6.99)$$

where P stands for the principal value integral. We first note that in the absence of a plasma, i.e., when the integrand in Equation (6.98) vanishes, we get the dispersion relation of the electromagnetic waves propagating in a vacuum ($\omega = kc$). The real and imaginary parts of the frequency ω are determined from Equation (6.97) by writing:

$$\varepsilon_{\pm R}(k, \omega_R + i\omega_I) + i\varepsilon_{\pm I}(k, \omega_R + i\omega_I) = \frac{k^2 c^2}{(\omega_R + i\omega_I)^2}, \qquad (6.100)$$

so that

$$\omega_R^2 \simeq \frac{k^2 c^2}{\varepsilon_{\pm R}(\omega_R)}, \tag{6.101}$$

and

$$\omega_I \simeq -\frac{\varepsilon_{\pm I}}{\left(\dfrac{\partial \varepsilon_{\pm R}}{\partial \omega_R} + \dfrac{2\varepsilon_{\pm R}}{\omega_R}\right)}. \tag{6.102}$$

We first discuss the excitation of electromagnetic waves in the absence of the magnetic field. For $B_0 = 0$. Equations (6.98) and (6.101) give:

$$\omega_R^2 = k^2 c^2 - \frac{4\pi e^2}{m_e} P \int \frac{(\omega_R - kV_z)V_\perp \dfrac{\partial f_{e0}}{\partial V_\perp} + kV_\perp^2 \dfrac{\partial f_{e0}}{\partial V_z}}{(\omega_R - kV_z)} d\vec{V}. \tag{6.103}$$

By performing partial integration and assuming $\omega_R \gg kV_z$, we find

$$\omega_R^2 = k^2 c^2 + \omega_{pe}^2 \left[1 + \frac{k^2 \left\langle V_\perp^2 \right\rangle}{\omega_R^2} + \frac{3k^4 \left\langle V_z^2 \right\rangle \left\langle V_\perp^2 \right\rangle}{\omega_R^4} \right]. \tag{6.104}$$

We solve for ω_R for an anisotropic velocity distribution function for which $< V_z^2 >= 0$ but $< V_\perp^2 > \neq 0$. The two roots of the quadratic Equation (6.104) are approximately given by

$$\omega_1^2 = \omega_{pe}^2 + k^2 c^2 + \frac{k^2 \left\langle V_\perp^2 \right\rangle \omega_{pe}^2}{\omega_{pe}^2 + k^2 c^2}, \tag{6.105}$$

and

$$\omega_2^2 = -\frac{k^2 \left\langle V_\perp^2 \right\rangle \omega_{pe}^2}{\omega_{pe}^2 + k^2 c^2}, \tag{6.106}$$

in the case where $\omega_{pe}^2 + k^2 c^2 \gg \omega_{pe} k < V_\perp^2 >^{1/2}$.

The root ω_2 corresponds to an instability. We must appreciate that this instability has arisen due to the anisotropy in the velocity distribution of electrons since $< V_\perp^2 > \neq < V_z^2 >$. A finite value of $< V_z^2 >$ reduces the growth rate of the instability. During the instability the electric and magnetic field perturbations propagate and grow with time. These are not electromagnetic waves as the real part ω_R of the frequency vanishes.

Problem 6.6: Show that for $< V_\perp^2 >= < V_z^2 >, \omega_2^2 > 0$ and the instability disappears.

From Equation (6.106), we find the growth rate

$$\omega_I = \frac{\left\langle V_\perp^2 \right\rangle^{1/2}}{c} \omega_{pe}, \tag{6.107}$$

for very short wavelengths i.e., for $k^2 c^2 \gg \omega_{pe}^2$. It is instructive to acknowledge that electromagnetic instabilities usually have smaller growth rates as compared to electrostatic instabilities which can have growth rates of the order of ω_{pe}.

What does $\epsilon_{\pm I}$ tell us? For $\omega_{ce} = 0$, Equation (6.99) shows that $\epsilon_{\pm I} \neq 0$ only if $\omega_R = kV_z$, i.e., the phase velocity of the electromagnetic wave is equal to the thermal velocity of the particles. For electromagnetic waves satisfying the dispersion relation

$$\omega_R^2 = \omega_{pe}^2 + k^2 c^2, \tag{6.108}$$

the phase velocity is greater than the speed of light c. Therefore the equality $\omega_R = kV_z$ cannot be satisfied. Electromagnetic waves cannot interact resonantly with free electrons. However, in a dielectric medium with a refractive index greater than unity, the phase speed of the electromagnetic waves is smaller than c and they can interact resonantly with free electrons. This is the essence of the well known **Cerenkov -Effect**. The refractive index of a plasma is less than unity. A plasma thus acts as a divergent lens since waves entering a plasma from air bend away from the normal. Waves with large amplitudes, can, however, change the properties of a plasma, such that ω_{pe} becomes a function of the wave amplitude, among other effects. Such a plasma exhibits very different properties and can be made to act like a converging lens (Figure 6.7).

These aspects fall in the domain of nonlinear plasma physics.

Coming back to the case with $\omega_{ce} \neq 0$, we have:

$$\varepsilon_{\pm R}(k, \omega_R) = 1 - \frac{\omega_{pe}^2}{n_0 \omega_R^2} \int \left[\frac{\omega_R - kV_z}{\omega_R - kV_z \pm \omega_{ce}} + \frac{k^2 V_\perp^2 / 2}{(\omega_R - kV_z \pm \omega_{ce})^2} \right] f_{e0} d\vec{V}, \tag{6.109}$$

and

$$\varepsilon_{\pm I}(k, \omega_R) = -\frac{\omega_{pe}^2 \pi}{2 n_0 \omega_R^2} \int \left[(\omega_R - kV_z)V_\perp \frac{\partial f_{e0}}{\partial V_\perp} + kV_\perp^2 \frac{\partial f_{e0}}{\partial V_z} \right] \times$$
$$\times \delta(\omega_R - kV_z \pm \omega_{ce}) d\vec{V}. \tag{6.110}$$

We can immediately check that in the absence of thermal effects, i.e., for $\omega_R \gg kV_{Te}$, Equations (6.109) and (6.101) respectively reduce to the dispersion relation of the right and the left circularly polarized waves, derived

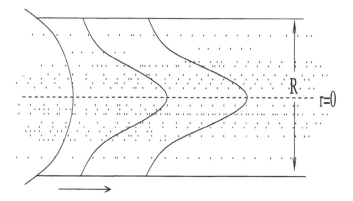

Figure 6.7. A Gaussian Radiation Beam Becomes Narrow as It Propagates Through a Plasma with a Density $n_e = n_0 \left(1 - r^2/R^2\right)$. The Refractive Index is Minimum on the Axis ($r = 0$). The Central Part of the Beam Has the Maximum Phase Speed, the Off-Centre Parts Lag Behind and the Beam Gets Narrow and Focused.

with the fluid treatment. The imaginary part $\epsilon_{\pm I}$ is the result of the kinetic treatment and describes the resonant interaction between the waves and particles, which could amplify or damp the wave amplitude depending upon the sign of the velocity gradient of the particle distribution function f_{e0}. From Equations (6.110) and (6.102), we see that the electromagnetic waves can grow in two ways:

$$(i) \qquad \text{when } \left. \frac{\partial f_{e0}}{\partial V_z} \right|_{V_z = (\omega_R \pm \omega_{ce}/k)} > 0, \qquad (6.111)$$

sometimes known as the **Parallel Driven Process** and

$$(ii) \qquad \text{when } \left. \frac{\partial f_{e0}}{\partial V_\perp} \right|_{V_z = (\omega_R \pm \omega_{ce}/k)} > 0, \qquad (6.112)$$

sometimes known as the **Perpendicular Driven Process**. We must carefully examine the resonance condition

$$\omega_R - kV_z \pm \omega_{ce} = 0, \qquad (6.113)$$

for all possible signs of k and V_z. Thus, for k and V_z both > 0, the Doppler shifted frequency of the wave $\omega_R - kV_z > 0$ can only satisfy the condition

$$\omega_R - kV_z - \omega_{ce} = 0, \qquad (6.114)$$

i.e., the electrons are in resonance with the right - circularly polarized wave. We may recall that electrons themselves undergo right - handed or anti-clockwise gyration in a uniform magnetic field. So, the important result is that for a resonant exchange of energy between the wave and the particles, the electric field of the wave must have the same sense of rotation as the particles themselves. This process is also known as the **Gyroresonance**. The resonance condition (6.114) for the right circularly polarized wave is referred to as the **Normal Doppler Resonance**, since for $k > 0$ and $V_z > 0, |\omega_R/k| > |V_z|$ and the wave continues to move in the positive z direction in the frame moving with electron velocity V_z.

In contrast, the resonance condition for the left circularly polarized wave is

$$\omega_R = -\omega_{ce} + kV_z, \qquad (6.115)$$

so that $|\omega_R/k| < |V_z|$ for $k > 0$ and $V_z > 0$. Therefore, in the frame moving with electron velocity V_z the phase velocity of the wave is negative and the wave propagates in the (-z) direction. The electric vector of the wave in the moving frame becomes counterclockwise, i.e., it acquires the same sense of rotation as the electrons. This type of resonance is known as the **Anomalous Doppler Resonance**. Thus the left-circularly polarized wave is in anomalous Doppler resonance with electrons whereas the right circularly polarized wave is in normal Doppler resonance with electrons.

Problem 6.7: Discuss the normal and the anomalous Doppler resonances for ion - cyclotron waves.

Electron velocity distributions with properties defined in Equation (6.111) and (6.112) occur in various space and astrophysical situations. Electrons trapped in planetary magnetic fields and in loop-like magnetic configurations which are present on the solar corona, in the galactic center, in planetary nebulae, in star forming regions, in accretion disks, in fact, in all magnetically dominated plasmas, acquire such anisotropic distributions. Some examples of anisotropic velocity distributions, such as for an electron beam - plasma system in the absence of magnetic field, have already been seen. We, here, give some other representations for f_{e0} commonly discussed in the context of cyclotron absorption and emission:

$$
\begin{aligned}
(i) \quad f_{e0} &= n_0 \left(\frac{m_e}{2\pi K_B T_\perp} \right) \delta(V_z) \exp\left[-\frac{m_e V_\perp^2}{2\pi K_B T_\perp} \right] \\
&\quad + n_b \left[\delta(V_z - V_0) + \delta(V_z + V_0) \right],
\end{aligned} \qquad (6.116)
$$

which represents two counter streaming beams of either both of electrons or one of electrons and the other of positrons propagating along the magnetic field in an ambient plasma with zero temperature along the magnetic field (i.e. $< V_z^2 >= 0$) and a finite temperature $T_\perp$ perpendicular to the magnetic field.

$$(ii) \quad f_{e0}(V_\perp, V_z) = n_0 \left(\frac{m_e}{2\pi K_B T_\perp} \right) \left(\frac{m_e}{2\pi K_B T_\parallel} \right)^{1/2}$$

$$\times \exp\left[-\frac{m_e V_\perp^2}{2K_B T_\perp} - \frac{m_e V_z^2}{2K_B T_\parallel} \right],$$

(6.117)

which describes a system with unequal average energies associated with the parallel and perpendicular degrees of freedom.

$$iii) \qquad f_{e0} = f_{e0}(V_\perp, V_\parallel)\Theta(\theta - \theta_M), \qquad \cdot \tag{6.118}$$

which is one representation of the so called loss-cone distribution. Here Θ is the unit step function, which has a value of unity for positive argument $(\theta - \theta_M)$ and has a zero value otherwise. This situation occurs in a magnetic mirror discussed in Chapter 3. The angle θ_M is the maximum value of the angle of the empty cone.

Thus, the choice of the zeroth order distribution function decides whether there is a resonant instability or not. The value of the real part of the frequency of the waves that go unstable or get absorbed is determined from the real part $\epsilon_{\pm R}$ of the dielectric function. The complete dispersion relation contains all the electromagnetic waves such as the ordinary and the extraordinary, the 'R' and the 'L' waves, the low frequency whistler waves discussed in the fluid treatment with the additional possibility of generating waves at high harmonies of the cyclotron frequency. The imaginary part $\epsilon_{\pm I}$ determines the growth or damping of these waves. The complete dispersion relation for all types of electromagnetic waves in a magnetized plasma can be found by substituting Equation (6.93) in the wave equation (6.84) and is given by the matrix equation:

$$\begin{pmatrix} \dfrac{k^2 c^2}{\omega^2} - \epsilon_{xx} & \epsilon_{xy} & \epsilon_{xz} \\[2mm] \epsilon_{yx} & \dfrac{k^2 c^2}{\omega^2} - \epsilon_{yy} & \epsilon_{yz} \\[2mm] \epsilon_{zx} & \epsilon_{zy} & \dfrac{k^2 c^2}{\omega^2} - \epsilon_{zz} \end{pmatrix} \begin{pmatrix} E_x \\ E_y \\ E_z \end{pmatrix} = 0, \tag{6.119}$$

where

$$\epsilon_{xx} = 1 + \sum_l \int \left(\frac{lV_\perp}{\lambda} \right) g_l F_{lx} d\vec{V},$$

$$\epsilon_{xy} = -\sum_l \int \left(\frac{lV_\perp}{\lambda} \right) g_l F_{ly} d\vec{V},$$

$$\epsilon_{xz} = -\sum_l \int \left(\frac{lV_\perp}{\lambda}\right) g_l F_{lz} d\vec{V},$$

$$\epsilon_{yx} = -\sum_l \int (iV_\perp) g_l F_{lx} d\vec{V}, \qquad (6.120)$$

$$\epsilon_{yy} = 1 + \sum_l \int (iV_\perp) g_l F_{ly} d\vec{V},$$

$$\epsilon_{yz} = -\sum_l \int (iV_\perp) g_l F_{lz} d\vec{V},$$

$$\epsilon_{zx} = -\sum_l \int V_z g_l F_{lx} d\vec{V},$$

$$\epsilon_{zy} = -\sum_l \int V_z g_l F_{ly} d\vec{V},$$

$$\epsilon_{zz} = 1 + \sum_l \int V_z g_l F_{lz} d\vec{V},$$

$$g_l = \frac{4\pi e^2}{m_e \omega} \frac{J_l(\lambda)}{(\omega - k_z V_z - l\omega_{ce})}, \qquad (6.121)$$

and l, λ, F_{lx}, F_{ly} and F_{lz} are defined in Equation (6.94). The imaginary part of the dielectric function for any specific electromagnetic wave can be found by giving a small positive imaginary part to ω and separating the integral over velocity into a principal value integral and Cauchy's integral as has been done several times by now. We could further venture (but won't!) into the quasi-linear relaxation and finally into the fully nonlinear regime.

Whatever has been done with electrons holds for protons too. The dielectric function for the protons can be determined by appropriately changing the sign of the electric charge and the mass of the particle. The study of the proton-associated waves can then be carried out following the procedures laid out for electron-associated waves.

We have omitted a discussion of the kinetic description of electrostatic waves in magnetized plasmas for two reasons. First, the obvious, is that the fluid description is simpler than the kinetic, and second, it is only after their conversion to electromagnetic waves that the electrostatic waves can be observed from astrophysical situations. The conversion involves nonlinear coupling and scattering processes. The coupling often occurs near the resonances and cutoffs. Thus, the resonance of the extraordinary wave which can be studied using Equations (6.84) and (6.93) occurs at the upper hybrid frequency, implying that some of the energy contained in the electromagnetic extraordinary wave is converted into the electrostatic upper hybrid waves. Anyhow, the real part $\epsilon_{PR}(\vec{k}, \omega)$ and the imaginary part $\epsilon_{PI}(\vec{k}, \omega)$ of the dielectric function $\epsilon_p(\vec{k}, \omega)$ for electrostatic waves in a magnetized plasma , including only the electronic contribution, can be found by combining the Vlasov and the Poisson equations and are given by

$$\epsilon_{PR}(\vec{k}, \omega_R) = 1 + \frac{\omega_{pe}^2}{n_0} \sum_l P \int \frac{J_l^2(\lambda) \left[\frac{l}{\lambda} \frac{\partial f_{e0}}{\partial V_\perp} \sin\theta + \frac{\partial f_{e0}}{\partial V_z} \cos\theta \right] d\vec{V}}{\omega_R - kV_z \cos\theta - l\omega_{ce}},$$

(6.122)

and

$$\epsilon_{PI}(\vec{k}, \omega_R) = -\pi \frac{\omega_{pe}^2}{n_0} \sum_l \int J_l^2(\lambda) \left[\frac{l}{\lambda} \frac{\partial f_{e0}}{\partial V_\perp} \sin\theta + \frac{\partial f_{e0}}{\partial V_z} \cos\theta \right] \times$$
$$\times \delta(\omega_R - kV_z \cos\theta - l\omega_{ce}) d\vec{V},$$

(6.123)

where $\lambda = \dfrac{k_x V_\perp}{\omega_{ce}}$ and $\vec{k} = (k \sin\theta, 0, k \cos\theta)$.

Problem 6.8: Determine the dispersion relations of the upper hybrid and the lower hybrid waves using Equations (6.122) and (6.123) in the hydrodynamic approximation $\lambda << 1$.

Problem 6.9: Find the dispersion relation of electrostatic waves in a magnetized plasma by solving the equation $\epsilon_{PR} = 0$ in the short wavelength limit $\lambda >> 1$.

In closing, we would like to make two qualifying remarks about the resonant instabilities driven by the positive velocity gradients of the equilibrium distribution functions. It was pointed out by Penrose (1960) that the criterion for instability as described by the positive velocity gradient of the distribution function is not Galilean invariant. The correct criterion is that the distribution function should have a minimum at the phase velocity of the wave. The second remark is about the parallel and perpendicular driven processes (Equations 6.111 and 6.112). Melrose (1986) has given a proof that in the nonrelativistic approximation, instabilities for $l > 0$ are essentially parallel driven. The resonance conditions (Equation 6.111 and 6.112) are independent of $V_\perp$, and therefore the integrals over V_z and $V_\perp$ decouple. Then it is shown that the term $\dfrac{\partial f_{e0}}{\partial V_\perp}$ does not contribute to the growth rate. In the relativistic case, however, the electron cyclotron frequency ω_{ce} is replaced by (ω_{ce}/γ) where γ is the Lorentz factor defined as

$$\gamma = \left[1 - \frac{V_z^2 + V_\perp^2}{c^2} \right]^{-1/2},$$

(6.124)

and therefore the resonance condition now becomes a function of both V_z and $V_\perp$. This gives rise to the perpendicular driven cyclotron emission - a point emphasized by Wu and Lee (1979) in the context of generation of kilometric radiation from the earth.

References

Akhiezer, A.I., Akhiezer, I.A., Polovin, P.V.,Sitenko, A.G., and Stepanov, K.N., 1975, Plasma Electrodynamics, Volumes 1 & 2., Pergamon Press.

Binney, J.J., Gravitational Plasmas in Plasma Physics, An Introductory Course, p291, Ed. Richard Dendy, Cambridge University Press.

Davidson, D.C., 1972, Methods in Nonlinear Plasma Theory, Academic Press.

Harris, E.G., 1975, Introduction to Modern Theoretical Physics Vol II, p731, A Wiley Interscience Publication.

Ichimaru, S., 1973, Basic Principles of Plasma Physics, A Statistical Approach, W.A. Benjamin, INC.

Kaplan, S.A. and Tsytovich, V.N., 1973, Plasma Astrophysics, Pergamon Press.

Landau,L., 1946, J.Phys. (USSR) **10**, 25.

Melrose, D.B., 1986, Instabilities in Space and Laboratory Plasmas, Cambridge University Press

Penrose, O.,1960, Phys. Fluid, **3**, 258.

Schmidt, G., 1966, Physics of High Temperature Plasmas, Academic Press.

Wu, C.S. and Lee, L.C., 1979, A theory of the terrestrial kilometric radiation. Astrophysical Journal **230**, 621.

Appendix

A. Plasma Dispersion Function

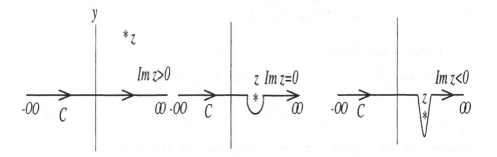

Figure A-1.   The Contour C for Integration Over t.

A function closely related to the plasma dielectric function ϵ is the plasma dispersion function $W(z)$ defined, for complex z, as

$$W(z) = \frac{1}{\sqrt{2\pi}} \int_C \frac{\exp(-t^2/2)}{t - z} dt \qquad (A.1)$$

The function $W(z)$ is an analytic function provided the contour of integration C passes always above the point $z = (x + iy)$ as shown in Figure(A-1).

For determining $W(z)$, we consider the case $Im\ z > 0$. For real values of t, we use the identity

$$\frac{1}{t - z} = i \int_0^\infty \exp\left[-i(t - z)t'\right] dt',\qquad (A.2)$$

in Equation (A.1) and carry out first t and then t' integration to find:

$$W(z) = 1 - z \exp(-z^2/2) \int_0^z dy \exp(y^2/2) + i\sqrt{\frac{\pi}{2}} z \exp(-z^2/2). \quad (A.3)$$

It is useful to find series expansion of $W(z)$:

$$W(z) = i\sqrt{\frac{\pi}{2}} z \exp\left(\frac{-z^2}{2}\right) + 1 - z^2 + \frac{z^4}{3} - \ldots l\frac{(-1)^{n+1}z^{2n+2}}{(2n+1)!!}\cdots$$

$$\text{for}|z| < 1,\qquad (A.4)$$

and

$$W(z) = i\sqrt{\frac{\pi}{2}} z \exp\left(\frac{-z^2}{2}\right) - \frac{1}{z^2} - \frac{3}{z^4} - \ldots - \frac{(2n-1)!!}{z^{2n}} \ldots \text{for}|z| >> 1. \quad (A.5)$$

The plasma dispersion function is also related to the error function of a complex argument. More on this topic can been seen in the book: The Plasma Dispersion Function by B.D. Fried and S.D. Conte, 1961, Academic Press, New York 1961, and in Basic Principles of Plasma Physics by S. Ichimaru, 1973, W.A. Benjamin, INC.

B. Bessel Functions

The differential equation satisfied by the Bessel functions is:

$$x^2\frac{d^2y}{dx^2} + x\frac{dy}{dx} + (x^2 - n^2)y = 0.\qquad (B.1)$$

For non–integral values of n, the two solutions of Equation (B.1) are $J_n(x)$ and $J_{-n}(x)$ where

$$J_n(x) = \sum_{k=0}^\infty \frac{(-1)^k}{k!\Gamma(n + k + 1)} \left(\frac{x}{2}\right)^{n+2k},\qquad (B.2)$$

and the Gamma function Γ is defined as:

$$\Gamma(n) = \int_0^\infty e^{-x} x^{n-1} dx,\qquad (B.3)$$

with

$$\Gamma(n+1) = n\Gamma(n). \tag{B.4}$$

For integral values of n and $x < 1$,

$$J_n(x) = (-1)^n J_{-n}(x), \tag{B.5}$$

and

$$J_n(x) = \frac{x^n}{2^n n!} \left[1 - \frac{x^2}{2^2 1!(n+1)} + \frac{x^4}{2^4 2!(n+1)(n+2)} + \ldots \right]. \tag{B.6}$$

For $x \gg 1$,

$$J_0(x) \simeq \left(\frac{2}{\pi x} \right)^{1/2} \cos(x - \pi/4),$$

$$J_1(x) \simeq \left(\frac{2}{\pi x} \right)^{1/2} \cos(x - 3\pi/4). \tag{B.7}$$

The recurrence formulae are:

$$J_{n-1}(x) + J_{n+1}(x) = \frac{2n}{x} J_n(x).$$

$$J_{n-1}(x) - J_{n+1}(x) = 2J_n'(x).$$

$$nJ_n(x) + xJ_n'(x) = xJ_{n-1}(x).$$

$$nJ_n(x) - xJ_n'(x) = xJ_{n+1}(x). \tag{B.8}$$

where prime represents the derivative with respect to the argument.

CHAPTER 7

NONCONDUCTING ASTROPHYSICAL FLUIDS

7.1. Whence Such Fluids?

If 99.99% of the universe is in the plasma state then for the remaining 0.01% must we learn a whole discipline of **Fluid Mechanics or Hydrodynamics**? The answer is No! More often than not, hydrodynamics provides a first level description of an astrophysical fluid. It is relatively easy to grasp gross features of any phenomenon governed predominantly by gravitational forces using gravitohydrodynamics. Magnetic fields are included later on in order to address additional issues. Thus early stages of star formation and evolution during which an interstellar cloud of very low density collapses under the action of its own gravity can be modeled by treating the cloud gas as a gravitating neutral or nonconducting fluid. But if we want to explain the discrepancy between the angular momentum of the cloud and that of the newly born star, we will have to include the effect of magnetic field. In the MHD approximation, the magnetic field is strongly coupled to the material. This system supports the excitation of Alfven waves, which, while propagating along the magnetic field can transport angular momentum from the body of the star to the external medium.

Several aspects of the structure of a galaxy can be understood by assuming that it is made up of a continuum fluid. In reality, a galaxy, especially its disk part, predominantly contains stars along with some interstellar gas. Therefore, such a system should be more appropriately studied using a kinetic treatment in which the evolution of the distribution function of stars in the phase space of positions and velocities is studied using Boltzmann or Vlasov type of equations. The justification for treating a galaxy as a fluid, apart from simplicity, is that we shall consider galactic processes on a spatial scale which is large enough to contain a large number of stars - the usual requirement of the continuum mechanics. But a fluid exerts pressure, and the pressure is due to molecular or atomic collisions. A typical star in a galaxy, however, rarely, if at all, collides with another star. The mean free path of a star in a galaxy is much larger than the spatial scale of the phenomenon, we wish to investigate using fluid dynamics. Therefore the stars form a collisionless system and do not exert pressure. Further, it is found that a pressureless selfgravitating system is unstable (Jean's Instability). Therefore a real galaxy must possess something akin to pressure to withstand the collapsing action of its gravity. This 'pressure' is taken to be associated with the random stellar velocities, so that the role of sound speed in the gaseous or fluid disk is assumed to be played by the root mean speed of the stars in the disk. Apart from describing the equilibrium of a galactic disk, the fluid properties are best put to use to investigate its observed spiral structure. Disks with real fluids occur around rotating compact objects such as pulsars and black holes. The stability of rotating fluids constitutes an important branch of nonconducting astrophysical fluids.

7.2. Equilibrium of Fluids

We write the momentum conservation law for a nonconducting fluid in a rotating frame of reference as (cf. Equation 4.14 with $\vec{B} = 0$):

$$\rho\left[\frac{\partial \vec{U}}{\partial t} + \left(\vec{U} \cdot \vec{\nabla}\right)\vec{U}\right] = -\vec{\nabla}p + \left(\varsigma + \frac{1}{3}\mu\right)\vec{\nabla}\left(\vec{\nabla} \cdot \vec{U}\right) + \mu\nabla^2\vec{U} + \rho\vec{g}$$

$$+2\rho\left(\vec{U} \times \vec{\Omega}_0\right) + \frac{\rho}{2}\vec{\nabla}\left(\left|\vec{\Omega}_0 \times \vec{r}\right|^2\right), \qquad (7.1)$$

where Ω_0 is the angular velocity of the rotating frame and $\vec{g}$ is the acceleration due to gravity. We see two new rotation dependent force densities — (i) $2\rho(\vec{U} \times \vec{\Omega}_0)$ is the coriolis force and (ii) $\frac{\rho}{2}\vec{\nabla}(\left|\vec{\Omega}_0 \times \vec{r}\right|^2)$ is the centrifugal force. The mass conservation, as usual, is contained in:

$$\frac{\partial \rho}{\partial t} + \vec{\nabla} \cdot \left[\rho\vec{U}\right] = 0. \qquad (7.2)$$

The energy conservation is nothing but the equation of state relating pressure p and mass density ρ through

$$\vec{\nabla}p = c_s^2\vec{\nabla}\rho,$$

where c_s is the adiabatic or isothermal sound speed, as the case may be.

The static equilibrium ($\vec{U} = 0, \vec{\Omega}_0 = 0$) of a fluid is given by the balance of pressure gradient forces and the gravitational forces. i.e.,

$$\vec{\nabla}p = \rho\vec{g}, \qquad (7.3)$$

when the pressure p includes radiation pressure in addition to thermal pressure, Equation (7.3) describes the hydrostatic equilibrium of a star like the sun. In highly compact stars such as, white dwarfs and pulsars, the material becomes degenerate, the equilibrium is then achieved by a balance of gravitational forces and degeneracy pressure gradient forces. We have already studied some examples of hydrostatic equilibrium in Chapter 4.

The coriolis force plays an important role in many astrophysical situations. We are familiar with the differential effect it produces on the trajectories of missiles shot from the northern and the southern hemispheres of the earth. There is a net displacement of a missile towards the east in the northern hemisphere and towards the west in the southern hemisphere. It is due to the coriolis force that the magnetic flux tubes appear on the surface of the sun at latitudes different from their place of generation in the solar convection zone. The coriolis force is also responsible for producing motions with a net handedness, called the kinetic helicity; this is absolutely essential for dynamo action to generate large scale magnetic fields, as we shall learn later.

The centrifugal force is usually combined with the gravitational force. We write:

$$\vec{g} = -\vec{\nabla}\varphi_g, \tag{7.4}$$

where φ_g is the gravitational potential and an effective acceleration due to gravity $\vec{g}'$ is defined as:

$$\vec{g}' = -\vec{\nabla}\varphi_g' = -\vec{\nabla}\left[\varphi_g - \frac{1}{2}\left|\vec{\Omega}_0 \times \vec{r}\right|^2\right], \tag{7.5}$$

with φ_g' as the effective gravitational potential. The centrifugal force essentially reduces the acceleration due to gravity.

Problem 7.1: Show that $\vec{g}' = \vec{g} - \vec{\Omega}_0 \times (\vec{\Omega}_0 \times \vec{r})$ and estimate g' for Earth, Jupiter, and a neutron star at their poles and equators.

We recognize that the balance of pressure gradient and the viscous forces for small velocities $\vec{U}$, such that the inertial nonlinear term $(\vec{U} \cdot \vec{\nabla})\vec{U}$ is negligible, gives what is known as the **Stokes Flow** described as (Cole 1962):

$$\vec{\nabla}p = \mu\nabla^2\vec{U} + \left(\zeta + \frac{1}{3}\mu\right)\vec{\nabla}(\vec{\nabla} \cdot \vec{U}). \tag{7.6}$$

For incompressible motions i.e., for $\vec{\nabla} \cdot \vec{U} = 0$, Equation (7.6) describes the well known parabolic velocity profile

$$U_z(r) = \frac{\delta p}{L_z}\left(\frac{R^2 - r^2}{4\mu}\right) + U_0, \tag{7.7}$$

where δp is the pressure difference over a length L_z in a cylinder of radius R and U_0 is the velocity at the surface $r = R$. This model could represent cylindrical flows such as in cometary tails or extragalactic jets if the magnetic effects are not too dominant.

In the absence of viscous forces Equation (7.1) can be rewritten as

$$\frac{\partial \vec{U}}{\partial t} - \vec{U} \times \left(\vec{\omega} + 2\vec{\Omega}_0 \right) = -\vec{\nabla}[P], \tag{7.8}$$

where $\vec{\omega} = \vec{\nabla} \times \vec{U}$ is the vorticity and

$$P = \frac{|\vec{U}|^2}{2} + h + \varphi'_g. \tag{7.9}$$

Here, the specific enthalpy h is defined as:

$$\rho\vec{\nabla}h = \vec{\nabla}p, \tag{7.10}$$

and the effective gravitational potential φ'_g is defined as:

$$\vec{g'} = -\vec{\nabla}\varphi'_g.$$

The curl of Equation (7.8) gives

$$\frac{\partial \vec{\omega}}{\partial t} - \vec{\nabla} \times \left[\vec{U} \times \left(\vec{\omega} + 2\vec{\Omega}_0 \right) \right] = 0. \tag{7.11}$$

In analogy to a streamline, which represents the instantaneous direction of velocity at any point, we can define a **Vortex Line** representing the instantaneous direction of vorticity $\vec{\omega}$ in a fluid. A vortex line is determined from the following equations:

$$\frac{dx}{\omega_x} = \frac{dy}{\omega_y} = \frac{dz}{\omega_z}. \tag{7.12}$$

Since

$$\vec{\nabla} \cdot \vec{\omega} = 0, \tag{7.13}$$

we find using Gauss's theorem that

$$\int_s \vec{\omega} \cdot d\vec{s} = 0, \tag{7.14}$$

where $d\vec{s}$ is an element of a closed surface s enclosed by the contour C. Now, integrating Equation (7.11) over the surface area s, we get:

$$\int_s \frac{\partial \vec{\omega}}{\partial t} \cdot d\vec{s} - \int_s \vec{\nabla} \times \left[\vec{U} \times \left(\vec{\omega} + 2\vec{\Omega}_0 \right) \right] \cdot d\vec{s} = 0. \tag{7.15}$$

For steady rotation, $2\vec{\Omega}_0$ can be added to $\vec{\omega}$ in the first integral. Using Stoke's theorem for the second integral, we find

$$\int_s \frac{\partial(\vec{\omega} + 2\vec{\Omega}_0)}{\partial t} \cdot d\vec{s} = \int_c \left[\vec{U} \times (\vec{\omega} + 2\vec{\Omega}_0)\right] \cdot d\vec{l}, \qquad (7.16)$$

where $d\vec{l}$ is an element of the contour C. Carrying out the steps taken to prove the constancy of the magnetic flux in Chapter 4 (Equation 4.19), we can show that

$$\frac{d}{dt} \int_s (\vec{\omega} + 2\vec{\Omega}_0) \cdot d\vec{s} = \frac{d\varphi_\omega}{dt}. \qquad (7.17)$$

Thus, the vorticity flux φ_ω remains constant as long as the surface enclosing it moves with the fluid velocity $\vec{U}$. This is known as the **Taylor's Theorem.** In the absence of rotation, Equation (7.17) can be recast as:

$$\int_s \vec{\omega} \cdot d\vec{s} = \int_C \vec{U} \cdot d\vec{l} = \Gamma = \text{ constant}, \qquad (7.18)$$

where Γ is called the **Circulation** along the closed curve C (Figure 7.1). The constancy of Γ is known as the **Kelvin-Helmholtz Theorem.**

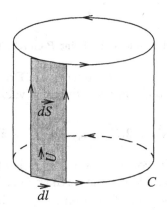

Figure 7.1. Conservation of Circulation Γ.

For steady, slow and incompressible motions, Equation (7.11) gives:

$$(\vec{\Omega}_0 \cdot \vec{\nabla})\vec{U} = 0 \qquad (7.19)$$

which implies that the velocity $\vec{U}$ cannot vary along the direction of $\vec{\Omega}_0$. Thus, if $\Omega_0 = (0, 0, \Omega_z)$ then U_x, U_y and U_z are not functions of (z), i.e., any fluid element, if initially on a line parallel to $\vec{\Omega}_0$, will always stay on that line and the distance between any two fluid elements on the z axis remains constant. Under the same conditions, Equation (7.8) tells us that the Bernoulli function P does not vary in the direction of $\vec{\Omega}_0$. Perpendicular to $\vec{\Omega}_0$, the equilibrium is given by

$$\vec{\nabla}_\perp P = 2(\vec{U} \times \vec{\Omega}_0). \tag{7.20}$$

A rotating fluid can thus support pressure gradients in a direction perpendicular to $\vec{\Omega}_0$.

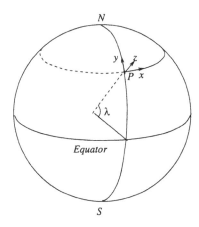

Figure 7.2. Latitude λ of a Point P on a Spherical Body

An equilibrium, known as the **Geostrophic Equilibrium**, results when gravitational forces and the nonlinear terms in velocity $\vec{U}$ are neglected. It is given by

$$\frac{1}{\rho}\vec{\nabla}p = 2(\vec{U} \times \vec{\Omega}_0) \equiv f(\vec{U}_g \times \hat{z}), \tag{7.21}$$

where $f = 2\Omega_0 \sin \lambda$ is the coriolis parameter, λ is the latitude of a point on the surface of a spherical body like the earth (Figure 7.2) and the geostrophic velocity $\vec{U}_g$ is determined from Equation (7.21). The sense of circular flow $\vec{U}_g$ can be determined from the vorticity

$$\omega_{gz} = \left(\vec{\nabla} \times \vec{U}_g\right)_z = \frac{1}{\rho f}\nabla^2 p, \tag{7.22}$$

which shows that in the northern hemisphere ($f > 0$), around a high pressure region ($\nabla^2 p < 0$) the fluid circulates in the clockwise direction causing **Anticyclones** and around a low pressure region ($\nabla^2 p > 0$), the fluid circulates in the anti-clock-wise direction causing **Cyclones**. We see from Equation (7.22) that the equatorial regions (small f) are prone to intense circulations. The flow velocity $\vec{U}_g$ is parallel to the isobars.

Problem 7.2: Find the equation of isobars and show that the fluid flows parallel to the isobars.

Problem 7.3: Estimate the geostrophic velocity U_g for $\lambda = 13.05°$ (latitude of Madras, highly prone to cyclones) for a pressure gradient of 10 millibar/100 km.

The equilibrium resulting from the balance of the inertial nonlinear force and the pressure gradient force, known as the **Cyclostrophic Equilibrium** is described as

$$\frac{1}{\rho}\vec{\nabla}p = -(\vec{U} \cdot \vec{\nabla})\vec{U}. \tag{7.23}$$

This equilibrium exists at low latitudes, where the coriolis parameter f is small. In the cylindrical coordinate system, the radial component of Equation (7.23) gives:

$$\frac{1}{\rho}\frac{\partial p}{\partial r} = \frac{U_\theta^2}{r}. \tag{7.24}$$

Since $(\partial p/\partial r) > 0$, the sense of circulation is always anticlockwise or cyclonic. A tropical cyclone usually have a pressure gradient of 30 millibars per hundred kilometers, an extent $r \simeq 100$ km and $U_\theta \simeq 50$ meters/sec.

The action of both coriolis and inertial forces gives rise to the flow known as the **Gradient Wind**. The relative importance of these two forces is described in terms of the **Rossby Number R_0** defined as:

$$R_0 = \frac{(\vec{U} \cdot \vec{\nabla})\vec{U}}{f\vec{U}} \simeq \frac{U}{fL}, \tag{7.25}$$

where L is a characteristic horizontal scale. It is straightforward to see that the geostrophic approximation is valid for small Rossby numbers.

There are several astrophysical situations in which matter settles in the form of disks around highly gravitating objects such as black holes, neutron stars and white dwarfs. In these so called accretion disks the matter usually executes Keplerian motion with angular velocity Ω_k around the central object. But it also acquires a radial velocity in the presence of viscous forces, due to which the matter spirals in, and may fall directly on to the surface of the gravitating object. The radial balance of the nonlinear inertial force and the viscous force in cylindrical geometry gives:

$$(\vec{U} \cdot \vec{\nabla})\vec{U}\Big|_r = \frac{\mu}{\rho}\nabla^2 U_r,$$

or

$$U_r = -\frac{\mu}{\rho r}. \tag{7.26}$$

Thus, the larger the viscosity $\nu = \mu/\rho$, the larger the infall velocity U_r, and the material spirals in with increasing velocity as it approaches the central object. In many a situations, such as active galactic nuclei harbouring a black hole or binary x-ray sources, the gravitational energy of the infalling matter is released in the form of high energy particles and radiation. Therefore, it is highly desirable that the infalling material possess large viscosity. The presence of turbulence has been found to enhance the viscosity. The classical or the so called molecular viscosity, ν_{mol} is given by

$$\nu_{mol} \simeq \frac{lV_T}{3}, \tag{7.27}$$

where l is the proton collisional mean free path and V_T is the proton thermal speed. Shakura and Sunyaev modeled an effective viscosity ν_{eff} as

$$\nu_{eff} = \alpha H c_s \tag{7.28}$$

where c_s is the sound speed, H is the disk half thickness, taken to be of the order of the vertical density scale height, for thin disks and α is the model parameter. Its value is determined from the observed light variations and is likely to be of the order of 0.01 - 1.

Problem 7.4: Show that with radial velocity given by equation (7.26), the specific angular momentum $(r^2\Omega_k)$ diffuses out with a diffusion coefficient $= 4\nu$.

The spiraling-in motion of the fluid results from the gravitational and the viscous forces. In the absence of viscous and rotational forces, the gravitational forces combined with pressure gradient forces give rise to what is known as **Spherical Accretion**. The steady state spherical accretion is governed by Bernoulli's equation:

$$P = \text{constant}. \tag{7.29}$$

The mass conservation law, in spherical symmetry, gives

$$\vec{\nabla} \cdot \left[\rho \vec{U} \right] = 0,$$

or

$$r^2 \rho U_r = \text{constant} \equiv -\dot{M}, \tag{7.30}$$

where $\dot{M}$ is the rate of change of mass, known as the **Accretion Rate**, if positive and **Mass Loss Rate**, if negative. In order to relate the pressure p and the mass density ρ, we use the polytropic equation of state

$$p = p_\infty \left(\frac{\rho}{\rho_\infty} \right)^\gamma, \tag{7.31}$$

where p_∞ and ρ_∞ are the values at $r = \infty$ and γ is the polytropic index. The specific enthalpy h is, then found to be:

$$h = \left(\frac{\gamma}{\gamma - 1}\right) c_\infty^2 \left[\left(\frac{\rho}{\rho_\infty}\right)^{\gamma - 1} - 1\right], \qquad (7.32)$$

where the integration over ρ has been carried out from ρ_∞ to ρ and c_∞ is the sound speed at $r = \infty$. The local sound speed c_s is determined, as usual from

$$c_s^2 = \frac{dp}{d\rho} = \gamma c_\infty^2 \left(\frac{\rho}{\rho_\infty}\right)^{\gamma - 1}. \qquad (7.33)$$

The constant in Bernoulli's equation (7.29) turns out to be zero as $h = U = \varphi_g = 0$ at $r = \infty$ and we have taken $\Omega_0 = 0$. We can express equation (7.29) in a dimensionless form by defining a characteristic radius $\mathbf{R_B}$, called the **Bondi Radius**, and a characteristic accretion rate $\dot{M}_B$, and then use them to normalize the radial distance r, and the accretion rate $\dot{M}$, respectively. Velocities are normalized with c_∞, so that equation (7.29) becomes

$$\frac{V^2}{2} + H - \frac{1}{R} = 0 \qquad (7.34)$$

and

$$\frac{\rho r^2 U_r}{\rho_\infty R_B^2 c_\infty} = \alpha R^2 V = \frac{\dot{M}}{\dot{M}_B} = \beta \qquad (7.35)$$

where

$$V = \frac{U_r}{c_\infty}, \quad \alpha = \frac{\rho}{\rho_\infty}, \quad H = \frac{h}{c_\infty^2}, \quad R = \frac{r}{R_B}, \quad R_B = \frac{GM}{c_\infty^2},$$

$$\text{and } \dot{M}_B = -R_B^2 \rho_\infty c_\infty$$

Substituting for α from Equation (7.35) in equation (7.34), we get:

$$\frac{V^2}{2} + \left(\frac{\gamma}{\gamma - 1}\right) \left[\left(\frac{\beta}{R^2 V}\right)^{\gamma - 1} - 1\right] - \frac{1}{R} = 0, \qquad (7.36)$$

which describes the flow pattern V vs R for different values of β (Figure 7.3) (Shu 1991).

Problem 7.5: By taking the limit $\gamma \to 1$, show that the flow pattern for the isothermal equation of state is given by $\dfrac{V^2}{2} + \ln \alpha - \dfrac{1}{R} = 0$. Find the extremum points of the flow.

Problem 7.6: Show that at $r = R_B$, the gravitational energy of a proton $= 2/3$ times its thermal energy. Estimate R_B for the sun.

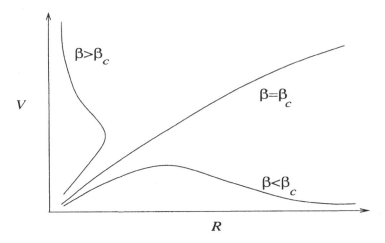

V

$\beta > \beta_c$

$\beta = \beta_c$

$\beta < \beta_c$

R

Figure 7.3. Flow Speed vs R for Normalized Accretion Rate $\beta \gtrless$ and Equal to a Critical Value β_c.

7.3. Waves in Fluids

We have already studied the excitation of, sound waves with the dispersion relation

$$\omega = kc_s,$$

where $(\omega, \vec{k})$ are the frequency and wave vector, and c_s is the sound speed, and gravity waves with the dispersion relation

$$\omega^2 = \omega_{BV}^2,$$

in a fluid. We now wish to study what new modes of oscillation, if any, rotation introduces. The most dramatic effect of rotation is the appearance of the coriolis force. In the presence of this force, an incompressible fluid supports what are called the **Rossby Waves**. These waves represent the exact solutions of the fluid equations. We need not go through the linearization process. The momentum conservation law for an incompressible rotating fluid, in the component form, is:

$$\frac{\partial U_i}{\partial t} + U_j \frac{\partial U_i}{\partial r_j} = -\frac{\partial}{\partial r_i} P + 2 \left(U_j \Omega_{ol} - U_l \Omega_{oj} \right) + \nu \nabla^2 U_i. \qquad (7.37)$$

The mass conservation law reduces to

$$\vec{\nabla} \cdot \vec{U} = 0. \qquad (7.38)$$

Assuming a plane-wave type variation for $\vec{U}$ and P such that

$$[P, \vec{U}] = \text{ constant } \exp \left[i\vec{k} \cdot \vec{r} - i\omega t \right], \qquad (7.39)$$

and substituting in Equations (7.37) and (7.38), we get:

$$(-i\omega + \nu k^2)U_i = -ik_i P + 2(U_j\Omega_{0l} - U_l\Omega_{0j}).$$ (7.40)

Equation (7.38) implies that the waves are transverse. Multiplying Equation (7.40) with U_i, k_i and Ω_{0i}, respectively, gives:

$$(-i\omega + \nu k^2)U_i^2 = 0,$$ (7.41)

$$k_i^2 P + 2ik_i(U_j\Omega_{0l} - U_l\Omega_{0j}) = 0,$$

and

$$(-i\omega + \nu k^2)U_i\Omega_{0i} + ik_i\Omega_{0i}P = 0.$$

For the coordinate system, such that $\vec{\Omega}_0 = (\Omega_{0x}, 0, \Omega_{0z})$ and $\vec{k} = (0, 0, k)$, we find:

$$U_z = 0; \quad U_y = \pm iU_x; \quad P = \pm\frac{2\Omega_{0x}U_x}{k};$$

and the dispersion relation of the Rossby waves is found to be

$$\omega = \pm 2\Omega_{0z} - i\nu k^2.$$ (7.42)

The ($\pm$) stands for the left and the right handed circular polarization respectively. The waves are damped, and the damping increases as the wavelength decreases. The frequency of the wave is a function of the angle between $\vec{\Omega}_0$ and $\vec{k}$; the highest value of ω is obtained for $\vec{\Omega}_0 \parallel \vec{k}$. These waves represent oscillations over the largest spatial scales ($k \to 0$) in a rotating fluid. For sufficiently large wavelengths, so that the coriolis force along with gravity needs to the retained in the discussion of waves, it is found that the dispersion relation is modified to:

$$\omega^2 = f^2 + \frac{k^2 r^2}{m^2}(\omega_{BV}^2 - \omega^2),$$ (7.43)

where ω_{BV} is the Brünt-Väisälä frequency, f is the coriolis parameter and (m/r) is the azimuthal wave number. Thus ω is always greater than f but Rossby waves are the waves of the lowest frequency and the largest wavelength.

We will now consider the excitation of waves in a system with a disk type configuration. Such disks occur in galaxies and around compact objects. We have already discussed how a galactic disk consisting of collisionless stars can be treated as a fluid. High temperature accretion disks around compact objects, though almost collisionless, could also be approximated as fluid disks for understanding their global features. The main motivation

Figure 7.4. The Galaxy M100 Exhibiting Spiral Pattern.

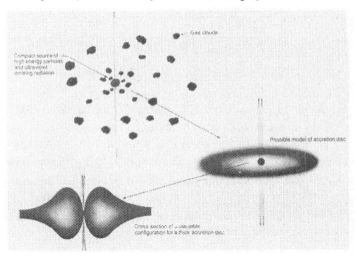

Figure 7.5. An Axisymmetric Accretion Disk Around a Compact Object.

for treating galactic disks as fluid is to understand their spiral structure (Figure 7.4) which, some believe to be **Spiral Density Waves**.

For this purpose, the disk is taken as a two dimensional axisymmetric structure (Figure 7.5). The mass and momentum conservation laws for this structure are obtained by integrating the 3-D forms of these laws over the vertical coordinate z. Thus, we define the surface mass density σ as

$$\sigma = \int_{-\delta}^{\delta} \rho dz, \qquad (7.44)$$

where 2δ represents the small vertical extent of the disk. Similarly, the enthalpy h is now defined as:

$$\vec{\nabla}h = \frac{1}{\sigma}\left[\int_{-\delta}^{\delta}\vec{\nabla}pdz\right]. \tag{7.45}$$

The fluid velocity $\vec{U}$ is two dimensional with the radial component U_r and the azimuthal component U_θ in a cylindrical coordinate system. After carrying out these operations, we get the disk equations:

$$\frac{\partial\vec{U}}{\partial t} + (\vec{U}\cdot\vec{\nabla})\vec{U} + 2\Omega_0(\hat{z}\times\vec{U}) - \Omega_0^2\vec{r} = -\vec{\nabla}h - \vec{\nabla}\varphi_g\Big|_{z=0} \tag{7.46}$$

$$\frac{\partial\sigma}{\partial t} + \nabla\cdot[\vec{U}\sigma] = 0,$$

and Poisson's equation:

$$\nabla^2\varphi_g = 4\pi G\rho. \tag{7.47}$$

The gravitational force is evaluated at $z = 0$. The equilibrium state of the disk is described by a rotation with a constant angular velocity $\Omega(r)$ around the z-axis under the balance of centrifugal, gravitational and pressure forces in a frame rotating with a constant angular velocity Ω_0. The equilibrium velocity $U_{0\theta}$ is found to be:

$$U_{0\theta} = (\Omega - \Omega_0)r, \tag{7.48}$$

along with

$$\Omega^2 r = \frac{\partial}{\partial r}(h_0 + \varphi_{g0}), \tag{7.49}$$

where the zeroth order enthalpy h_0 is given by

$$\frac{\partial h_0}{\partial r} = \frac{c_{s0}^2(r)}{\sigma_0(r)}\frac{d\sigma_0(r)}{dr}, \tag{7.50}$$

and

$$c_{s0}^2 = \left[\frac{d}{d\sigma}\int pdz\right]_{\sigma=\sigma_0},$$

with $c_{s0}(r)$ and $\sigma_0(r)$ as the equilibrium sound speed and surface mass density respectively. The zeroth order gravitational potential φ_{g0} satisfies Poisson's equation

$$\nabla^2\varphi_{g0} = 4\pi G\left[\rho_0(r,z) + \rho_{ext}(r,z)\right], \tag{7.51}$$

where ρ_{ext} is the mass density associated with objects other than the disk itself, such as the bulge and halo of a galaxy for a galactic disk and a compact object for an accretion disk.

We, now, perturb the equilibrium state and represent all first order quantities by the subscript one, Thus, we write:

$$
\begin{aligned}
\sigma &= \sigma_0(r) + \sigma_1(r,\theta,t), \\
U_r &= U_{1r}(r,\theta,t), \\
U_\theta &= U_{0\theta} + U_{1\theta}(r,\theta,t), \\
\varphi_g &= \varphi_{go}(r.z) + \varphi_{g1}(r,\theta,z,t), \\
h &= h_0 + h_1.
\end{aligned}
\tag{7.52}
$$

The linearized form of equations (7.46), (7.47) and (7.51) are:

$$
\frac{\partial U_{1r}}{\partial t} + \frac{U_{0\theta}}{r}\frac{\partial U_{1r}}{\partial \theta} - \frac{U_{0\theta}U_{1\theta}}{r} - 2\Omega_0 U_{1\theta} = -\frac{\partial h_1}{\partial r} - \frac{\partial \varphi_{g1}}{\partial r}\bigg|_{z=0},
\tag{7.53}
$$

$$
\frac{\partial U_{1\theta}}{\partial t} + U_{1r}\frac{\partial U_{0\theta}}{\partial r} + \frac{U_{0\theta}}{r}\frac{\partial U_{1\theta}}{\partial \theta} + \frac{U_{0\theta}U_{1r}}{r} + 2\Omega_0 U_{1r} = -\frac{1}{r}\frac{\partial h_1}{\partial \theta} - \frac{1}{r}\frac{\partial \varphi_{g1}}{\partial \theta}\bigg|_{z=0},
\tag{7.54}
$$

$$
\frac{\partial \sigma_1}{\partial t} + \frac{1}{r}\frac{\partial}{\partial r}(r\sigma_0 U_{1r}) + \frac{1}{r}\frac{\partial}{\partial \theta}(\sigma_0 U_{10} + \sigma_1 U_{0\theta}) = 0,
\tag{7.55}
$$

and

$$
\frac{1}{r}\frac{\partial}{\partial r}(r\frac{\partial \varphi_{g1}}{\partial r}) + \frac{1}{r^2}\frac{\partial^2 \varphi_{g1}}{\partial \theta^2} + \frac{\partial^2 \varphi_{g1}}{\partial z^2} = 4\pi G \rho_1,
\tag{7.56}
$$

with

$$
h_1 = \frac{\partial h}{\partial \sigma}\bigg|_{\sigma=\sigma_0} (\sigma_1) = c_{s0}^2\frac{\sigma_1}{\sigma_0}.
\tag{7.57}
$$

Assuming a plane-wave type variation for all the first order quantities, i.e.,

$$
U_1 = U_1(r)\exp[i(kr + m\theta - \omega t)]
\tag{7.58}
$$

and substituting in the linearized Equations (7.53) - (7.57) except the Poisson Equation (7.56) which will be discussed soon, we find:

$$
\begin{aligned}
i\omega' U_{1r} + (\Omega + \Omega_0)U_{1\theta} &= ik(h_1 + \varphi_{g1}), \\
\frac{\Omega_E^2}{2\Omega}U_{1r} - i\omega' U_{1\theta} &= -\frac{im}{r}(h_1 + \varphi_{g1}), \\
\omega' \sigma_1 - \sigma_0(kU_{1r} + \frac{m}{r}U_{1\theta}) &= 0,
\end{aligned}
\tag{7.59}
$$

with

$$
\omega' \equiv \omega - m(\Omega - \Omega_0),
$$

and

$$
\Omega_E^2 \equiv 2\Omega\left(2\Omega + r\frac{d\Omega}{dr}\right),
$$

where the short wavelength approximation $kr \gg 1$ and the local approximation $\left(\dfrac{1}{k\sigma_0}\dfrac{d\sigma_0}{dr}\right) \ll 1$ have been used. This implies that we are investigating phenomena on spatial scales much smaller than the spatial scales over which the equilibrium configuration changes. The dispersion relation obtained under these conditions is known as the **Local Dispersion Relation**. Here, Ω_E is known as the **Epicyclic Frequency**. It can be shown that Ω_E is the frequency with which a single particle of mass m oscillates about its circular orbit in a central force of magnitude $mr\Omega^2$.

Problem 7.7: Reduce the fluid Equations (7.59) to equations for a single particle ($h_1 = 0, \varphi_{g1} = 0, \sigma_0 = 0, \sigma_1 = 0$) and show that the particle executes oscillations with a frequency $\omega = \Omega_E$ for $\Omega = \Omega_0$.

The radial variation of the rotation frequency Ω has important implications in the dynamics of galaxies and stability of rotating fluids. It is expressed in the form

$$A = -\frac{r}{2}\frac{d\Omega}{dr} = \Omega - \frac{\Omega_E^2}{4\Omega}, \tag{7.60}$$

and is known as the **Oort's Constant A**. We now, solve Poisson's equation (7.56) in order to relate φ_{g1} and σ_1. We write:

$$\rho_1(r, \theta, z) = \sigma_1(r, \theta)\delta(z), \tag{7.61}$$

so that the definition of the surface density σ_1 (Equation 7.44) is recovered. Integrating Poisson's equation over z, we get

$$\int_{-\delta}^{\delta} \nabla^2 \varphi_{g1}(r, \theta, z)dz = 4\pi G\sigma_1,$$

or

$$\frac{\partial \varphi_{g1}}{dz}\Bigg|_{z=-\delta}^{z=\delta} = 4\pi G\sigma_1, \tag{7.62}$$

where we have used the property of the gravitational potential that it is continuous in r and θ but can suffer a jump across the plane $z = 0$. From the symmetry of the problem, it is clear that φ_{g1} must be a function of $|z|$, therefore

$$\frac{\partial \varphi_{g1}}{\partial z}\Bigg|_{z=-\delta}^{z=\delta} = 2\frac{\partial \varphi_{g1}}{d|z|}\Bigg|_{|z|=0} = 4\pi G\sigma_1. \tag{7.63}$$

For $|z| \neq 0$, Poisson's equation becomes a Laplacian equation:

$$\nabla^2 \varphi_{g1} = \frac{\partial^2 \varphi_{g1}}{\partial r^2} + \frac{1}{r}\frac{\partial \varphi_{g1}}{\partial r} + \frac{1}{r^2}\frac{\partial^2 \varphi_{g1}}{\partial \theta^2} + \frac{\partial^2 \varphi_{g1}}{\partial |z|^2} = 0. \tag{7.64}$$

Let us try a WKB type solution

$$\varphi_{g1} = r^{-1/2} \exp\left[iS(r, |z|)\right] e^{im\theta}. \tag{7.65}$$

We find:

$$\left(\frac{\partial S}{\partial |z|}\right)^2 + \left(\frac{\partial S}{\partial r}\right)^2 = i\left(\frac{\partial^2 S}{\partial r^2} + \frac{\partial^2 S}{\partial |z|^2}\right) - \left(m^2 - \frac{1}{4}\right)\frac{1}{r^2} = 0. \tag{7.66}$$

For short wavelengths ($kr \gg 1$) and small values of m, the right hand side of equation (7.66) can be put equal to zero, so that

$$\left(\frac{\partial S}{\partial |z|}\right) = \pm i\left(\frac{\partial S}{\partial r}\right), \tag{7.67}$$

or

$$S = F(r \pm i|z|),$$

where F is any function of $(r \pm i|z|)$. The potential φ_{g1} is now given by

$$\varphi_{g1} = r^{-1/2} \exp\left[iF(r \pm iz)\right] e^{im\theta}. \tag{7.68}$$

The sign (+) or (−) is decided by the requirement that φ_{g1} must diminish as $|z| \to \infty$ and depends on the radial derivative of the function F; for $\dfrac{\partial F}{\partial r} > 0$, the plus sign is the right choice for $k > 0$. The surface density σ_1 can be expressed as:

$$\begin{aligned}
\sigma_1 &= \frac{i}{2\pi G}\varphi_{g1}(z = 0), \frac{\partial S}{\partial z}\bigg|_{z=0}, \\
&= -\frac{k}{2\pi G}\varphi_{g1}(z = 0) \tag{7.69}
\end{aligned}$$

since $\dfrac{\partial S}{\partial z} = i\dfrac{\partial S}{\partial r} = ik$.

A more accurate treatment of Poisson's equation can be seen in Lebovitz (1983). Having determined σ_1 and φ_{g1}, we can express the first order enthalpy

$$h_1 = c_s^2 \frac{\sigma_1}{\sigma_0} \quad \text{and} \quad \varphi_{g1} = -\frac{2\pi G\sigma_1}{k},$$

so that

$$h_1 + \varphi_{g1} = \frac{\sigma_1}{\sigma_0}\left[c_s^2 - \frac{2\pi G\sigma_0}{k}\right]. \tag{7.70}$$

We are now ready to eliminate the first order quantities from Equations (7.59) and obtain the dispersion relation (for $\dfrac{m}{r} \ll k$),

$$[\omega - m(\Omega - \Omega_0)]^2 = \Omega_E^2 + k^2\left(c_s^2 - \frac{2\pi G\sigma_0}{k}\right), \tag{7.71}$$

of the spiral density waves for $k > 0$. Why are they called spiral? The surface density

$$\sigma_1 = r^{-1/2} \exp\left[i(m\theta + F(r) - \omega t)\right] \tag{7.72}$$

at any given time, say $t = 0$, has the maxima when

$$m\theta + F(r) = 2\pi n \tag{7.73}$$

where $n = 0, 1, 2, \ldots$ so that

$$\left(\theta - \frac{2\pi n}{m}\right) = -\frac{F(r)}{m} \tag{7.74}$$

which is the equation of a spiral rotated through an angle (n/m). The spiral leads or trails the disk rotation depending upon whether $\dfrac{d\theta}{dr} > 0$ or < 0. Thus, we see that the spiral trails for $\dfrac{\partial F(r)}{\partial r} = k > 0$, and vice-versa. There is a spiral for each value of m. The inclination of the spiral to the disk is expressed through an angle θ_I where

$$\tan \theta_I = \frac{1}{r}\frac{\partial r}{\partial \theta} = -\frac{m}{kr},$$

at a constant value of the phase $(m\theta + F(r) - \omega t)$ at a position r. The condition $kr \gg 1$ gives $\tan \theta_I \ll 1$ which corresponds to the tightly wound spiral density waves. The spiral pattern rotates with a constant frequency Ω_s given by

$$\Omega_s = \left(\frac{\partial \theta}{\partial t}\right) = \frac{\omega}{m},$$

again at constant phase. We have four frequencies in the system: (1) Ω_0 of the rotating coordinate system, (2) Ω of the rotating disk, (3) ω of the spiral density waves and (4) Ω_s of the spiral pattern.

Problem 7.8: Show that we have tacitly assumed that $F = kr$.

Let us examine the dispersion relation (Equation 7.71) for axisymmetric perturbations; i.e. for $m = 0$, we get

$$\omega^2 = \Omega_E^2 + k^2 c_s^2 - 2\pi k G \sigma_0. \tag{7.75}$$

We see that the spiral density waves exist only when

$$\omega^2 > 0. \tag{7.76}$$

There are three contributions to ω^2 : Ω_E^2 from the rotation of the disk, $k^2 c_s^2$ from pressure and $2\pi k G \sigma_0$ from the self gravity of the disk. For $\omega^2 < 0$, the density perturbation grows at an exponential rate and the disk becomes unstable. This can happen either due to self gravity and or due to the differential rotation $\Omega(r)$. The pressure term, counteracting gravity, is a stabilizing term. The condition of marginal stability ($\omega^2 = 0$) is obtained for

$$k_0 = \frac{\pi G \sigma_0}{c_s^2} \pm \frac{1}{c_s^2} \left(\pi^2 G^2 \sigma_0^2 - c_s^2 \Omega_E^2 \right)^{1/2}. \qquad (7.77)$$

Remember that Equation (7.75) is a local dispersion relation. All the zeroth order quantities, σ_0 and c_s^2 and Ω_E^2 have been taken to vary over spatial scales much larger than the wavelength $\lambda = (2\pi)/k$. Therefore k_0 given by Equation (7.77) is not an explicit function of r but it is to be determined using the local values of the physical parameters. The spiral density waves have, typically, periods of the order of the rotation period of the disk (Problem 7.9).

The thickness of the disk can be estimated from the balance between the pressure force and the gravitational force in the vertical direction. The acceleration due to gravity of a mass point M at a position (r, z) is given by:

$$\vec{g} = -\frac{GM\vec{r}}{(r^2 + z^2)^{3/2}}. \qquad (7.78)$$

The hydrostatic balance in the z direction

$$-\frac{\partial p_0}{\partial z} + \rho_0 g_z = 0, \qquad (7.79)$$

shows that

$$\rho(r, z) = \rho_0(r) \exp\left[-\frac{z^2}{H^2} \right],$$

where the scale height H is given by:

$$H = \frac{\sqrt{2} c_s r^{3/2}}{(GM)^{1/2}} = \frac{\sqrt{2} c_s}{\Omega_k} = \frac{\sqrt{2} c_s r}{V_k(R)}. \qquad (7.80)$$

Thus, the thickness of the disk can approximately be taken equal to the scale height H. The disk is called thin, if

$$\frac{H}{r} \simeq \frac{H}{R} = \frac{\sqrt{2} c_s}{V_k(R)} \ll 1, \qquad (7.81)$$

where $V_k = \Omega_k R$ is the Keplerian velocity at the outer edge of the disk at $r = R$.

Problem 7.9: A galactic disk has typically: $\rho_0 \simeq 0.18 M_\odot/\text{pc}^3$, $\sigma_0 \simeq 75 M_\odot/\text{pc}^2$, angular speed $\Omega_0 \simeq 26$ km/sec/kpc, the radius $R \simeq 10$ kpc, the root mean square velocity of stars, which we take to represent the sound speed $c_s, \simeq 40$ km/sec. Estimate Ω_E, k_0 and the thickness H and compare these numbers with those for an accretion disk around a black hole of mass $10^8 M_\odot$, taking the temperature $T \simeq 10^6$ K at a position five Schwartzchild radii away.

Problem 7.10: Show that the dispersion relation of the spiral density waves is given by $\omega^2 = \Omega_E^2 + k^2 c_s^2 - 2\pi|k|G\sigma_0$ irrespective of the sign of k.

7.4. Instabilities in Fluids

In this section, we shall dwell upon the circumstances which lead to an enhancement of matter density and or motion under the influence of gravitational and rotational forces. Such studies have very important bearing on the distribution of matter in the universe, as a result of which, perhaps, we see regions differing in mass density by many many orders of magnitude co-existing side by side.

First of all, we can discuss the case of spiral density waves becoming a **Spiral Density Instability**. Let us recall the dispersion relation of the spiral density waves

$$\omega^2 = \Omega_E^2 + k^2 c_s^2 - 2\pi|k|G\sigma_0. \qquad (7.82)$$

In a non-rotating frame ($\Omega_0 = 0$) and for a pressureless medium ($c_s = 0$), ω^2 becomes negative if

$$\Omega_E^2 < 0, \qquad (7.83)$$

and the disk goes unstable. This corresponds to the **Rotational Instability** in which the circular orbit of a particle no longer oscillates with the epicyclic frequency Ω_E, but there is an exponentially growing departure from the circular orbit, the growth rate being given by Ω_E. In terms of the radial variation of the angular velocity Ω, the condition (Equation 7.83) for rotational instability becomes:

$$\frac{d\Omega}{dr} < 0 \text{ and } \left|\frac{d\Omega}{dr}\right| > \frac{2\Omega}{r}. \qquad (7.84)$$

These conditions are also expressed as:

$$\frac{d}{dr}\left[(r^2\Omega)^2\right] < 0,$$

and in this form are called **Rayleigh's Criterion**.

Problem 7.11: Show that a disk is stable under Keplerian rotation.

In the absence of rotation ($\Omega = 0$), Equation (7.82) shows that the disk experiences a gravitational instability, known as the **Jean's Instability** when

$$2\pi |k| G\sigma_0 > k^2 c_s^2. \tag{7.85}$$

The pressure term ($k^2 c_s^2$) defines the range of wavelengths

$$\frac{2\pi}{k} = \lambda > \frac{c_s^2}{G\sigma_0}, \tag{7.86}$$

at which the disk undergoes gravitational collapse, i.e., the density increases at an exponential rate at these scales. Is this the way to form condensed objects such as stars and galaxies?

The **Jean's Instability** is usually investigated in a 3-dimensional fluid in an expanding frame of reference in order to appreciate the difficulties of forming condensed objects in a medium which is undergoing expansion since its beginning - **the Hubble Expansion**. The other reason for including a zeroth order motion (e.g. expansion or rotation) is that a gravitating medium cannot be static. This difficulty raises its head while satisfying Poisson's equation in the zeroth order. For a homogeneous and isotropic fluid or universe, the gravitational field

$$\vec{E}_g = -\vec{\nabla}\varphi_g, \tag{7.87}$$

must be zero as there is no preferred direction in which $\vec{E}_g$ could point. This implies that the zero order mass density ρ_0 must vanish too, since

$$\vec{\nabla} \cdot \vec{E}_g = 4\pi G\rho_0. \tag{7.88}$$

So, how do condensations form in vacuum (classical, not quantum mechanical which has quantum fluctuations)? In Jean's analysis, this point was glossed over and the static equilibrium with a nonzero ρ_0 has been indulgently described as the Jean's Swindle.

The expanding model of the universe is based on the **Cosmological Principle** which states that the universe is homogeneous and isotropic. This is an assumption and is supported only approximately by observations, for such a universe cannot have mass condensations such as galaxies and clusters of galaxies. This implies that the behaviour of the universe depends upon the scale on which it is observed, very much akin to a fluid, which has large scale and small scale properties. In an isotropic and homogeneous universe, the only allowed large scale velocity field could be an expansion or a contraction since any other motion like rotation or shear would have a preferred direction. In the expanding model, the relative velocity $\vec{U}$ between

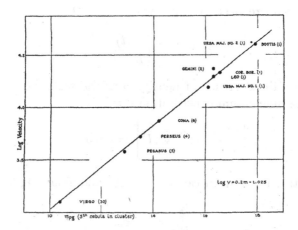

Figure 7.6. The Hubble Law of Velocity – Distance Relation for Clusters of Galaxies. Logarithms of Velocities in km sec^{-1} Are Plotted Against Apparent Magnitudes of the Fifth Brightest Nebulae in Clusters. Each Cluster Velocity is the Mean of the Various Individual Velocities Observed in the Cluster, the Number Being Indicated by the Figure in Brackets. The Apparent Magnitude m, The Absolute Magnitude M and the Distance d in Light-years are Related as $\log d = 0.2(m - M) + 1.513$.

any two points is a function of the relative distance $\vec{r}$ between these two points or in other words, the velocity of expansion

$$\vec{U} = H(t)\vec{r}. \tag{7.89}$$

This is the famous **Hubble's Law of Expansion** (Figure 7.6) and H is known as **Hubble's Constant**, though it is a function of time. The separation $\vec{r}$ increases accordingly as:

$$\vec{r}(t) = \frac{R(t)}{R(t_0)}\vec{r}_0(t_0), \tag{7.90}$$

so that

$$\vec{U}(t) = \frac{d\vec{r}}{dt} = \frac{\dot{R}}{R(t_0)}\vec{r}_0(t_0) = \frac{\dot{R}}{R}\vec{r}(t).$$

Thus

$$H(t) = \frac{\dot{R}}{R}. \tag{7.91}$$

The factor $R(t)$, a function only of time, is called the **Scale Factor**. Let us see what the mass conservation of an expanding fluid gives us. The continuity equation

$$\frac{\partial \rho}{\partial t} = -\vec{\nabla} \cdot \left[\rho \vec{U}\right] = -\vec{\nabla} \cdot [\rho H(t)\vec{r}]$$

$$= -3\rho H(t) \text{ for } \vec{\nabla}\rho = 0, \tag{7.92}$$

on integrating with respect to time gives:

$$\rho(t) = \rho(t_0) \left[\frac{R(t)}{R(t_0)} \right]^{-3}, \tag{7.93}$$

which describes the decrease of density ρ as the scale factor R increases with time. The momentum conservation law

$$\rho \left[\frac{\partial \vec{U}}{\partial t} + \left(\vec{U} \cdot \vec{\nabla} \right) \vec{U} \right] = -\vec{\nabla} p + \vec{F}_g, \tag{7.94}$$

after substituting for $\vec{U}$ from Equation (7.89) and assuming $\vec{\nabla} p = 0$, gives:

$$\frac{\ddot{R}}{R} \vec{r} = -\vec{\nabla} \varphi_g,$$

or

$$\frac{\ddot{R} r}{R} = -\frac{\partial \varphi_g}{\partial r}. \tag{7.95}$$

We have defined the gravitational force $\vec{F}_g$ as:

$$\vec{F}_g = -\rho \vec{\nabla} \varphi_g, \tag{7.96}$$

and φ_g satisfies Poisson's equation:

$$\frac{1}{r^2} \frac{\partial}{\partial r} \left(r^2 \frac{\partial \varphi_g}{\partial r} \right) = 4\pi G \rho,$$

or

$$\frac{\partial \varphi_g}{\partial r} = \frac{4\pi G \rho}{3} r. \tag{7.97}$$

Substituting Equations (7.93) and (7.97) in Equation (7.95), we get

$$\frac{\ddot{R}}{R} = -\frac{4\pi G \rho(t_0) R^3(t_0)}{3R^3}, \tag{7.98}$$

which, on integration becomes:

$$(\dot{R})^2 = \frac{8\pi G \rho(t_0) R^3(t_0)}{3R} - \frac{\kappa}{T^2}, \tag{7.99}$$

where the integration constant has been written as $(-\kappa/T^2)$ with κ dimensionless and T having the dimensions of time. Equation (7.99) is solved for three cases: (1) for $\kappa = 0$, (ii) $\kappa > 0$ and (iii) $\kappa < 0$. For $\kappa = 0$, we find

$$R = R_0 \left(\frac{t}{t_0} \right)^{2/3}, \quad H = \frac{\dot{R}}{R} = \frac{2}{3t} \quad \text{and} \quad \rho = \frac{3H^2}{8\pi G} \equiv \rho_c, \tag{7.100}$$

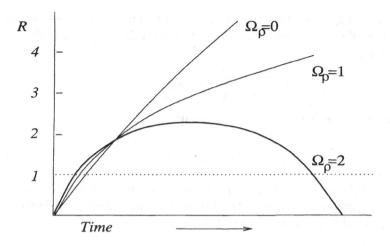

Figure 7.7. The Three Models of the Expanding Universe.

where $t_0 = \dfrac{2}{3}\left(\dfrac{3}{8\pi G\rho(t_0)}\right)^{1/2}$ and ρ_c is called the critical density. This model is known as the **Einstein - de Sitter Model**. In this model the scale factor R increases with time for all time. Two parameters, the **Deceleration Parameter** q defined as:

$$q = -\frac{1}{H^2}\frac{\ddot{R}}{R},\qquad (7.101)$$

and the density parameter Ω_ρ defined as

$$\Omega_\rho = \frac{\rho}{\rho_c},\qquad (7.102)$$

are used to describe different cosmological models. We find that the Einstein-de Sitter model is characterized by $\Omega_\rho = 1$ and $q = 1/2$.

A similar analysis can be carried out for positive and negative values of the constant κ. We shall only give the results here. For $\kappa > 0$, it is found that the scale factor R has a maximum value given by

$$R_{max} = \frac{8\pi G\rho_0 R_0^3 T^2}{3\kappa},\qquad (7.103)$$

after which R decreases with time and the universe begins to contract. In this model, the parameters Ω_ρ and q are functions of time with $\Omega_\rho > 1$ and $q > 1/2$. For $\kappa < 0$, R increases with time for all time with $\Omega_\rho < 1$ and $q < 1/2$. Thus, $\Omega_\rho = 1$ represents the transition between the ever expanding ($\kappa < 0$) and those undergoing contraction after an initial expansion ($\kappa > 0$) models of the universe (Figure 7.7). This is as far as we need to venture into cosmology. We, now, know the equilibrium state, which is, though, time dependent. We describe it with the mass density ρ_0 and the velocity $\vec{V}_0$ varying as

$$\rho_0 = \frac{3}{8\pi G}\left(\frac{\dot{R}}{R}\right)^2 \quad \text{and} \quad \vec{V}_0 = \frac{\dot{R}}{R}\vec{r}, \tag{7.104}$$

and the zeroth order gravitational potential φ_{g0} is given by

$$\vec{\nabla}\varphi_{g0} = \frac{4\pi G\rho_0}{3}\vec{r}. \tag{7.105}$$

We shall now perturb this equilibrium with first order perturbations ρ_1, U_1 and φ_{g1} satisfying the linearized mass and momentum conservation laws, and φ_{g1} the linear Poisson's equation:

$$\frac{\partial\rho_1}{\partial t} + \nabla\cdot\left[\rho_0\vec{U}_1 + \rho_1\vec{V}_0\right] = 0, \tag{7.106}$$

$$\frac{\partial\vec{U}_1}{\partial t} + \left(\vec{V}_0\cdot\vec{\nabla}\right)\vec{U}_1 + (\vec{U}_1\cdot\vec{\nabla})\vec{V}_0 = -\frac{c_s^2}{\rho_0}\vec{\nabla}\rho_1 - \vec{\nabla}\varphi_{g1}, \tag{7.107}$$

and

$$\nabla^2\varphi_{g1} = 4\pi G\rho_1 \tag{7.108}$$

where the pressure perturbation p_1 is expressed in terms of ρ_1 using

$$\vec{\nabla}p_1 = c_s^2\vec{\nabla}\rho_1, \tag{7.109}$$

and the self gravity of the perturbed density $(\rho_1\vec{\nabla}\varphi_{g0})$ has been neglected. Defining the density contrast

$$\delta(\vec{r},t) = \frac{\rho_1(\vec{r},t)}{\rho_0(t)}, \tag{7.110}$$

and expanding the perturbations in Fourier integrals as:

$$D(\vec{r},t) = \frac{1}{(2\pi)^3}\int D_k(t)\exp\left[-\frac{i\vec{k}\cdot\vec{r}}{R(t)}\right]d\vec{k}, \tag{7.111}$$

where $D(\vec{r},t)$ stands for δ, $\vec{U}_1$ and φ_{g1}, we obtain, on substituting equation (7.111) in Equations (7.106) – (7.108):

$$\frac{\partial\delta_k}{\partial t} - \frac{i\vec{k}\cdot\vec{U}_k}{R} = 0, \tag{7.112}$$

$$\frac{d}{dt}(R\vec{U}_k) - ic_s^2\delta_k\vec{k} - i\vec{k}\varphi_k = 0, \tag{7.113}$$

$$\varphi_k = -\frac{4\pi G \rho_0 R^2}{k^2} \delta_k,$$ (7.114)

with

$$\frac{d}{dt} \equiv \frac{\partial}{\partial t} + (\vec{V}_0 \cdot \vec{\nabla}) \ .$$

Here, $\vec{k}$ is the comoving wavenumber and $\dfrac{\vec{r}}{R}$ is the comoving coordinate. The proper wavelength λ is given as

$$\lambda = \frac{2\pi R}{k}.$$ (7.115)

We now split the perturbed velocity $\vec{U}_1$ into rotational $\vec{U}_r$ and irrotational $\vec{U}_{ir}$ parts as:

$$\vec{U}_1 = \vec{U}_{ir} + \vec{U}_r,$$ (7.116)

such that

$$\vec{\nabla} \cdot \vec{U}_r = 0 \ \text{and} \ \vec{\nabla} \times \vec{U}_{ir} = 0.$$

This means that $\vec{U}_r$ can be written as the curl of a vector and $\vec{U}_{ir}$ as the gradient of a scalar. Equation (7.112) becomes

$$\dot{\delta}_k - \frac{i\vec{k} \cdot \vec{U}_{ir}(\vec{k})}{R} = 0.$$ (7.117)

Equation (7.113) gives

$$\frac{d}{dt}(R\vec{U}_r(\vec{k})) = 0,$$ (7.118)

or

$$\vec{U}_r \propto R^{-1},$$

i.e., the rotational component of the velocity perturbation decreases with time since R increases with time. Further, the rotational part decouples from the irrotational part and therefore from the density, i.e., we cannot transfer energy from one to the other. But caution is required as this is the result of a linear analysis. If we venture into the nonlinear regime, the terms $(\vec{U}_1 \cdot \nabla)\vec{U}_1$ couple the rotational and irrotational flows with density and show us a way of enhancing one at the cost of the other. The irrotational component $\vec{U}_{ir}$ is given by

$$\vec{U}_{ir}(\vec{k}) = -\frac{iR}{k}\dot{\delta}_k + \frac{\text{constant}}{R}.$$ (7.119)

Now, using Equations (7.117) and (7.119), we find the evolution equation for the density contrast as:

$$\frac{\partial^2 \delta_k}{\partial t^2} + \frac{2\dot{R}}{R}\frac{\partial \delta_k}{\partial t} + \left(\frac{c_s^2 k^2}{R^2} - 4\pi G\rho_0\right)\delta_k = 0.$$ (7.120)

We observe that the pressure and gravity terms balance each other at a wavenumber k_J given by

$$k_J^2 = \frac{4\pi G\rho_0 R^2}{c_s^2}. \tag{7.121}$$

The corresponding wavelength λ_J, called the **Jean's Length**, is determined from Equation (7.115) to be:

$$\lambda_J = \left(\frac{\pi c_s^2}{G\rho_0}\right)^{1/2}. \tag{7.122}$$

For $\lambda > \lambda_J$, the gravity term dominates the pressure term and vice-versa.

Problem 7.12: Show that in the absence of expansion, Equation (7.120) describes the sound waves of frequency $\omega = \dfrac{kc_s}{R}$.

In order to see the effect of expansion for $\lambda < \lambda_J$ we assume a solution of the form

$$\delta_k = A(t)e^{-i\alpha(t)}, \tag{7.123}$$

substitute in Equation (7.120) and separate the real and imaginary parts. We find from the imaginary part, that

$$A\ddot{\alpha} + 2\dot{A}\dot{\alpha} + 2A\dot{\alpha}\frac{\dot{R}}{R} = 0,$$

or

$$\frac{\partial}{\partial t}(R^2 A^2 \dot{\alpha}) = 0,$$

so that

$$\dot{\alpha} = \frac{\text{constant}}{R^2 A^2}. \tag{7.124}$$

The real part reads:

$$\ddot{A} + \frac{2\dot{R}}{R}\dot{A} - A\dot{\alpha}^2 + \left(\frac{c_s^2 k^2}{R^2} - 4\pi G\rho_0\right)A = 0. \tag{7.125}$$

We realize that:

$$\frac{\dot{\alpha}}{\alpha} >> \frac{\dot{A}}{A}, \tag{7.126}$$

i.e., the density oscillation period is much smaller than the characteristic expansion time. Neglecting $\ddot{A}$ and $\dot{A}$ in Equation (7.125), we get:

$$A = \frac{\text{constant}}{(c_s^2 k^2 R^2 - 4\pi G\rho_0 R^4)^{1/4}}. \tag{7.127}$$

Therefore, the density contrast is given by:

$$\delta_k = \frac{\text{constant}}{(c_s^2 k^2 R^2 - 4\pi G \rho_0 R^4)^{1/4}} \exp\left[-i \int \dot{\alpha} dt\right], \qquad (7.128)$$

and shows the effect of expansion on the propagation of sound waves for adiabatic perturbations since the sound speed c_s is a function of time. The isothermal conditions could also obtain if there is a source of background radiation which could offset the cooling due to expansion.

Problem 7.13: Show that the temperature T varies as $R^{3(1-\gamma)}$ where γ is the adiabatic index.

For $\lambda \gg \lambda_J$, we find that equation (7.120) supports solutions corresponding to an increase of δ_k with time, i.e., the system is unstable, though not in an exponential way as we would expect from a linearly unstable system. This is due to the presence of expansion. For $\lambda \gg \lambda_J$ we neglect the pressure term and Equation (7.120) becomes:

$$\frac{\partial^2 \delta_k}{\partial t^2} + \frac{2\dot{R}}{R} \frac{\partial \delta_k}{\partial t} - 4\pi G \rho_0 \delta_k = 0. \qquad (7.129)$$

Using the Einstein-de Sitter model for R and ρ_0, this becomes:

$$\frac{\partial^2 \delta_k}{\partial t^2} + \frac{4}{3t} \frac{\partial \delta_k}{\partial t} - \frac{2\delta_k}{3t^2} = 0. \qquad (7.130)$$

The two solutions of Equation (7.130) are:

$$(i) \quad \delta_1 = \text{constant } (t)^{-1} \quad \text{and} \quad (ii) \quad \delta_2 = \text{constant } (t)^{2/3}. \qquad (7.131)$$

Thus, the second solution δ_2, corresponds to a growing density contrast. The growth is not an exponential, but rather a power law in time. If we had accepted the Jean's swindle, we would have found an exponentially growing density perturbation. Thus, the expansion has slowed down the growth rate. The variation of ρ_1 with time is found from

$$\rho_1 = \rho_0 \delta_2 \propto t^{-4/3}. \qquad (7.132)$$

Thus, ρ_1 decreases with time, so, is there really an instability? It is fortunate that the density contrast δ increases with time and we thus have some hope of forming mass condensations. The problem of creation of dense structures in the universe is no where near its solution and is a topic of active research at present. Nevertheless, we could estimate the Jean's length λ_J and the mass contained in sphere of radius λ_J – Jean's mass at any epoch t, in order to have a feel for the numbers involved. The time t is chosen as some fraction of the age of the universe which is taken as a few billion years, Hubble's constant is parameterized as $H = 100h$ km/sec/Mpc where $1 \geq h \geq 1/2$, the mass density ρ_0 is expressed in terms of the density parameter as $\rho_0 = \Omega_\rho \rho_c$ where Ω_ρ could be less than, equal to or greater than unity, the temperature T is a function of t through its R dependence, and finally the mass m appearing in the expression for sound speed is the mass of a hydrogen atom.

Problem 7.14: Estimate the Jean's mass at the epoch when the universe was half its present age.

It can be shown that the Jean's criterion for instability remains unchanged in the presence of uniform rotation $\vec{\Omega}$ and uniform magnetic field $\vec{B}$ acting either singly or jointly (Chandrasekhar 1961) except for waves propagating at right angles to $\vec{\Omega}$ or $\vec{B}$. Perhaps, this is the reason for accepting Jean's swindle. For, the equilibrium motion of a fluid of uniform density ρ due to the gravitational force turns out to be a uniform rotation with velocity $U_\theta = \left(\dfrac{4\pi G\rho}{3}\right)^{1/2} r$. We can go ahead and apply Jean's criterion for instability for a non-expanding medium too, and study the gravitational collapse of, say, interstellar clouds to form stars. The condition for a cloud to become unstable and undergo gravitational collapse is that the gravitational energy E_g must be greater than the sum of thermal E_T and any other energy, for example, those associated with turbulence, magnetic fields and rotation. A cloud of uniform mass density ρ, mass M, radius R_M and temperature T has

$$E_g = \frac{3}{5}\frac{GM^2}{R_M}, \tag{7.133}$$

and

$$E_T = \frac{3}{2}\frac{R_g TM}{\mu_m}, \tag{7.134}$$

where R_g is the gas constant and μ_m the mean molecular weight per particle. Therefore, for collapse:

$$M > \frac{5}{2}\frac{R_g T R_M}{\mu_m G}. \tag{7.135}$$

For a typical diffuse cloud of neutral hydrogen: $T = 100$K, $\rho = 10^{-23}$ gm cm^{-3}, $\mu_m = 1$ and we find that $M > 4 \times 10^3 M_\odot$. Thus, less massive clouds cannot directly undergo gravitational collapse and additional modes of compression must be invoked.

A star is made up of several different layers distinguished, by their opacity to radiation, among other characteristics. If a region is opaque to radiation, it absorbs the radiation and heats up. Due to the increase in temperature, it expands, which leads to cooling. The cooling results in a decrease of opacity, in a reduced absorption of radiation and the region contracts back. And so, the cycle repeats. The star is said to be **Pulsating** (Figure 7.8). The over-all pulsations of a star reveal themselves in the form of the light curve of the star. Definite relations have been inferred to exist between the periods of pulsation and the luminosities of certain classes of stars. The pulsation period is a characteristic of a star and gives us clues about the stellar interior (Aller 1954).

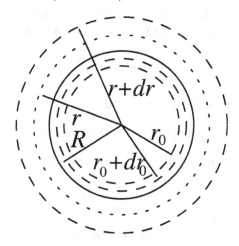

Figure 7.8. A Pulsating Star

The model of a star contains a knowledge of pressure, density and temperature variations which produce the observed luminosity at its surface. For a pulsating star, in addition, we must know these physical parameters as a function of time too. Let $p(r_0, t)$, $T(r_0, t)$ and $\rho(r_0, t)$, be the pressure, the temperature and the density of a layer of the star situated at a distance r_0 from the center ($r_0 = 0$) of the star at a given time t. We would like to know how these quantities vary with the radial coordinate r. Let M_r be the mass within this shell of radius r_0. Let dM_{r_0} be the mass contained in the region between $(r_0 + dr_0)$ and r_0. Due to pulsation, the surfaces at r_0 and $(r_0 + dr_0)$ move to new positions at r and $(r + dr)$. The mass dM_r contained in the region between $(r + dr)$ and r must be equal to dM_{r_0} in the frame moving with the material, so that:

$$dM_r = 4\pi\rho r^2 dr = 4\pi\rho_0 r_0^2 dr_0 = dM_{r_0},$$

or

$$\left(\frac{r}{r_0}\right)^2 \frac{\rho}{\rho_0} \frac{\partial r}{\partial r_0} = 1, \tag{7.136}$$

where the subscript 0 describes the unperturbed quantities. The radial equation of motion is:

$$\rho \frac{\partial^2 r}{\partial t^2} = -\frac{\partial p}{\partial r} - \frac{GM_r}{r^2}\rho, \tag{7.137}$$

where M_r is the mass of the star lying within the radial position r. The energy equation or the equation of state is taken for adiabatic changes and therefore,

$$\frac{p}{p_0} = \left(\frac{\rho}{\rho_0}\right)^{\gamma}. \tag{7.138}$$

The mass conservation law gives $M_r = M_{r0}$. The equilibrium described by the unperturbed quantities is contained in the hydrostatic balance:

$$\frac{dp_0}{dr} = -\frac{GM_{r0}}{r_0^2}\rho_0. \tag{7.139}$$

Let us perturb this equilibrium such that

$$p = p_0(1 + p_1), \quad \rho = \rho_0(1 + \rho_1) \text{ and } r = r_0(1 + r_1). \tag{7.140}$$

Substitution of Equation (7.140) in Equation (7.136) and linearization gives:

$$\rho_1 + 3r_1 + \frac{\partial r_1}{\partial r_0}r_0 = 0. \tag{7.141}$$

Also using Equation (7.136), we determine the pressure gradient:

$$\frac{\partial p}{\partial r} = \frac{\partial p}{\partial r_0}\frac{\partial r_0}{\partial r} = \frac{\partial p}{\partial r_0}\left(\frac{r}{r_0}\right)^2\frac{\rho}{\rho_0}. \tag{7.142}$$

The linearization of Equation (7.137) gives:

$$\frac{\partial^2 r_1}{\partial r_0^2} + \frac{\partial r_1}{\partial r_0}\left(\frac{4}{r_0} - \frac{GM_{r0}}{r_0^2}\frac{\rho_0}{p_0}\right) + r_1\left(\frac{4 - 3\gamma}{\gamma}\right)\frac{GM_{r0}}{r_0^3}\frac{\rho_0}{p_0} - \frac{\rho_0}{\gamma p_0}\frac{\partial^2 r_1}{\partial t^2} = 0. \tag{7.143}$$

Equation (7.143) describes acoustic waves in a stratified medium. We need boundary conditions in order to solve it. The first is that the displacement $r_1 = 0$ at the center $r_0 = 0$. All coefficients in Equation (7.143) remain finite at $r_0 = 0$ except the term $(4/r_0)$ which blows up. Therefore, we must put

$$\frac{\partial r_1}{\partial r_0} = 0 \text{ at } r_0 = 0. \tag{7.144}$$

We should also examine the surface of the star at $r_0 = R$. Since,

$$\frac{\rho}{p} = \frac{\mu m}{R_g T}, \tag{7.145}$$

the coefficient of (ρ_0/p_0) in Equation (7.143) must vanish at the surface where $T \to 0$, so that at $r_0 = R$,

$$-\frac{\partial r_1}{\partial r_0}\frac{GM_{r0}}{r_0^2} + r_1\left(\frac{4-3\gamma}{\gamma}\right)\frac{GM_{r0}}{r_0^3} - \frac{1}{\gamma}\frac{\partial^2 r_1}{\partial t^2} = 0. \qquad (7.146)$$

These circumstances are analogous to a long pipe closed at one end in which a standing wave pattern forms through most of its length. The open end corresponds to the vanishing density at the surface of a star. Let us try a solution of the form

$$r_1(r_0, t) = f(r_0)\cos\omega t, \qquad (7.147)$$

and substitute in Equation (7.143). We get

$$f'' + f'\left(\frac{4}{r_0} - \frac{GM_{r0}\rho_0}{r_0^2 p_0}\right) + f\left[\left(\frac{4-3\gamma}{\gamma}\right)\frac{GM_{r0}}{r_0^3}\frac{\rho_0}{p_0} + \frac{\rho_0\omega^2}{\gamma p_0}\right] = 0. \quad (7.148)$$

This equation can only be solved by numerical integration to determine f and ω. First, $f(r_0)$ is determined for $\omega = 0$ such that it satisfies the boundary conditions. Then, for a trial value of ω, Equation (7.148) is solved by a power series expansion near $r_0 = 0$ and $r_0 = R$ and the solutions are extrapolated from these points outward and inward respectively. The value of ω for which the outward and the inward solutions meet at some point, say half-way, in the star is taken to be the correct value. The longest period represents the fundamental mode of oscillation of the star. It is found that

$$\omega\rho_0^{1/2} = \text{constant.} \qquad (7.149)$$

For all homologous stars, the relation (7.149) must be satisfied. The periods of the stars are observed to be of the order of a day to a few weeks. The constant has a value in the range 0.3 - 0.5. This standing wave theory, although, it accounted well for the period, gave a wrong phase relationship between the light variation and the radial velocity $\left(\dfrac{\partial r_1}{\partial t}\right)$. This flaw was removed by allowing the standing waves to become running waves at the surface of the star. The star is observed to be the brightest and hottest under the maximum expansion and coolest and faintest under the maximum contraction. The relation between the frequency ω and the mass density ρ_0 (Equation 7.149) can be converted into the **Period–Luminosity** relation by using the radial dependences of the luminosity L of a star and mass density ρ_0. We know from the Stefan – Boltzmann law that $L \propto R^2$ and using $\rho_0 = M/R^3$, we find

$$L \propto P^{-4/3} \qquad (7.149')$$

for a constant value of M (Cox 1985). The period – luminosity relation (Equation 7.149′) has been amply ratified by observations of pulsating stars such as Cepheids.

With this rather incomplete but, we trust, instructive account of a few types of instabilities that the astrophysical fluids support, we move on to the subject of fluid turbulence.

7.5. Turbulence in Fluids

We have seen that fluids support waves and instabilities under small departures from their equilibria. During its instability phase, a fluid develops flows on several different spatial scales defined by the range of wavevectors k and on several different time scales defined by the range of frequencies ω determined from the dispersion relations. We could ask two questions: How do the various flow scales evolve with time? and what happens if departures from the equilibria are large? The answer to both these questions can be obtained only if we give up the luxury of linearization and dare to venture into the nonlinear regime.

How do we enter the nonlinear regime? From Equation (7.1), we find that we must retain the inertial term $(\vec{U} \cdot \vec{\nabla})\vec{U}$, which we have been neglecting until now. This term produces coupling among flows on different spatial scales and it influences the dynamics at large flow velocities or at large departures from an equilibrium. So, we can begin with the 'Kindergarten' definition of a **Turbulent Fluid** or more accurately a **Turbulent Flow** as a sheared flow (to produce instabilities) with several interacting flow scales.

Turbulence is predominantly a property of a flow and not as much of a fluid. The molecular properties of fluids in a state of turbulence play less important roles than do their macroscopic characteristics. This is quantitatively expressed in terms of **Reynold's Number** R_Y which is the ratio of the inertial force $\rho(\vec{U} \cdot \vec{\nabla})\vec{U}$ and the viscous force $\nu\nabla^2\vec{U}$. Turbulent flows are characterized by very large values of R_Y. In most of the astrophysical situations, $R_Y \sim 10^{12}$ or larger, essentially due to the large spatial scales of the flows. Turbulent flows are ubiquitous. Flows, oceanic or atmospheric, planetary or stellar, interstellar or galactic, intergalactic or protogalactic, have been fascinating scientists of all generations and genres. However, not all the consequences of the inherently turbulent nature of flows are known, inspite of prolonged studies of them spanning centuries. The novelties in a nonlinear medium, which a fluid essentially is, arise due to the multiplicity of ways in which the interactions among the large number of spatial and temporal scales of motion are specified and systemized.

The presence of turbulence in cosmic objects is best inferred from measurements of widths of atomic and molecular spectral lines. If the spectral

lines have Doppler widths much larger than those due to the thermal motion
of atoms and molecules, they are said to originate in a turbulent medium.
Further, a pure thermal motion gives rise to Gaussian line profile. The observed line profiles are seen to possess broad wings indicating the magnitude
and the scale of the turbulent flow. Whereas the thermal width depends on
the mass of the line emitting atom or molecule, the turbulent widths arising due to the motion of a macroscopic fluid element, consisting of a very
very large number of atoms or molecules are independent of the mass of the
constituent atoms or molecules.

Before, embarking upon a quantitative account of turbulent flows, let us
keep in mind that turbulence is not always randomness. Large scale coherent
flows are seen to develop and sustain themselves for finite durations of time
under specific conditions in an otherwise turbulent flow. Specifically, two-
dimensional flows offer the best examples of creation of ordered structures.
The situation in three-dimensional systems is not yet clear. We shall give a
brief summary of the latest developments in this mysterious facet of flows
towards the end of this chapter.

7.6. General Characteristics of Turbulent Flows

Turbulent flows are described by **Continuum Mechanics** since even the
smallest identifiable scale is much bigger than any molecular scale. The dynamics of the flow is contained in the Euler equations or the Navier-Stokes
equations which are nothing but a statement of the momentum conservation
law (Equation 7.1). Turbulent flows are **Diffusive** in nature. They enhance
the rates of all transport processes resulting in efficient and fast transfer of
mass, momentum and energy. Turbulent flows are **Dissipative**. They lose
energy due to viscous processes and need to be maintained by an external
supply of energy. Turbulent flows are essentially **Rotational**, their **Vorticity** and **Helicity** play decisive roles in their dynamics. However, due to
nonlinearity, it is not possible to arrive at any general features of different
types of flows. Boundary conditions, buoyancy and dissipation determine
the nature of flows in a very specific manner.

The presence of a large number of scales makes it necessary to use
statistical methods and characterize turbulence in terms of correlation functions of velocity and density in a fluid. As we shall learn later on, the
method of dimensional analysis has been put to the best use in discerning
some universal properties of turbulence. The hope is that the solutions of
the Navier-Stokes equations would ratify the predictions of the dimensional
analysis, also known as the **Kolmogorov Approach** after its proposer,
A.N.Kolmogorov. We cannot even pretend to be exhaustive while listing
the general characteristics of turbulent flows, since no one claims to know

them all. Times were, when ignorance of astrophysical conditions lead astrophysicists to suggest higher and higher values of the magnetic field in order to understand a particular piece of observational data. In fact the comment that the magnitude of the needed magnetic field was directly proportional to our ignorance remained popular for quite sometime. We hope that an understanding of astrophysical turbulence will, similarly, diminish our demand of the existence of exotic matter in the making of the universe and its constituents. Let us move on!

7.7. Quantification of Turbulence

The velocity field $\vec{U}(\vec{r}, t)$ in a turbulent flow can take any value in a completely random fashion. Since $\vec{U}(\vec{r}, t)$ is a random function its values must be distributed according to some definite probability laws which could be ascertained from the experimental data of the problem. A knowledge of $\vec{U}(\vec{r}, t)$ at every point $(\vec{r}, t)$ constitutes a realization of the turbulent field. In general there exists a statistical connection among the values of $\vec{U}(\vec{r}, t)$. The probability laws describe this statistical connection and enable us to determine the average values of the various quantities of interest. The laboratory measurements of the velocity components at a fixed space point have been found to follow, very closely, the **Gaussian Probability Density Function**. This result holds at different values of R_γ as well as at different stages of turbulence. The criteria for a good fit to a Gaussian are contained in the **Flatness Factor** defined as:

$$\left\langle U_i^4 \right\rangle \left(\left\langle U_i^2 \right\rangle \right)^{-2},$$

and the **Skewness Factor** defined as:

$$\left\langle U_i^3 \right\rangle \left(\left\langle U_i^2 \right\rangle \right)^{-3/2}.$$

For a perfect Gaussian function, the flatness factor has a value of 3 and the skewness factor is zero. Experiments have confirmed these values within the limits of experimental errors. For spatially homogeneous turbulence, all regions of space are alike. Thus, averaging over a large number of realizations or ensembles is same as averaging over a large region of space for one realization. But, what is measured in a laboratory experiment is the time variation of $\vec{U}(\vec{r}, t)$ at a fixed space point. A time average is then taken.

With the help of the **Ergodic Principle**, according to which, given enough time, all realizations of the turbulent velocity field are realized, we can equate the time average and the probability average — the latter being further equal to the space average for homogeneous turbulence. Mean values of the products of velocity fields and their derivatives form the fabric of a turbulent flow. Out of these functions the most important is the two-point **Velocity Correlation Function**, $R_{ij}(\vec{r}, t)$ defined as (Batchelor 1953; Shore 1992):

$$R_{ij}(\vec{r}, t) = \langle U_i(\vec{x}, t) U_j(\vec{x} + \vec{r}, t) \rangle, \qquad (7.150)$$

where the angular brackets represent the space average. The function $R_{ij}(\vec{r})$ describes correlation between the ith and the jth components of the three dimensional velocity fields separated by a displacement vector $\vec{r}$ at a given instant of time t. For homogeneous turbulence, we see that

$$R_{ij}(\vec{r}) = R_{ij}(-\vec{r}), \qquad (7.151)$$

i.e., R_{ij} is **Reflectionally Symmetric**. We omit writing the time coordinate t with the understanding that all quantities are taken at the same instant of time. We shall also assume that viscous forces have smoothened the velocity field and therefore it is a continuous function of $\vec{x}$, which implies that

$$\lim_{\vec{\delta} \to 0} \left\langle U_i(\vec{x}) U_j(\vec{x} + \vec{r} + \vec{\delta}) \right\rangle = \langle U_i(\vec{x}) U_j(\vec{x} + \vec{r}) \rangle,$$

or

$$\lim_{\vec{\delta} \to 0} R_{ij}(\vec{r} + \vec{\delta}) = R_{ij}(\vec{r}), \qquad (7.152)$$

i.e., the two-point correlation function is a **continuous** function of $\vec{r}$. The necessary and the sufficient condition for $R_{ij}(\vec{r})$ to be the correlation tensor of a continuous and stationary random process is that we can define its **Fourier Transform** $Q_{ij}(\vec{k})$ as:

$$R_{ij}(\vec{r}) = \int Q_{ij}(\vec{k}) e^{i\vec{k} \cdot \vec{r}} d\vec{k}, \qquad (7.153)$$

and

$$Q_{ij}(\vec{k}) = \frac{1}{(2\pi)^3} \int R_{ij}(\vec{r}) e^{-i\vec{k} \cdot \vec{r}} d\vec{r}.$$

Problem 7.15: show that $Q_{ij}(\vec{k}) = Q_{ji}(-\vec{k}) = Q_{ij}^*(\vec{k})$ where the symbol $*$ represents the complex - conjugate.

For zero separation ($|\vec{r}| = 0$), we get

$$R_{ij}(|\vec{r}| = 0) = \langle U_i(\vec{x}) U_j(\vec{x}) \rangle = \int Q_{ij}(\vec{k}) d\vec{k}. \qquad (7.154)$$

Since $R_{ij}(\vec{r})$ has the dimensions of energy per unit mass, the quantity $Q_{ij}(\vec{k})$ is the energy per unit mass per unit volume in the wave vector space and is

known as the **Energy Spectrum Tensor**. For isotropic conditions, therefore, we can define the **Energy Spectrum Function** $E(k)$ as

$$\int_0^\infty E(k)dk = \frac{1}{2}\int \sum_i Q_{ii}(\vec{k})4\pi k^2 dk$$

$$= \frac{1}{2}\sum_i \langle U_i(\vec{x})U_i(\vec{x})\rangle = \frac{3}{2}\langle U^2 \rangle, \qquad (7.155)$$

where U is the mean square velocity on some spatial scale. In anisotropic
conditions, the off diagonal components of $R_{ij}(|\vec{r}| = 0)$ may also exist,
Therefore, $R_{ij}(|\vec{r}| = 0)$ is generally called the **Energy Tensor**. The values
of the wave-vector $\vec{k}$ describes the various flow scales or eddy sizes present
in turbulence and the energy spectrum tensor $Q_{ij}(\vec{k})$ describes the distribution of energy associated with each component of the velocity over various
spatial scales (k^{-1}). The energy tensor and the energy spectrum tensor can
be integrated over angular coordinates and can define average correlation
tensors which are functions only of the magnitude r of the separation vector
$\vec{r}$.

Problem 7.16: Show that the angle averaged velocity correlation tensor
$S_{ij}(r)$ and the angle averaged energy spectrum tensor $\psi_{ij}(k)$ are related as:

$$\psi_{ij}(k) = \frac{2}{\pi}\int_0^\infty S_{ij}(r)kr\sin kr\ dr, \qquad (7.156)$$

where

$$\psi_{ij}(k) = \int Q_{ij}(\vec{k})k^2 d\overline{\Omega_k}$$

and

$$S_{ij} = \int \frac{R_{ij}(\vec{r})}{4\pi}d\overline{\Omega_r}$$

and $\overline{\Omega_k}$ and $\overline{\Omega_r}$ are the solid angles in k and r spaces respectively.

The energy tensor $R_{ij}(\vec{r})$ can be expressed in terms of the general tensors
of the second order as:

$$R_{ij}(\vec{r}) = G(r)\delta_{ij} + F(r)r_i r_j + h(r)\epsilon_{ijk}r_k, \qquad (7.157)$$

where $G(r), F(r)$ and $h(r)$ are arbitrary scalar functions of r^2. The term
with $h(r)$ allows the possibility of reflectionally asymmetric turbulence and
is zero for isotropic turbulence for which R_{ij} must be symmetric in the
indices i and j. If the flows are **Incompressible** then it follows that

$$\sum_i \frac{\partial R_{ij}}{\partial r_i} = \sum_j \frac{\partial R_{ij}}{\partial r_j} = 0. \qquad (7.158)$$

Therefore, from Equations (7.157) and (7.158) we find

$$\frac{G'}{r} + rF' + 4F = 0, \qquad (7.159)$$

where the prime denotes the derivative with respect to r. Equation (7.159) implies that we need only one scalar function F or G to describe isotropic and incompressible turbulence. The actual measurements of the velocity correlation function are done by measuring velocity components, $\vec{U}_{\parallel}$ in the direction of the separation vector $\vec{r}$, and $\vec{U}_{\perp}$, perpendicular to it (Figure 7.9); we, therefore, define the longitudinal velocity correlation function $f(r)$ as:

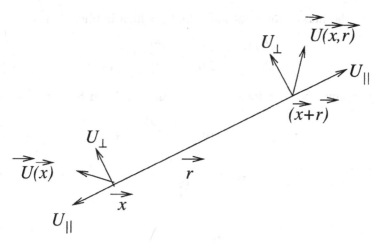

Figure 7.9. Velocities Parallel and Perpendicular to the Separation Vector $\vec{r}$.

$$f(r) = \frac{\left\langle \vec{U}_{\parallel}(\vec{x})\vec{U}_{\parallel}(\vec{x}+\vec{r})\right\rangle}{U^2}, \qquad (7.160)$$

and the transverse velocity correlation function $g(r)$ as:

$$g(r) = \frac{\left\langle \vec{U}_{\perp}(\vec{x})\vec{U}_{\perp}(\vec{x}+\vec{r})\right\rangle}{U^2}, \qquad (7.161)$$

where

$$U^2 = \left\langle \vec{U}_{\parallel}^2(\vec{x})\right\rangle = \left\langle \vec{U}_{\perp}^2(\vec{x})\right\rangle = \frac{1}{3}\sum_i \langle U_i(\vec{x})U_i(\vec{x})\rangle .$$

If we take $\vec{r}$ along the x-axis, say $\vec{r} = (r, 0, 0)$ then R_{xx} is $R_{\parallel}$ and $R_{\perp}$ stands for R_{yy} and R_{zz}. we find

$$R_{\parallel} = r^2 F(r) + G(r) = U^2 f(r), \qquad (7.162)$$

and

$$R_\perp = G(r) = U^2 g(r),$$

or

$$R_{ij} = U^2 \left[\frac{(f-g)}{r^2} r_i r_j + g \delta_{ij} \right], \tag{7.163}$$

and

$$g = f + \frac{r f'}{2}. \tag{7.164}$$

We observe that for zero separation, i.e., for $r = 0$,

$$f(0) = g(0) = 1. \tag{7.165}$$

These are the maximum values of f and g which implies that

$$\begin{aligned}
f'(r)|_{r=0} &= 0, & f''|_{r=0} &< 0, \\
g'(r)|_{r=0} &= 0, & g''|_{r=0} &< 0.
\end{aligned} \tag{7.166}$$

We can expand the functions f and g in the neighbourhood of $r = 0$ as:

$$\begin{aligned}
f(r) &= f(0) + \frac{r^2}{2} f'' \bigg|_{r=0} + \dots \\
&= 1 - \frac{r^2}{2\lambda_T^2}, \tag{7.167}
\end{aligned}$$

and

$$g(r) = 1 - \frac{r^2}{\lambda_T^2},$$

where $f''(r)|_{r=0} \equiv \dfrac{1}{\lambda_T^2}$.

The spatial scale λ_T is known as the **Taylor Microscale** and represents the curvature of the two-point velocity correlation function at zero separation. We shall describe the physical significance of λ_T, a little while later. We expect that the correlation should vanish as $r \to \infty$ so that $f \to 0$. We can define the moments of f and g and find:

$$\int_0^\infty r^m f(r) dr = \frac{2}{1-m} \int_0^\infty r^m g(r) dr. \tag{7.168}$$

For $m = 0$, we get an estimate of the linear spatial scale over which the velocities are correlated. Thus, $L_\parallel$, known as the longitudinal integral scale and $L_\perp$, known as the transverse integral scale are defined as:

$$L_\parallel = \int_0^\infty f(r) dr \quad \text{and} \quad L_\perp = \int_0^\infty g(r) dr. \tag{7.169}$$

Problem 7.17: Prove the equality given in Equation (7.168). Show that $L_\perp = \frac{1}{2}L_\parallel$.

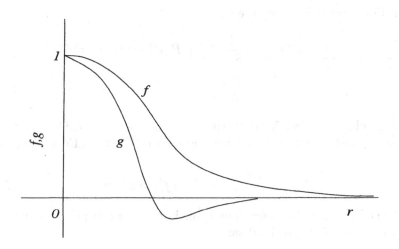

Figure 7.10. Variation of Longitudinal and Transverse Velocity Correlation Functions f and g with Separation r for Isotropic and Incompressible Turbulence.

For $m = 1$,

$$\int_0^\infty rG(r)dr = 0, \tag{7.170}$$

and for $m > 1$, the moments of f and g have opposite signs which suggests that for large r, $f(r) > 0$ but $g(r) < 0$. This suggestion has been ratified by experiments on incompressible turbulence. The approximate variation of f and g with r built on the above considerations is shown in Figure (7.10). Using the properties of the second order tensor, we expressed $R_{ij}(\vec{r})$ by Equation (7.157). Similarly, we express the Fourier transform $Q_{ij}(\vec{k})$ as:

$$Q_{ij}(\vec{k}) = A(k)k_ik_j + B(k)\delta_{ij}, \tag{7.171}$$

and the incompressibility condition requires that

$$B(k) = -k^2A(k),$$

so that

$$Q_{ij}(\vec{k}) = \left(k_ik_j - k^2\delta_{ij}\right)A(k). \tag{7.172}$$

The energy spectrum $E(k)$ is related to (Equation 7.155) $Q_{ii}(\vec{R})$ as:

$$E(k) = \frac{1}{2}\sum_i 4\pi k^2 Q_{ii}(k) = -4\pi k^4 A(k),$$

or

$$Q_{ij}(\vec{k}) = \frac{E(k)}{4\pi k^4}\left(k^2\delta_{ij} - k_ik_j\right).$$ (7.173)

From Equation (7.153), we see that

$$
\begin{aligned}
\sum_i Q_{ii}(\vec{k}) &= \frac{1}{8\pi^3}\sum_i \int R_{ii}(\vec{r})\exp(-i\vec{k}\cdot\vec{r})d\vec{r} \\
&= \frac{E(k)}{2\pi k^2}.
\end{aligned}
$$ (7.174)

We can relate the energy spectrum $E(k)$ and the measurable quantities f and g by substituting for $R_{ii}(\vec{r})$ from Equation (7.163) in Equation (7.174). We get:

$$E(k) = \frac{U^2}{\pi}\int_0^\infty (3f + rf')\,kr\sin kr\,dr.$$ (7.175)

The other important correlation function is the two-point vorticity correlation tensor $T_{ij}(\vec{r})$ defined as:

$$T_{ij}(\vec{r}) = \langle \vec{\omega}_i(\vec{x})\vec{\omega}_j(\vec{x}+\vec{r})\rangle$$ (7.176)

where $\vec{\omega} = \vec{\nabla}\times\vec{U}$ is the vorticity. It is a simple task to express $T_{ij}(\vec{r})$ in terms of $R_{ij}(\vec{r})$. We find, using the incompressibility conditions:

$$T_{ij}(\vec{r}) = \sum_m \left[-\delta_{ij}\nabla^2 R_{mm} + \partial_i\partial_j R_{mm}\right] + \nabla^2 R_{ij},$$ (7.177)

$$T_{ii}(\vec{r}) = \nabla^2 R_{ii},$$ (7.178)

and

$$\sum_i T_{ii}(\vec{r}) = -\frac{15}{\lambda_T^2}U^2 \equiv T_{EN}.$$ (7.179)

We must remember that the result in Equation (7.179) is valid near $r \to 0$. The quantity T_{EN} is the mean square vorticity and is known as the **Enstrophy**. We find the ratio:

$$\lim_{r\to 0}\frac{\sum_i R_{ii}}{\sum_i T_{ii}} = -\frac{\lambda_T^2}{5}$$ (7.180)

i.e., the Taylor microscale λ_T is determined from the ratio of energy to enstrophy in turbulence. This is the scale which contains most of the energy and is generally a function of r.

We can now be a little more specific about a turbulent flow. Instead of describing it as a random velocity field, we can describe it as a random collection of vortices, i.e., the orientation of the vorticity vector $\vec{\omega}$ is a random variable in three dimensions. These vortices may merge to generate larger or smaller vorticity then that existing initially. The decrease of vorticity is associated with dissipation of the associated energy. Thus, enstrophy is a measure of the energy contained in the vortical field.

Problem 7.18: Obtain equations (7.177) – (7.180). Show that the Fourier Transform of $T_{ij}(\vec{r})$ is equal to k^2 times the Fourier Transform of $R_{ij}(\vec{r})$.

It is well known that the properties of inhomogeneous or anisotropic turbulence vary drastically from that of homogeneous or isotropic turbulence. The most important new feature of the anisotropic turbulence is that it is reflectionally asymmetric i.e., its two point velocity correlation tensor $R_{ij}(\vec{r})$ is not invariant under space reflection. This is expressed as

$$R_{ij}(\vec{r}) = -R_{ji}(-\vec{r}). \tag{7.181}$$

One important consequence of reflectional asymmetry is that such a flow possesses what is known as **Helicity**, i.e., it has handedness. The helicity H is defined as:

$$H(\vec{x}, t) = \vec{U}(\vec{x}, t) \cdot \vec{\omega}(\vec{x}, t), \tag{7.182}$$

and represents the projection of the vorticity vector $\vec{\omega}$ along the velocity vector $\vec{U}$.

Problem 7.19: Show that for a two-dimensional flow $U_x(x, y)$ and $U_y(x, y)$, $H = 0$.

Therefore, a turbulent flow with helicity is necessarily three dimensional. We can show that the average value of the helicity in an incompressible and isotropic turbulence vanishes. We write:

$$
\langle H(\vec{r}, t) \rangle = \left\langle \sum_{ijk} \epsilon_{ijk} U_i(\vec{x}, t) \frac{\partial}{\partial r_j} U_k(\vec{x} + \vec{r}, t) \right\rangle,
$$

$$
= \sum_{ijk} \epsilon_{ijk} \frac{\partial}{\partial r_j} R_{ik}(\vec{r}). \tag{7.183}
$$

Substituting for $R_{ik}(\vec{r})$ from Equation (7.163) and carrying out the summation, we get $< H >= 0$. However, if we use the anisotropic part of $R_{ij}(\vec{r})$ from Equation (7.157), we find:

$$
\langle H \rangle = \sum_{i,j,k,l} \epsilon_{ijk} \epsilon_{ilk} \frac{\partial}{\partial r_j} (r_l h(r))
$$

$$
\propto h(r). \tag{7.184}
$$

Thus, the antisymmetric part of the velocity correlation function is a measure of the average helicity. We could also form the helicity correlation function, which can be expressed in terms of the velocity and vorticity correlation functions, but the algebra is lengthy and can be performed only under strong motivation!

7.8. Invariants of a Nonlinear Flow

It is of immense value to know what remains invariant with time in a complex system such as a nonlinear fluid flow. We already know that the mass, the momentum and the energy are conserved. We can use these conservation laws to see if there are other invariants. In fact it is well known that the search for invariants remains an ongoing programme. Let us begin with the momentum conservation for a nonrotating ideal incompressible fluid (Equation 7.1)

$$\left[\frac{\partial \vec{U}}{\partial t} - \vec{U} \times \vec{\omega}\right] = -\vec{\nabla}\left(\frac{p}{\rho} + \frac{U^2}{2}\right). \tag{7.185}$$

Taking a dot product of Equation (7.185) with $\vec{U}$ and integrating over volume, we find:

$$\frac{\partial E}{\partial t} = \frac{\partial}{\partial t}\int \frac{U^2}{2}d\vec{r} = -\int \nabla \cdot \left[\vec{U}\left(\frac{p}{\rho} + \frac{U^2}{2}\right)\right]d\vec{r}$$

$$= -\int \left(\frac{p}{\rho} + \frac{U^2}{2}\right)\vec{U} \cdot d\vec{S}, \tag{7.186}$$

where the volume integral has been converted into the surface integral. If the fluid is surrounded by either a periodic boundary or a rigid boundary so that the component of velocity field along the normal to the surface vanishes on the boundary, the surface integral vanishes and we get the constancy of the total energy E. So this is the first invariant of a 3-dimensional system.

We shall now show that the total helicity

$$H(t) = \int \vec{U} \cdot \vec{\omega} \, d\vec{r}, \tag{7.187}$$

is also an invariant for an ideal incompressible and barotropic fluid under the action of conservative forces. We first derive an equation for the vorticity $\vec{\omega}$ by taking the curl of the momentum equation (7.185), to find:

$$\frac{\partial \vec{\omega}}{\partial t} + \left(\vec{U} \cdot \vec{\nabla}\right)\vec{\omega} = \left(\vec{\omega} \cdot \vec{\nabla}\right)\vec{U}, \tag{7.188}$$

where we have made use of the incompressibility condition $\vec{\nabla} \cdot \vec{U} = 0$. Using Equations (7.185) and (7.188) we find that the time derivative of $h = \vec{U} \cdot \vec{\omega}$ satisfies the relation:

$$\frac{\partial h}{\partial t} + \vec{\nabla} \cdot \left[h\vec{U} + \left(\frac{p}{\rho} - \frac{U^2}{2} \right) \vec{\omega} \right] = 0, \qquad (7.189)$$

which has the standard form of a conservation law if the quantity

$$h\vec{U} + \left(\frac{p}{\rho} - \frac{U^2}{2} \right) \vec{\omega}, \qquad (7.190)$$

is identified as the helicity flux. Integration of Equation (7.189) over volume for the rigid or periodic boundary conditions shows that

$$\frac{\partial H}{\partial t} = 0, \qquad (7.191)$$

or the total helicity is an invariant of the system under the circumstances which ensure constancy of flux of vorticity discussed earlier.

Problem 7.20: Show that for a compressible flow

$$\frac{\partial H}{\partial t} = \frac{\partial}{\partial t} \int \left(\frac{h}{\rho} \right) \rho \, d\vec{r} = \int \rho \left(\frac{\partial}{\partial t} + \vec{U} \cdot \vec{\nabla} \right) \left(\frac{h}{\rho} \right) d\vec{r}$$

and

$$\left(\frac{\partial}{\partial t} + \vec{U} \cdot \vec{\nabla} \right) \left(\frac{h}{\rho} \right) = \frac{1}{\rho} \vec{\nabla} \cdot \left[\vec{\omega} \left(\frac{p}{\rho} - \frac{U^2}{2} \right) \right]. \qquad (7.192)$$

Thus, we find that for compressible flows the total time derivative (D/Dt) of helicity density per unit mass density is zero if the divergence on the right hand side of Equation (7.192) vanishes. This is possible if

$$\vec{\omega} \cdot \vec{\nabla} \left[\frac{p}{\rho} - \frac{U^2}{2} \right] = 0. \qquad (7.193)$$

Under these conditions, the helicity density h and the mass density ρ change at the same rate, or

$$\frac{1}{h} \frac{Dh}{Dt} = \frac{1}{\rho} \frac{D\rho}{Dt}. \qquad (7.194)$$

Now, h is a pseudoscalar and takes positive as well as negative values. If $h \gtrless 0$ and $\frac{Dh}{Dt} \gtrless 0$, we see that $\frac{D\rho}{Dt} > 0$ i.e., an increase of helicity is concomitant with an increase of mass density. Does it have anything to do with formation of helical structures, whether circulating winds or spiraling galaxies? For $h \gtrless 0$ and $\frac{Dh}{Dt} \lessgtr 0$, we find $\frac{D\rho}{Dt} < 0$ i.e., a decrease of helicity is concomitant with evacuation. The role of helicity has not been yet fully explored in the formation of structures in the universe. These ideas are just beginning to draw some attention and the results are awaited ! We shall say more on this topic a while later.

Continuing our discussion of invariants, we now show that **Total Enstrophy EN** defined as:

$$EN = \int \frac{(\vec{\omega})^2}{2} d\vec{r} \tag{7.195}$$

is an invariant of a two-dimensional fluid. Take a dot product of Equation (7.188) with $\vec{\omega}$ and integrate over volume. We get:

$$\frac{\partial}{\partial t} \int \frac{(\vec{\omega} \cdot \vec{\omega})}{2} d\vec{r} = \int \vec{\omega} \cdot \left[(\vec{\omega} \cdot \vec{\nabla})\vec{U} - (\vec{U} \cdot \vec{\nabla})\vec{\omega} \right] d\vec{r}. \tag{7.196}$$

It is easy to see that the first term on the right hand side vanishes for a two-dimensional flow (U_x, U_y) because then $\vec{\omega} = (0, 0, \omega_z)$. The second term can be simplified as:

$$\vec{\omega} \cdot \left(\vec{U} \cdot \vec{\nabla} \right) \vec{\omega} = \vec{U} \cdot \vec{\nabla} \left(\omega^2 / 2 \right) = \vec{\nabla} \cdot \left[\frac{\vec{U}\omega^2}{2} \right] - \frac{\omega^2}{2} \vec{\nabla} \cdot \vec{U}. \tag{7.197}$$

Therefore, for incompressible two dimensional ideal flows with periodic or rigid boundary conditions

$$\frac{\partial}{\partial t} \int \frac{(\vec{\omega})^2}{2} d\vec{r} = 0. \tag{7.198}$$

In three dimensions, the presence of helicity invalidates enstrophy conservation. In fact, it is this term that is responsible for generation or reduction of vorticity.

Problem 7.21: Show that the vorticity Equation (7.188) can be recast as:

$$\frac{\partial \omega_i}{\partial t} + U_j \frac{\partial \omega_i}{\partial x_j} = \omega_j S_{ij} \tag{7.199}$$

where $S_{ij} = \frac{1}{2} \left[\frac{\partial U_i}{\partial x_j} + \frac{\partial U_j}{\partial x_i} \right]$ is the rate of strain.

The term $\omega_j S_{ij}$ represents amplification and rotation of the vorticity vector by the strain rate.

This is as far as we shall go in the discussion of the invariants. We will soon study their consequences.

7.9. Spectral Representation of the Fluid Equations

Turbulent flows consist of flows on several spatial scales. Each scale is known as an eddy. A fully developed turbulent flow contains a wide range of eddy sizes. In analogy to the kinetic theory of gases, where molecular collisions give rise to pressure, a turbulent fluid exerts turbulent pressure caused by the interactions among the eddies. The turbulent energy spectrum, already defined, describes the distribution of energy among eddies of different sizes. This spectral behaviour of a turbulent fluid can be studied by decomposing the turbulent velocity into the Fourier components as:

$$U_i(\vec{r}, t) = \sum_{\vec{k}} U_{ik}(\vec{k}, t) \exp(i\vec{k} \cdot \vec{r}), \tag{7.200}$$

where U_{ik} are the components of the amplitude vector $\vec{U}_k$. Since velocity is a measurable quantity and therefore must be real, this implies

$$U_{ik}(\vec{k}, t) = U_{ik}(-\vec{k}, t) = U_{ik}^*(\vec{k}, t). \tag{7.201}$$

Problem 7.21: Show that the turbulent kinetic energy E_k per unit mass is equal to $\sum_i \frac{1}{2} \left| U_{ik}(\vec{k}, t) \right|^2$.

The Navier-Stokes equation for an incompressible and viscous flow is:

$$\frac{\partial U_i}{\partial t} + \frac{\partial}{\partial r_l}(U_l U_i) = -\frac{\partial P}{\partial x_i} + \nu \nabla^2 U_i, \tag{7.202}$$

where for constant ρ, we have written for the pressure $p = \rho P$. Let us substitute Equation (7.200) in Equation (7.202) as well as use

$$P = \sum_k P_k(\vec{k}, t) \exp(i\vec{k} \cdot \vec{r}), \tag{7.203}$$

to get

$$\left(\frac{\partial}{\partial t} + \nu k^2\right) U_i(\vec{k}) = -ik_i P_k - \sum_{l,k''} i\left(k_l - k_l''\right) U_i\left(\vec{k} - \vec{k}''\right) U_l(\vec{k}''). \tag{7.204}$$

Multiply equation (7.204) by k_i and sum over i. The use of the incompressibility condition $\sum_i k_i U_{ik} = 0$ then gives

$$P_k = -\sum_{l,j,k''} \frac{k_j}{k^2}\left(k_l - k_l''\right) U_j\left(\vec{k} - \vec{k}''\right) U_l(\vec{k}''). \tag{7.205}$$

Substitution for P_k in Equation (7.204), and a couple of simple manipulations, get us to:

$$\left(\frac{\partial}{\partial t} + \nu k^2\right) U_i(\vec{k}) = -i \sum_{l,j,k''} k_l Q_{ij} U_l(\vec{k}'') U_j(\vec{k} - \vec{k}''), \qquad (7.206)$$

where $Q_{ij} = \delta_{ij} - \dfrac{k_i k_j}{k^2}$.

We have now been able to express pressure term in terms of velocities. This has been possible due to the incompressibility condition. We can, now, derive the energy equation by multiplying Equation (7.206) by $U_i(\vec{k})$ and summing over i. we find:

$$\frac{\partial E_k}{\partial t} = -2\nu k^2 E_k + \sum_{i,j,l,k'} \Lambda_{ijl}\left(\vec{k}, \vec{k}'\right), \qquad (7.207)$$

where

$$E_k = \sum_i \frac{1}{2} U_i(\vec{k}) U_i(\vec{k}), \qquad (7.208)$$

has already been identified as the turbulent energy spectrum and

$$\Lambda_{ijl}(\vec{k}, \vec{k}') = -i Q_{ij}(\vec{k}) U_i(\vec{k}) k_l U_l(\vec{k}') U_j(\vec{k} - \vec{k}'). \qquad (7.209)$$

Equation (7.207) describes the rate of change of energy of an eddy of wavevector $\vec{k}$. The first term on the right hand side is the viscous dissipation of E_k. The second term (Λ) represents the net gain in E_k due to interaction of the eddy k with all other eddies of wave vector k'. For steady nonviscous flows, $\Lambda = 0$. For steady and viscous flows:

$$\sum \Lambda = 2\nu k^2 E_k. \qquad (7.210)$$

We notice that $\Lambda_{ijl}(\vec{k}, \vec{k}')$ is antisymmetric in $\vec{k}$ and $\vec{k}'$ i.e.,

$$\Lambda_{ijl}(\vec{k}, \vec{k}') = -\Lambda_{ijl}(\vec{k}', \vec{k}). \qquad (7.211)$$

We also realize that Q_{ij} is either zero or has a negative value. Therefore, it can be shown that:

$$\Lambda > 0 \text{ for } \vec{k} > \vec{k}' \text{ and } \Lambda < 0 \text{ for } \vec{k} < \vec{k}'.$$

We rewrite Equation (7.207), splitting Λ into its positive part Λ_1 and negative part Λ_2 as:

$$\frac{\partial E_k}{\partial t} = \sum_{\substack{\vec{k} > \vec{k}' \\ \text{or } L < L'}} \Lambda_1(\vec{k}, \vec{k}') - \sum_{\substack{\vec{k} < \vec{k}' \\ \text{or } L > L'}} \Lambda_2(\vec{k}', \vec{k}) - 2\nu k^2 E_k. \qquad (7.212)$$

Thus, we see that an eddy of size $L = 2\pi/k$ gains energy from eddies of larger sizes ($L' > L$) through Λ_1 and loses energy to eddies of smaller sizes ($L' < L$) through Λ_2. The viscous term dominates at small spatial scales. it could happen that for some large (small) value of $k(L)$, the gains overwhelm the losses i.e., $\Lambda_1 \gg \Lambda_2$. In this case a steady state obtains when the gain of energy is equal to viscous loss or

$$\sum_{\vec{k} > \vec{k}'} \Lambda_1(\vec{k}, \vec{k}') = 2\nu k^2 E_k. \tag{7.213}$$

For large enough volume V, the summation over $\vec{k}'$ can be converted into an integral. For isotropic conditions, we write:

$$E_k = |U_k|^2 = kF_k,$$

where F_k is the energy per unit gram per unit wave number k. Equation (7.212) becomes

$$\left(\frac{\partial}{\partial t} + 2\nu k^2\right) F_k = \frac{1}{k^3} \int Y(k, k')dk', \tag{7.214}$$

where $Y(k, k') = (4\pi k^2)\dfrac{V}{(2\pi)^3}4\pi k'^2\Lambda(k, k')$.

In order to determine the spectral energy distribution function F_k, we must know $Y(k, k')$. Several attempts have been made to arrive at an expression for $Y(k, k')$. The most successful is the one by Heisenberg (1948), which is based on the concept of **Turbulent Viscosity**. It is assumed that $Y \propto F$ and Y is expressed as:

$$\begin{aligned} Y(k, k') &= h_k F_{k'} k'^5 \quad \text{for } k > k' \\ &= -h_{k'} F_k k^5 \quad \text{for } k < k', \end{aligned} \tag{7.215}$$

where h_k, determined from dimensional considerations, is given by

$$h_k = 2Ak^{-3/2}F_k^{1/2}, \tag{7.216}$$

and A is a constant. The energy equation (7.214) describing the total energy loss of large eddies with wavevectors, say in the range $k = 0$ to $k = k_1$ due to interaction with small eddies in the range $k' = k_1$ to $k' = \infty$, on substitution for Y, becomes:

$$\int_0^{k_1} \frac{\partial F_k}{\partial t} dk = -2\nu \int_0^{k_1} F_k k^2 dk - 2A \int_0^{k_1} F_k k^2 dk \int_{k_1}^{\infty} \frac{F_{k'}^{1/2}}{k'^{3/2}} dk'. \tag{7.217}$$

An effective eddy viscosity ν_T is defined as

$$\nu_T = A \int_{k_1}^{\infty} \frac{F_{k'}^{1/2}}{k'^{3/2}} dk', \tag{7.218}$$

using which equation (7.217) takes the form:

$$\int_0^{k_1} \frac{\partial F_k}{\partial t} dk = -2(\nu + \nu_T) \int_0^{k_1} F_k k^2 dk. \tag{7.219}$$

Thus the energy transfer or cascade from large scales to small scales has been represented as a viscous loss. We may check that if $F_k \propto k^{-n}$, Equation (7.218) gives $\nu_T \propto (U_k/k)$, which is correct dimensionally.

Problem 7.22: Show that the effective, also called, the turbulent eddy viscosity $\nu_T \simeq \nu R_Y$. Thus turbulent diffusion is enhanced by the factor R_Y.

For stationary conditions, the left hand side of Equation (7.219) vanishes. Heisenberg (1948) and Chandrasekhar (1949) have shown us the way to obtain the stationary energy spectrum F_k, by solving equation (7.219). We give, some key steps here. First, differentiate Equation (7.219) with respect to k to get:

$$\frac{\nu}{A} + \int_k^{\infty} \frac{F_k^{1/2} dk}{k^{3/2}} = \frac{1}{k^2 [k^3 F_k]^{1/2}} \int_0^k F_k k^2 dk. \tag{7.220}$$

We define

$$\alpha = k^3 F_k \text{ and } y = \int_0^k F_k k^2 dk, \tag{7.221}$$

and find:

$$\frac{dk}{dy} = \frac{k}{\alpha}, \tag{7.222}$$

or

$$\log k = \text{constant} + \int_0^y \frac{dy}{\alpha}. \tag{7.223}$$

Using Equations (7.221) and (7.222), Equation (7.220) takes the form:

$$\frac{\nu}{A} + \int_y^{\infty} \frac{dy}{\sqrt{\alpha}k^2} - \frac{y}{\sqrt{\alpha}k^2} = 0. \tag{7.224}$$

Now, differentiating Equation (7.224) with respect to y and using (7.222), we get,

$$\frac{d\alpha}{dy} - \frac{4}{y}\alpha + 4 = 0. \tag{7.225}$$

This equation can be easily solved by the series method. The solution is:

$$\alpha = \frac{4}{3}y\left[1 - ay^3\right],$$
(7.226)

where a is a constant. Substituting for α into Equation (7.223) and performing the integration, we get:

$$k = b\left(\frac{y^3}{1 - ay^3}\right)^{1/4}.$$
(7.227)

Using equations (7.226) and (7.227), we find the energy spectrum F_k to be:

$$F_k = F_0\left(\frac{k_0}{k}\right)^{5/3}\left[1 + \left(\frac{k}{k_s}\right)^4\right]^{4/3},$$
(7.228)

where $F_0 = \frac{4}{3}(b)^{-\frac{4}{3}}(k_0)^{-\frac{5}{3}}$ and $k_s = b(a)^{-\frac{1}{4}}$.

Problem 7.23: Using Equation (7.220), show that the constants a and b are related to ν and A as:

$$\frac{\nu}{A} = \frac{\sqrt{3}}{2}\frac{a^{1/2}}{b^2} \quad \text{and} \quad k_s = \frac{1}{\sqrt{b}}\left(\frac{\sqrt{3}A}{2\nu}\right)^{3/4}.$$
(7.229)

We observe that when $\nu \to 0$ or the Reynolds number becomes infinitely large, $k_s \to \infty$ and the energy spectrum $F_k \propto k^{-\frac{5}{3}}$. So, in equilibrium, this is how the energy is distributed among the various wavevectors k. For a finite value of ν

$$F_k \propto k^{-7} \quad \text{for} k \gg k_s.$$
(7.230)

Thus, $k = k_s$ represents the scale at which the rate of viscous dissipation becomes equal to rate of transfer of energy from large scales to small scales. A complete account of the energy loss due to nonlinear eddy interactions as well as viscous losses can be found in Chandrasekhar (1949). We, now, go on to give another, a rather simple and intuitive way of arriving at the form of the energy spectra for two– and three –dimensional turbulent flows.

7.10. The Kolmogorov–Oboukhov Way

The celebrated energy spectrum $F_k \propto k^{-5/3}$ was first derived by A. N. Kolmogorov and A.M. Oboukhov in 1941. They used an intuitive approach based on the method of dimensional analysis. A flow with a large value of the Reynolds number R_Y is said to be fully turbulent. The energy contained in these large number of eddies is ultimately converted into heat due to viscous dissipation. This, we observe, every day, when we stop stirring our cup of tea! The eddies, initially nearly the size of the cup become smaller and smaller and finally disappear. Let L be the largest scale, often equal to the dimensions of the container, also known as the fundamental eddy size, and (ΔU) be the magnitude of the variation of the mean turbulent velocity over the distance L. Then, from the mass density ρ, the eddy size L and ΔU, we can construct three quantities important for describing turbulence. The quantity $\rho(\Delta U)^2$ has the dimensions of pressure, therefore, the variation of pressure Δp over the scale L can be written as

$$\Delta p = \rho \left(\Delta U \right)^2 . \tag{7.231}$$

Let ϵ be the mean energy input into an eddy of size d per unit time per unit gram. In the stationary state this is also equal to the rate of loss of energy of this eddy due to viscous dissipation. We see that the quantity $(\Delta U)^3 L^{-1}$ has the right dimensions to represent ϵ. Therefore, we write:

$$\epsilon = \frac{(\Delta U)^3}{L} \tag{7.232}$$

The quantity $(L\Delta U)$ has the dimensions of viscosity and we write the turbulent viscosity ν_T as:

$$\nu_T = L\Delta U = \nu R_Y . \tag{7.233}$$

This shows that turbulent viscosity is large at large scales, contrary to shear viscosity which is dominant at small scales. We further note that the quantities (ν^3/ϵ) and $(\nu\epsilon)$ have, respectively, the dimensions of the fourth power of length and velocity.

We, now consider eddies with sizes d much smaller than the fundamental scale but much larger than the **Internal Scale** d_0, which is the scale at which the value of the Reynolds number becomes unity. At this scale the viscous dissipation begins to compete with nonlinear transfer or cascade due to eddy interactions. Thus, the range of scales $L >> d >> d_0$ is known as the **Inertial Range** which lies in the middle, far separated both from source (L) and the sink (d_0) scales. The turbulence in the inertial range is called the **Local Turbulence** and qualifies to be isotropic, so that the turbulent flow properties do not depend upon the mean velocity direction. Since in the inertial range, viscous forces are negligible, the cascade of energy from large scales to small scales occurs through the inertial term $\Lambda(kk')$. Therefore the turbulent properties in the inertial range may depend only on $(\rho, L, \Delta U, \epsilon, d)$. From dimensional arguments, Kolmogorov and Oboukhov proposed that (Kolmogrov 1941)

$$\Delta U_d \simeq (\epsilon d)^{1/3} , \tag{7.234}$$

and interpreted (ΔU_d) to be the velocity of an eddy of size d. The variation in velocity ΔU_τ, at a fixed space-point over a time interval τ is expressed as:

$$\Delta U_\tau \simeq (\epsilon l)^{1/3} = (\epsilon U \tau)^{1/3}, \tag{7.235}$$

where $l = U\tau$ is the distance traveled by the turbulent region if the entire fluid has a mean velocity U, and $l << L$ in order to avoid the effect of boundaries.

Kolmogorov and Oboukhov proposed **Two Similarity Principles** in order to derive the energy spectrum F_k. The **First Principle** states that the statistical distribution of energy in the eddies can depend only upon the rate of energy input per unit gram ϵ and on the kinematical viscosity ν.

Now, F_k has the dimensions of [(velocity)^2 (length)]. Therefore F_k can be expressed as:

$$F_k = (\nu\epsilon)^{1/2} \left(\nu^3/\epsilon\right)^{1/4} f_k \left(k\left(\nu^3/\epsilon\right)^{1/4}\right), \tag{7.236}$$

where f_k is a function of the dimensionless quantity $k(\nu^3/\epsilon)^{1/4}$. The **Second Principle** states that in the inertial range, the energy distribution should be such that as $R_Y \to \infty$, F_k must be independent of ν, or as $R_Y \to \infty$, $f_k \propto \nu^{-5/4}$. Since f_k is dimensionless, we must have

$$F_k \propto \nu^{5/4}\epsilon^{1/4} \left(\frac{k\nu^{3/4}}{\epsilon^{1/4}}\right)^n, \tag{7.237}$$

and n should be determined from the requirement that F_k must be independent of ν. We find $n = -5/3$ and

$$F_k = \text{constant } \epsilon^{2/3} k^{-5/3}. \tag{7.238}$$

This is the **Kolmogorov Energy Spectrum** often called the $k^{-5/3}$ universal spectrum, independent of the viscosity ν. The energy content of the eddy is proportional to its size. The spectrum (7.238) is identical to that derived by Heisenberg (Equation (7.228) in the limit $k_s \to \infty$) and has enjoyed a good experimental support.

Having learnt the thoughts behind the Kolmogorov spectrum, we now show that it can be derived very simply, again using dimensional arguments, as:

$$(kU_k) U_k^2 = (kU_k) kF_k \simeq \epsilon,$$

or

$$F_k \simeq \epsilon^{2/3} k^{-5/3}. \tag{7.239}$$

The constant in Equation (7.237) has been determined experimentally to be of the order of unity. The energy per unit gram contained in an eddy of size $d \simeq \frac{1}{k}$ is

$$U_d^2 = \epsilon^{2/3} d^{2/3}. \tag{7.240}$$

Thus, in isotropic turbulence in the inertial range, the velocity U_d scales as $\sqrt[3]{d}$. The energy cascades from large eddies to small eddies. This phenomenon is known as the **Direct Cascade** in contrast to the transfer of energy from small eddies to large eddies, known as the **Inverse Cascade**. Does the inverse cascade really take place?

7.11. Two-Dimensional Turbulent Flows

We have seen that there are two invariants in a two-dimensional turbulent flow, viz., the total energy and the enstrophy. Are there two inertial ranges corresponding to these two invariants? The energy spectrum in the inertial range is still given by Equation (7.239). We can carry out the dimensional treatment for enstrophy too. Treating enstrophy on par with energy, we demand that:

$$(kU_k)\left(k^2 U_k^2\right) = k^{9/2} F_k^{3/2} = \epsilon'$$

or

$$F_k = \left(\epsilon'\right)^{2/3} k^{-3} \tag{7.241}$$

where in analogy with ϵ, ϵ' is the enstrophy injection rate. So, we, now, have two energy spectra, one varying as $k^{-5/3}$ and the other as k^{-3}. And what is the direction of energy cascade? It has been confirmed by more ways than one (Hasegawa 1978) that the energy cascades from small eddies to large eddies following the $k^{-5/3}$ law and enstrophy cascades from large scales to small scales following the k^{-3} law. If we inject energy and enstrophy in a two-dimensional fluid at a given wavevector k_e, then it is found that the k^{-3} law operates in the region $k > k_e$ and the $k^{-5/3}$ law operates in the region $k < k_e$ (Figure 7.11). This implies that energy cascades towards large spatial scales ($k < k_e$). Thus, the phenomenon of Inverse cascade has been established for a two-dimensional turbulent flow. The flow of energy from small scales to large scales is seen as order arising from disorder. Does it mean that the inverse cascade is accompanied by a decrease of entropy? We must remember that this decrease of entropy occurs only in a part of the system; in the inertial range, and that the energy and enstrophy have to be injected in the system at a constant rate. Therefore, we are not violating the laws of thermodynamics as we are not dealing with equilibrium systems!

The real world is, however, a three dimensional system. Can an Inverse Cascade of energy occur in a 3-D system?

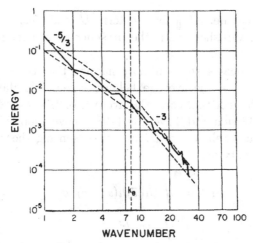

Figure 7.11. Energy Spectrum of a 2–D Turbulent Fluid

7.12. Inverse Cascade in a 3–D System

We learn from the 2-D example that the inverse cascade is perhaps an outcome of the existence of the two invariants as well as the anisotropy. We have shown earlier that energy and helicity are the two invariants of a 3-D turbulent flow and such a system is by definition anisotropic. Therefore, we do expect the inverse cascade of energy to take place in a 3-D anisotropic turbulent flow. The energy spectrum corresponding to the helicity invariant can be derived from

$$\left(kU_k\right)\left(kU_k^2\right) = k^{7/2}F_k^{3/2} = \epsilon'', \tag{7.242}$$

to be

$$F_k = \left(\epsilon''\right)^{2/3}k^{-7/3},$$

where ϵ'' is the helicity injection rate. Again, based on our experience with a 2-D system, we may conjecture that like enstrophy, helicity cascades from large scales to small scales and the energy does the reverse. However an inverse cascade of energy in 3-D systems has not yet been studied enough to unequivocally establish it or otherwise. Besides, the mechanism of inverse cascade in 2-D and 3-D must be different, as helicity vanishes for a 2-D flow. Nevertheless, there is an attractive feature of a helically turbulent flow which might facilitate an inverse cascade of energy. We discussed earlier on how the inertial term $(\vec{U} \cdot \vec{\nabla})\vec{U}$ brings about the direct cascade of energy in an isotropic turbulent flow. We can express

$$\left(\vec{U} \cdot \vec{\nabla}\right)\vec{U} = \vec{U} \times \vec{\omega} - \frac{\vec{\nabla}U^2}{2}. \tag{7.243}$$

If the flow is helical i.e., $\vec{U} \parallel \vec{\omega}$, the term $\vec{U} \times \vec{\omega}$, responsible for direct cascade of energy, vanishes. Thus the presence of helicity retards the flow of energy to small scales.

In an astrophysical system, it may happen that the system possesses helical fluctuations but the average helicity is zero (Moffat and Tsinober 1992). Under such circumstances, the average of the square of the helicity need not vanish. Levich and Tzvetkov (1985) have used this realization to define the statistical helicity invariant I representing the conserved mean square helicity density as:

$$
\begin{aligned}
I &= \int \langle h(\vec{x}) h(\vec{x} + \vec{r}) \rangle \, d\vec{r}, \\
&= \int I(r) d\vec{r}.
\end{aligned}
\tag{7.244}
$$

For a quasi-normal distribution of helicities, the invariant I, essentially from dimensional arguments, can be expressed in terms of the energy spectrum function F_k as:

$$
I = I_0 \int F_k^2 dk = \int I(k) dk,
\tag{7.245}
$$

where I_0 is a constant.

Problem 7.24: Using dimensional analysis, show that equation (7.245) follows from Equation (7.244).

Again, assuming that I cascades at a constant rate, we have

$$
(kU_k)\left(kF_k^2\right) = \epsilon'',
\tag{7.246}
$$

or

$$
F_k = \left(\epsilon''\right)^{2/5} k^{-1},
$$

where ϵ'' is the average mean square helicity density exchange rate among the scales. The total energy is obtained by integrating over k. Thus, the $k^{-5/3}$ spectrum, when integrated over k, yields $k^{-2/3}$, which translates as $L^{2/3}$ in the real space. The spectrum in equation (7.246) corresponds to a logarithmic dependence on the spatial scale L. Again due to stronger dependence on k of $I(k)$, proportional to $(k^{-5/3})^2$, compared to the $k^{-5/3}$ dependence of energy, it is speculated that I dominates at large spatial scales (small k) and the energy at small spatial scales. This is taken to imply that I cascades to large spatial scales. As I cascades to large spatial scales, the helicity-helicity correlation scale increases, but due to the logarithmic dependence, the increase in energy is very small. But, the increase of correlation length implies that the velocity and the vorticity become more and more aligned, as a consequence of which the nonlinear term $(\vec{U} \cdot \vec{\nabla})\vec{U}$ decreases. This retards the flow of energy to small spatial scales. How long will this continue? This cannot go on indefinitely, especially if the medium is restricted in one direction, say vertical, by gravity or buoyancy as is true of atmospheres of celestial objects, may it be a planet or a star. Under such circumstances, the helicity-helicity correlation reaches a maximum constant value in the vertical direction, but continues to increase in the horizontal plane, and the system becomes more and more anisotropic. From a 3-D isotropic system, it has become a quasi 2-D system, with horizontal scale much larger than the vertical scale, and the vertical velocity U_z much smaller than the horizontal velocities (U_x, U_y). Further the vertical and the horizontal motions decouple, and U_z becomes independent of z, leading to

$$\omega_{x,y} = \left(\vec{\nabla} \times \vec{U}\right)_{x,y} = 0. \qquad (7.247)$$

The invariant I acquires the form:

$$
\begin{aligned}
I &= \int \left\langle (U_z \omega_z)^2 \right\rangle dx\,dy\,dz, \\
&= L_z \left\langle U_z^2 \right\rangle \int \omega_z^2 dx\,dy, \\
&= L_z \left\langle U_z^2 \right\rangle \left(k^2 U_k^2\right) k^{-2}, \\
&\propto U_k^2,
\end{aligned}
$$

or

$$I(k) \propto k^{-5/3}, \qquad (7.248)$$

using dimensional analysis.

Thus, in this quasi-2D regime $I(k)$ and $E(k)$ become identical and it is expected that a large fraction of energy is now transferred to large spatial scales. This stage continues until, the coriolis and centrifugal forces, if present, begin to compete with the nonlinear inertial force. The spatial scale L_c at which this happens can be determined from the balance:

$$\left(\vec{U} \cdot \vec{\nabla}\right)\vec{U} = 2\left(\vec{U} \times \vec{\Omega}_0\right) - \vec{\Omega}_0 \times (\Omega_0 \times \vec{r}). \qquad (7.249)$$

Again, dimensionally speaking, we find,

$$L_c \simeq \frac{U}{\Omega_0}, \qquad (7.250)$$

where Ω_0 is the angular speed of the rotating turbulent fluid. Since the coriolis force usually begins to exert its influence on large spatial scales, we expect that large scale $\sim L_c$ size helical eddies are formed, given a sufficient supply of energy. Beyond these scales, the system may begin to behave more like a 2-D system and observe enstrophy conservation. We may consider scales $L > L_c$ as a source of vorticity injection into the system which may now exhibit the two spectral (k^{-3} and $k^{-5/3}$) branches of the 2-D system.

The complete energy spectrum of a helically turbulent medium derived on the basis of dimensional arguments given above, is shown in Figure (7.12). Beginning at the smallest scales, the energy cascades from large scales to small scales with a $L^{2/3}$ law. The $\ln L$ branch corresponds to a very slow inverse cascade of energy, corresponding to the invariant I. The first break in the spectrum at $L = L_z$ signals the onset of anisotropy. The magnitude of L_z depends upon the magnitude of buoyancy forces. Further on for $L > L_z$, a quasi-2D regime with energy spectrum following a $L^{2/3}$ variation represents the predominant action of the inverse cascade of energy until the break at $L = L_c$. Beyond L_c, the system is more akin to a 2-D system. In this region enstrophy decays from large to small scales. The energy can continue to cascade from small scales to large scales, provided enough energy is available in the system. Otherwise, the large scale structure begins to decay. Thus, we see that the inverse cascade occurs in the intermediate range of scales. The smallest and the largest scales undergo direct cascade. In this picture, the energy in large structures has been inverse cascaded from small structures. If it is so, then the energy per unit gram $E(L)$ in a large scale L should not exceed that in the small scale l. This should be true for the spectral region where the phenomenon of inverse cascade takes place. Thus, considering the quasi-2D region it follows that

$$E(L) = \left[\frac{E_0(l)}{\tau}\right]^{2/3} L^{2/3} \leq E_0(l). \tag{7.251}$$

Here, $E_0(l)$ is the energy per unit gram in the small scale l, and τ is the time for which $E_0(l)$ is available; $E_0(l)/\tau$ is the energy injection rate ϵ. Thus, given $E_0(l)$ and L we can estimate the duration τ for which the source $E_0(l)$ must be kept on.

The energy spectrum shown in Figure (7.12) and Equation (7.251) have found their preliminary applications in rather disparate situations. Levich and Tzvetkov (1985) have used the ideas of inverse cascade of energy in a helically turbulent medium to account for cloud formation and large scale flows in the earth's atmosphere. Krishan (1991) found that the observed energy spectrum of the cellular velocity fields on the solar atmosphere with a $k^{-5/3}$ branch joining on to a $k^{-0.7}$ variation towards small k (Figure 7.13) could be accounted well by the $k^{-5/3}$ and k^{-1} branches of the theoretically predicted energy spectrum (Figure 7.12). The corresponding spatial scales

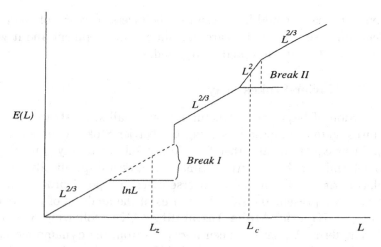

Figure 7.12. Energy Spectrum of a Helically Turbulent Fluid, Break *I* is due to Anisotropy and Break *II* is due to Coriolis Force.

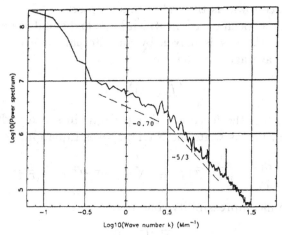

Figure 7.13. Power Spectrum of the Solar Photospheric Motions (Zahn 1987).

of granulation and supergranulation patterns along with their lifetimes were also estimated using Equation (7.251).

On galactic scales, the issue of the flat nature of the rotation curves of galaxies has also been addressed in the light of the turbulent energy spectrum. Prabhu and Krishan (1994) have tried to model the rotation curves combining keplerian and turbulent motions, specifically the $k^{-5/3}$ and k^{-1} branches of the turbulent energy spectrum. Krishan and Sivaram (1991) and Krishan (1996a) suggested that the whole hierarchy of mass distribution in the form of galaxies, clusters of galaxies and superclusters of galaxies, perhaps, could be a result of the inverse cascade of energy in a turbulent universe. These ideas are in a state of development and it will be a while before they are accepted or rejected.

7.13. The Navier-Stokes Way

The formation of large scale structures from small scale structures is also being investigated by actually solving the Navier-Stokes equations. These are nonlinear equations and therefore do not submit to any general procedure of determining their solutions. Since, here, we are specifically interested in exploring the possibility of an inverse cascade type of process in a turbulent flow, we present the salient features of the analysis of the so–called **Anisotropic Kinetic Alpha Instability** (AKA) following Sulem et al (1989). The term 'Alpha' has been borrowed from the dynamo mechanism of generation of large scale magnetic field from small scale magnetic field in MHD turbulent flows. The magnetic alpha effect enjoys a longer history than the AKA effect. We shall discuss some aspects of magnetohydrodynamic turbulence towards the end of this chapter.

We consider a specific case in which large and small scale flows coexist. The small scale (l) flow is driven by an external force density $\vec{f}_E$ whose space and time average vanishes i.e.,

$$\left\langle \vec{f}_E(\vec{r}, t) \right\rangle = 0. \tag{7.252}$$

This can happen if the force is either periodic in space and time or completely random. Let us write the Navier-Stokes Equation:

$$\frac{\partial \vec{U}}{\partial t} + \left(\vec{U} \cdot \vec{\nabla} \right) \vec{U} = -\vec{\nabla} h + \nu \nabla^2 \vec{U} + \vec{f}_E(\vec{r}, t), \tag{7.253}$$

along with the incompressibility condition:

$$\vec{\nabla} \cdot \vec{U} = 0. \tag{7.254}$$

Let us write $\vec{U}$ as the sum of the small scale velocity $\vec{V}$ and the large scale velocity $\vec{W}$:

$$\vec{U} = \vec{V} + \vec{W}, \tag{7.255}$$

where

$$\left\langle \vec{V} \right\rangle = 0, \ \left\langle \vec{U} \right\rangle = \vec{W}, \ \vec{\nabla} \cdot \vec{V} = 0 \text{ and } \vec{\nabla} \cdot \vec{W} = 0.$$

Substitute equation (7.255) in Equation (7.253) and take the average (space-time or ensemble) to get:

$$\frac{\partial \vec{W}}{\partial t} + \left\langle \left(\vec{V} \cdot \vec{\nabla} \right) \vec{V} \right\rangle + \left(\vec{W} \cdot \vec{\nabla} \right) \vec{W} = -\vec{\nabla} h_w + \nu \nabla^2 \vec{W}, \tag{7.256}$$

where we have used $h = h_w + h_V$ with $\langle h_V \rangle = 0$. We must appreciate that though $\langle \vec{V} \rangle = 0, \left\langle (\vec{V} \cdot \vec{\nabla})\vec{V} \right\rangle$ need not vanish. Equation (7.256) governs the large scale flow $\vec{W}$ and

$$\left\langle \left(\vec{V} \cdot \vec{\nabla} \right) \vec{V} \right\rangle_i = \sum_j \partial_j \langle V_i V_j \rangle = \sum_j \partial_j R_{ij}, \qquad (7.257)$$

acts as a driving term. It represents the force due to stresses R_{ij} associated with small scale flows. These are known as the **Reynolds Stresses** (it is nothing but the two–point velocity correlation function at zero separation). We can find the equation governing the small scale flow by subtracting Equation (7.256) from Equation (7.253) as:

$$\frac{\partial \vec{V}}{\partial t} + \left(\vec{W} \cdot \vec{\nabla} \right) \vec{V} + \left(\vec{V} \cdot \vec{\nabla} \right) \vec{V} \; + \; \left\langle \left(\vec{V} \cdot \vec{\nabla} \right) \vec{V} \right\rangle + \left(\vec{V} \cdot \vec{\nabla} \right) \vec{W} =$$
$$-\vec{\nabla} h_V + \nu \nabla^2 \vec{V} + \vec{f}_E. \qquad (7.258)$$

We note that the Reynolds stresses are a function of the large scale velocity $\vec{W}$. In order to understand the interrelationship between $\vec{V}$ and $\vec{W}$, we simplify Equation (7.258) by linearizing it in $\vec{V}$ and assuming $\vec{W}$ to be independent of space and time on the scale of variation of $\vec{V}$. For the sake of illustration, we choose the following form of $\vec{f}_E$:

$$\begin{aligned} f_x &= f_0 \cos \left(\frac{y}{l_0} + \frac{\nu t}{l_0^2} \right), \\ f_y &= f_0 \cos \left(\frac{x}{l_0} - \frac{\nu t}{l_0^2} \right), \qquad (7.259) \\ f_z &= \beta \left(f_x + f_y \right), \end{aligned}$$

where l_0, f_0 and β are constants. This functional form of the force represents a parity violating or an anisotropic force. We neglect pressure related forces. The solution of Equation (7.258) under these conditions is found to be:

$$V_x = \frac{-f_0 \nu}{l_0^2} \frac{\cos \left[\frac{y}{l_0} - \frac{\nu t}{l_0^2} + a \right]}{\left[1 + \left(1 + \frac{W_y l_0}{\nu} \right)^2 \right]^{1/2}}, \quad \tan a = - \left(1 + \frac{W_y l_0}{\nu} \right),$$

$$V_y = \frac{f_0 \nu}{l_0^2} \frac{\cos \left[\frac{x}{l_0} - \frac{\nu t}{l_0^2} + b \right]}{\left[1 + \left(1 - \frac{W_x l_0}{\nu} \right)^2 \right]^{1/2}}, \quad \tan b = \left(1 - \frac{W_x l_0}{\nu} \right), \qquad (7.260)$$

$$V_z = \beta \left(V_x + V_y \right).$$

We can now determine Reynolds stresses as:

$$R_{xx} = \langle V_x^2 \rangle, \; R_{yy} = \langle V_y^2 \rangle, \; R_{xy} = 0,$$
$$R_{xz} = \beta R_{xx}, \; R_{yz} = \beta R_{yy} \; \text{and} R_{zz} = \beta^2 \left(R_{xx} + R_{yy} \right). \quad (7.261)$$

We linearize equation (7.256) by retaining only the linear dependence of the Reynolds stresses on the large scale velocity $\vec{W}$. We find:

$$\frac{\partial W_x}{\partial t} - \alpha_0 \frac{\partial W_y}{\partial x} = \nu \left(\frac{\partial^2}{\partial x^2} + \frac{\partial^2}{\partial y^2} \right) W_x,$$

$$\frac{\partial W_y}{\partial t} + \alpha_0 \frac{\partial W_x}{\partial y} = \nu \left(\frac{\partial^2}{\partial x^2} + \frac{\partial^2}{\partial y^2} \right) W_y, \qquad (7.262)$$

where $\alpha_0 = \dfrac{(1+\beta) f_0^2 \nu}{4 l_0}$,
and

$$\frac{\partial W_z}{\partial t} = \nu \left(\frac{\partial^2}{\partial x^2} + \frac{\partial^2}{\partial y^2} \right) W_z.$$

We see that the horizontal and the vertical components of the large scale flow decouple. The vertical flow is just a viscous flow. We can easily solve for the horizontal flow and find that:

$$W_x = W_1 \exp\left[i(k_x x + k_y y) \right] \exp\left[\pm \alpha \sqrt{k_x k_y} t - \nu(k_x^2 + k_y^2) t \right],$$

$$W_y = W_2 \exp\left[i(k_x x + k_y y) \right] \exp\left[\pm \alpha \sqrt{k_x k_y} t - \nu(k_x^2 + k_y^2) t \right], \quad (7.263)$$

and

$$\frac{W_2}{W_1} = \pm i \left(\frac{k_y}{k_x} \right)^{1/2}.$$

The large scale flow is an unstable elliptically polarized helical flow. The growth rate ω_I of this instability is given by

$$\omega_I = \alpha_0 k - 2\nu k^2, \qquad (7.264)$$

for $k_x = k_y = k$.

The energy for the growth of the large scale flow comes from the Reynolds stresses, which derive their sustenance from the driving force $\vec{f}_E$. The important point to note is that the anisotropy of the force $\vec{f}_E$ introduced anisotropy in the small scale flow which in turn produced anisotropic Reynolds stresses causing the asymmetry in Equation (7.262) through the sign of the α_0 dependent terms. This is the essence of the **AKA Instability**. There are several issues related to the nature of relationship between inverse cascade and the AKA instability. The complete nonlinear Navier-Stokes equations can only be solved using numerical methods which need large computational facilities. Further, the role of AKA instability in astrophysical fluids awaits exploration. A maiden attempt in this direction can be seen in Krishan (1993) where AKA has been studied in an expanding fluid to see the effect of Hubble's expansion on the growth rate ω_I. There is some work on inverse cascade as well as AKA in compressible turbulent flows too. This field is rich with possibilities and has not yet received the attention it deserves.

Problem 7.25: Calculate the vorticity and helicity of the small scale $\vec{V}$ and the large scale flow $\vec{W}$.

7.14. Magnetohydrodynamic Turbulence

At the end of the chapter on 'Magnetohydrodynamics of Conducting Fluids' we promised that we will discuss MHD turbulence after getting some familiarity with fluid turbulence. We can imagine that MHD turbulence contains magnetic fields on several spatial and time scales in addition to the velocity fields. There is a magnetic field associated with each eddy, so, when they interact, an amplification or reduction or a redistribution of both velocity and magnetic fields could result. We are familiar with MHD equations. By using these equations and following the procedures outlined in previous sections, we can easily check that the MHD turbulent flow possesses at least three invariants. These are:

$$(i) \quad \text{the total energy, } E = \int \frac{1}{2} \left(\rho U^2 + \frac{B^2}{4\pi} \right) d\vec{r},$$

$$(ii) \quad \text{the magnetic helicity , } H_B = \int \vec{A}_B \cdot \vec{B} d\vec{r}, \qquad (7.265)$$

and

$$(iii) \quad \text{the cross helicity , } H_c = \int \vec{U} \cdot \vec{B} d\vec{r}.$$

Here, ρ is the mass density, $\vec{B}$ is the magnetic field with $\vec{A}_B$ the associated vector potential. The magnetic helicity H_B is a measure of the tangled nature of the magnetic field in MHD turbulence. The vector potential $\vec{A}_B$ is to magnetic helicity what the velocity $\vec{U}$ is to the kinetic helicity. The cross helicity, by definition, characterizes the alignment of the velocity and the magnetic field.

By using dimensional arguments and the Kolmogorovic hypotheses of the constancy of the decay rates of the invariants, we could derive the corresponding spectral energy distributions. We consider a specific state of equipartition of energy between magnetic and velocity fields i.e., $\frac{1}{2}\rho U^2 = B^2/8\pi$. The turbulence is then in the form of an ensemble of MHD waves propagating near the Alfven velocity and the spectral distributions, B_k of the magnetic energy and F_k of the kinetic energy, become nearly identical. Therefore, we expect that in the inertial range, the decay rates of magnetic and kinetic energy are constants and equal i.e.,

$$\frac{kB_k}{\tau_m} = \frac{kF_k}{\tau_m} = \epsilon, \qquad (7.266)$$

but τ_m is no longer given by the eddy turn over time $(kU_k)^{-1}$. It is now associated with the wave-wave interaction processes. We should construct a quantity with dimensions of $(\text{time})^{-1}$ by using the flow velocity U_k, the Alfven velocity V_A and the eddy size $(k)^{-1}$. We find that τ_m^{-1} can be expressed as:

$$\tau_m^{-1} \simeq \frac{kU_k^2}{V_A}. \qquad (7.267)$$

The energy spectrum is then obtained, as before, from:

$$\left(\frac{kU_k^2}{V_A}\right)\left(U_k^2\right) = \text{constant} = \epsilon,$$

or

$$F_k = (\epsilon V_A)^{1/2} k^{-3/2} = B_k. \qquad (7.268)$$

Since τ_m is a characteristic of the wave-wave interaction processes, it must be much larger than the wave period $(kV_A)^{-1}$. This condition defines the minimum value of k for which the $k^{-3/2}$ spectrum remains valid.

The decay of magnetic helicity can also be described as:

$$\left(\frac{kU_k^2}{V_A}\right)\left(\frac{kB_k}{k}\right) = \frac{k}{V_A}(kF_k)F_k = \text{constant}, \qquad (7.269)$$

or

$$F_k \propto k^{-1}.$$

The decay of cross helicity

$$\left(\frac{kU_k^2}{V_A}\right)(U_kB_k) = \left(\frac{kU_k^2}{V_A}\right)(4\pi\rho)^{1/2} U_k^2 \propto k^3 F_k^2 = \text{constant}, \qquad (7.270)$$

gives the same $k^{-3/2}$ spectrum. So again we have two branches of the energy spectrum F_k given by Equations (7.268) and (7.269). Obviously, both cannot operate in the same inertial range. How is this incompatibility resolved?

Again, taking a cue from the 2-D hydrodynamic turbulence, it is expected that the k^{-1} spectrum represents transfer of energy from small to large scales and $k^{-3/2}$ from large to small scales. This expectation is partially supported by the observation that $(H_B/E) \propto k^{-1}$ implying thereby that the magnetic helicity dominates at large spatial scales, compared to energy. This further points out that the magnetic helicity suffers less dissipation compared to energy. This is one of the reasons to postulate the **Selective Decay Hypothesis**. According to which, in a turbulent MHD flow, the energy E, decays to a minimum value, while the magnetic helicity H_B remains more or less a constant (Montgomery et al.1978). Using this hypothesis, a variational principle is set up to learn the credentials of the resulting state. Thus, if $\delta E, \delta H_B$ and δH_c are the variations, then the variational principle reads:

$$\delta E - \lambda_B \delta H_B - \lambda_c \delta H_c = 0, \qquad (7.271)$$

where λ_B and λ_c are the Lagrange multipliers. Substituting for the variations from the defining equations (7.265), we find:

$$\int \left[\delta \left(\vec{U} - \lambda_c \vec{B} \right) + \delta \vec{A}_B \cdot \left(\vec{\nabla} \times \vec{B} - 2\lambda_B \vec{B} - \lambda_c \vec{\nabla} \times \vec{U} \right) \right] d\vec{r} = 0,$$

or

$$\vec{\nabla} \times \vec{B} = \alpha_c \vec{B}, \qquad (7.272)$$

$$\vec{\nabla} \times \vec{U} = \alpha_c \vec{U},$$

where

$$\alpha_c = \frac{2\lambda_B}{1 - \lambda_c^2} \text{ and } \vec{U} = \lambda_c \vec{B}.$$

Equation (7.272) represents a **Force-Free** magnetic field $\vec{B}$ and a **Beltrami Flow** velocity $\vec{U}$. The force-free nature of magnetic fields has been invoked in many astrophysical situations, the most discussed being the solar photospheric magnetic fields. The observations of these fields using the Zeeman effect have provided favourable support to the existence of the force-free magnetic fields. The solar coronal fields are believed to be extrapolations of the photospheric fields and manifest themselves in the form of coronal loops. Modeling of coronal loops using the force free magnetic fields and the Beltrami flows has proved to be reasonably successful in accounting for the observed pressure profiles (Krishan 1996b).

Perhaps the most important and the best considered role of astrophysical MHD turbulence is in the generation of large scale magnetic field from small scale turbulent MHD fluctuations. There are more than one underlying mechanisms and all of them are collectively known as the **Dynamo Mechanism**. We give here a beginner's description of a dynamo. It goes

along the lines of "the Navier-Stokes way" for generating large scale flows
from small scale flows and is known as the **Alpha Effect**. We begin with
Faraday's Induction law:

$$\frac{\partial \vec{B}}{\partial t} = \nabla \times \left(\vec{U} \times \vec{B} \right) + \eta \nabla^2 \vec{B}, \tag{7.273}$$

where η is the resistivity. We split the fields into large and small scales as:

$$\begin{aligned} \vec{U} &= \vec{U}_0 + \vec{U}', \\ \vec{B} &= \vec{B}_0 + \vec{B}', \end{aligned} \tag{7.274}$$

such that $\left\langle \vec{U} \right\rangle = \vec{U}_0$, $\left\langle \vec{B} \right\rangle = \vec{B}_0$, $\left\langle \vec{U}' \right\rangle = \left\langle \vec{B}' \right\rangle = 0$.
Substituting equations (7.274) into Equation (7.273), and performing the
averaging process, we get:

$$\frac{\partial \vec{B}_0}{\partial t} = \nabla \times \left(\vec{U}_0 \times \vec{B}_0 + \left\langle \vec{U}' \times \vec{B}' \right\rangle \right) + \eta \nabla^2 \vec{B}_0. \tag{7.275}$$

The equation for $\vec{B}'$ is obtained by subtracting the mean field Equation
(7.275) from Equation (7.273) as:

$$\frac{\partial \vec{B}'}{\partial t} = \nabla \times \left(\vec{U}_0 \times \vec{B}' + \vec{U}' \times \vec{B}_0 \right) + \vec{\nabla} \times \vec{S} + \eta \nabla^2 \vec{B}', \tag{7.276}$$

where $\vec{S} = \vec{U}' \times \vec{B}' - \left\langle \vec{U}' \times \vec{B}' \right\rangle$.
The fluctuating velocity $\vec{U}'$ and magnetic field $\vec{B}'$ produce an electric field
$\vec{E}$ given by

$$\vec{E} = \frac{1}{c} \left\langle \vec{U}' \times \vec{B}' \right\rangle. \tag{7.277}$$

The force due to this electric field is the equivalent of the Reynolds stresses,
and it is this correlation between $\vec{U}'$ and $\vec{B}'$ that acts as a source term for
the enhancement of the mean field $\vec{B}_0$. Equation (7.276) shows that $\vec{B}'$ and
$\vec{B}_0$ bear a linear relationship. Therefore, we expect that $\vec{E}$ is linearly related
to $\vec{B}_0$. Since $\vec{U}'$ and $\vec{B}'$ exist on scales much smaller than that of $\vec{B}_0$, we can
express $\vec{E}$ as a function of the local value of $\vec{B}_0$. Thus, we write:

$$c E_i = \sum_j \alpha_{ij} B_{0j} + \sum_{j,k} \beta_{ijk} \frac{\partial B_{0j}}{\partial x_k}. \tag{7.278}$$

The tensors α_{ij} and β_{ijk} are determined from the statistical properties of
the turbulence. Further, it is assumed that:

$$\left| \vec{B}' \right| << \left| \vec{B}_0 \right|, \tag{7.279}$$

so that the nonlinear terms (quadratic in fluctuating quantities) in Equation (7.276) can be ignored. Again, as in the AKA instability, we must determine $\vec{B}'$ in terms of $\vec{B}_0$ and $\vec{U}'$ for a prescribed $\vec{U}_0$ and substitute in $\vec{E}$. For weakly isotropic turbulence, i.e., the one that is invariant under rotations of the frame of reference but not under reflection, we can write:

$$\alpha_{ij} = \alpha_m \delta_{ij} \text{ and } \beta_{ijk} = \beta \epsilon_{ijk},$$

and it is found that

$$\alpha_m = -\frac{1}{3}\tau_{cor}\left\langle \vec{U}' \cdot \vec{\nabla} \times \vec{U}' \right\rangle,$$
$$\beta = \frac{1}{3}\tau_{cor}\left\langle \vec{U}' \cdot \vec{U}' \right\rangle. \tag{7.280}$$

Problem 7.26: By integrating equation (7.276) to determine B' and than forming its cross product with U', obtain Equation (7.280).

We, immediately recognize that α_m is proportional to the mean helicity and τ_{cor}, the correlation time of the fluctuations. The coefficient β is proportional to the energy tensor. This term is identified with turbulent resistivity $\eta_T = \beta$ as we did for turbulent viscosity ν_T. Thus, the electric field $\vec{E}$ becomes:

$$c\vec{E} = \alpha_m \vec{B}_0 - \eta_T \vec{\nabla} \times \vec{B}_0 \tag{7.281}$$

and the dynamo Equation (7.275) acquires the form:

$$\frac{\partial \vec{B}_0}{\partial t} = \nabla \times \left(\alpha_m \vec{B}_0\right) + \nabla \times \left(\vec{U}_0 \times \vec{B}_0\right) - \vec{\nabla} \times \left(\eta_t \vec{\nabla} \times \vec{B}_0\right), \tag{7.282}$$

where η_t stands for the total resistivity. The magnetic field $\vec{B}(\equiv \vec{B}_0)$ and the velocity $\vec{U}(\equiv \vec{U}_0)$ are split into poloidal B_p and toroidal B_θ components in a cylindrical coordinate system (r, θ, z) as

$$\vec{B} = \vec{B}_p + B_\theta\hat{\theta}, \quad \vec{B}_p = \vec{\nabla} \times \left(A_p\hat{\theta}\right),$$
$$\vec{U} = \vec{U}_p + U_\theta\hat{\theta}, \quad U_\theta = \Omega r, \tag{7.283}$$

where A_p is the vector potential and Ω is the angular velocity. Taking a scalar product of Equation (7.282) with $\hat{\theta}/r$, we find the evolution equations for the toroidal component as:

$$\frac{\partial B_\theta}{\partial t} + r\vec{U}_p \cdot \vec{\nabla}\left(\frac{B_\theta}{r}\right) = r\vec{B}_p \cdot \vec{\nabla}\Omega - \alpha_m\left(\nabla^2 - \frac{1}{r^2}\right)A_p - \frac{1}{r}\vec{\nabla}\alpha_m \cdot \vec{\nabla}\left(rA_p\right)$$
$$- B_\theta\vec{\nabla} \cdot \vec{U}_p + \eta_t\left(\nabla^2 - \frac{1}{r^2}\right)B_\theta, \tag{7.284}$$

and for the vector potential A_p as:

$$\frac{\partial A_p}{\partial t} + \frac{1}{r}\vec{U}_p \cdot \vec{\nabla}\left(rA_p\right) = \alpha_m B_\theta + \eta_t \left(\nabla^2 - \frac{1}{r^2}\right) A_p. \qquad (7.285)$$

The left-hand side of Equation (7.284) is the total time derivative of B_θ.

The differential rotation (Ω) stretches the poloidal field B_p to produce a B_θ component. This is known as the ω – **Effect**. The α- dependent terms produce both $\vec{B}_p$ (in Equation 7.285) and B_θ (in Equation 7.284) and this is known as the $\boldsymbol{\alpha}$ – **Effect**. The compressibility $(\vec{\nabla} \cdot \vec{U}_p)$ also enhances B_θ due to the crowding together of field lines. The resistive terms cause diffusion or dissipation of the field. In the absence of the α – effect, the poloidal component B_p (or A_p) suffers dissipation without any enhancement. The differential rotation, though, enhances B_θ but since B_p decays so does B_θ. This is the essence of the **Cowling Theorem** which forbids a steady axisymmetric magnetic field being maintained by dynamo action. So, in addition to differential rotation, motions possessing net helicity are needed for dynamo action. In sun, for example, the convective motions acquire rotation due to coriolis force, so that the rising and the falling elements have opposite sense of helicities. The α - effect operates when there is an asymmetry between the up flows and down flows leading to a nonzero helicity. A rather picturesque (Figure 7.14) description of the α - effect has been given by an eminent scientist as: 'stretch B_p (through $\vec{\nabla}\Omega$), twist (through coriolis force), fold and reconnect " (Colgate 1994).

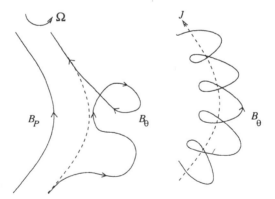

Figure 7.14. Differential Rotation of the Conducting Fluid Stretches the Poloidal Field B_p into the Toroidal field B_0. Helical Motions Lift, Fold and Reconnect the Loops. The Associated Current J Produces B_p.

We can assess the relative contributions of the α - and the ω - effects through the ratio $R_{\alpha\omega}$ of these terms which is found to be, on dimensional grounds:

$$R_{\alpha\omega} = \frac{(\alpha_m A_p/L^2)}{(LB_p\Omega/L)} = \frac{\alpha_m}{\Omega L}. \tag{7.286}$$

We see that $R_{\alpha\omega}$ is approximately the ratio of turbulent velocity ($|U'|$) to rotation velocity $U_{0\theta}$ at the scale L. When $\alpha_m >> \Omega L$, the α – **Effect** produces B_θ from B_p and B_p from B_θ and is referred to as α^2 - **Effect**. Let us estimate the growth time

$$T_g \simeq \frac{L}{\alpha_m}, \tag{7.287}$$

of a magnetic structure of size L, say a sunspot on the solar atmosphere. We know that cellular velocity patterns, known as granulation exist on the solar photosphere. They are a manifestation of the convective turbulence with typical velocity $U' \simeq 5 \times 10^3$ cm sec $^{-1}$, spatial scale $\simeq 10^8$ cm and a lifetime $\tau_{cor} \simeq 300$sec. Therefore

$$\alpha_m \simeq \frac{1}{3}\frac{(5 \times 10^3)^2 \times 300}{10^8} \text{ cm sec}^{-1} \tag{7.288}$$

For a large sunspot of size $L \simeq 10^{10}$ cm., we find $T_g \simeq 13$ years. This is fairly close to the 11 year decay and growth period of sunspot magnetic fields. The turbulent resistivity

$$\eta_T \simeq \frac{1}{3}U'^2\tau_{cor} \simeq 2.5 \times 10^9 \text{ cm}^2\text{sec}^{-1} \tag{7.289}$$

and the corresponding decay time of the sunspot magnetic field

$$T_d \simeq \frac{L^2}{\eta_T} \simeq 10^3 \text{ years}$$

is much larger than the growth time T_g. The 22 year cycle of solar magnetic field is explained in terms of periodic dynamo waves. Of course, neither is the sun so simple as to exhibit only one periodicity, nor is this simple minded treatment constituting various combinations of α - and ω - effect adequate to account for the complex temporal behaviour of solar magnetic fields. Then, there are issues of the relative rates at which the small scale magnetic field $\vec{B}'$, and the large scale magnetic field $\vec{B}_0$, grow. It may happen that the growth of $\vec{B}'$ overwhelms that of $\vec{B}_0$. The system is essentially nonlinear and easily susceptible to catastrophe or chaos. The search is on for dynamo actions capable of producing multiperiodic and multiscale iden- tifiable magnetic structure.

References

Aller, L.H., 1954, Nuclear Transformations, Stellar Interiors and Nebulae, The Ronald Press Company

Batchelor, G.K., 1953, The Theory of Homogeneous Turbulence, Cambridge University Press.

Chandrasekhar, S., 1949, Proc. Roy. Soc. A., **200**, 20.

Chandrasekhar, S., 1961, Hydrodynamic and Hydromagnetic Stabilty, Oxford University Press.

Cole, G.H.A., 1962, Fluid Dynamics, Methuen & Co. Ltd., London.

Colgate, S.A., 1994, Physica Scripta, **T52**, 96.

Cox, J.P., 1985, in Cephids: Theory and Observations, p126, Ed: B.F. Madore, Cambridge University Press.

Hasegawa, A., 1985, Adv. Phys., **34**, 1.

Heisenberg, W., 1948, Proc. Roy. Soc. A., **195**, 402.

Kolmogrov, A.N., 1941, Dokl. Akad. Nauk. SSR., **30**, 301.

Krishan, V., 1991, Mon. Not. R. Astr. Soc., **250**,50.

Krishan, V. and Sivaram, C., 1991, Mon. Not. R. Soc., **250**, 157.

Krishan, V., 1993, Mon. Not. R. Astr. Soc., **264**, 257.

Krishan, V., 1996a, Current Science, **71**, 541.

Krishan, V., 1996b, J. Plasma Physics, **56**, 427.

Lebovitz, N.R., 1983, Fluid Dynamics in Astrophysics & Geophysics, Vol. 20.

Levich, E. and Tzvetkov, E., 1985, Phy. Rep., **128**, 1.

Moffat, H.K. and Tsinobar, A., 1992, Ann. Rev. Fluid Mech., **24**, 281.

Montgomery, D., Turner, L. and Vahala, G., 1978, Phys. Fluids, **21**, 757.

Prabhu, R.D. and Krishan, V., 1994, Ap. J., **428**, 483.

Shakura, N.I. and Sunyaev, R.A., 1973, A & A, **24**, 337.

Shore, N.S., 1992, An Introduction to Astrophysical Hydrodynamics, Academic Press, Inc.

Shu, F.H., 1991, The Physics of Astrophysics, Vol 2, University Science Books.

Sulem, P.L., She, Z.S., Scholl, H. and Frisch, U., 1989, J. Fluid Mech., **205**, 341.

Zahn, J.P., 1987, in Solar and Stellar Physics, p55

8. Physical Constants

Boltzmann constant $K_B = 1.38 \times 10^{-16}$ erg deg^{-1} (K)

Electron charge $e = 4.8 \times 10^{-10}$ statcoulomb

Electron mass $m_e = 9.11 \times 10^{-28}$ gm

Proton mass $m_p = 1.67 \times 10^{-24}$ gm

Gravitational constant $G = 6.67 \times 10^{-8}$ dyne cm^2 gm^{-2}

Planck constant $h = 6.62 \times 10^{-27}$ erg sec

$\hbar = 1.05 \times 10^{-27}$ erg sec

Speed of light in vacuum $c = 3 \times 10^{10}$ cm sec^{-1}

Rydberg constant $R = \dfrac{2\pi^2 m e^4}{ch^3} = 1.09 \times 10^5$ cm^{-1}

Bohr radius $a_0 = \dfrac{\hbar^2}{m_e e^2} = 5.29 \times 10^{-9}$ cm

Classical electron radius $r_e = \dfrac{e^2}{m_e c^2} = 2.82 \times 10^{-13}$ cm

Thomson cross-section $\sigma_T = \dfrac{8\pi r_e^2}{3} = 6.65 \times 10^{-25}$ cm^2

Compton wavelength of electron $\dfrac{h}{m_e c} = 2.42 \times 10^{-10}$ cm

Fine structure constant $\dfrac{e^2}{\hbar c} = 7.3 \times 10^{-3} \simeq 1/137$

Stefan-Boltzmann constant $\sigma = 5.67 \times 10^{-5}$ erg cm^{-2} deg^{-4} (K) sec^{-1}

Temperature corresponding to 1eV $= 1.16 \times 10^4$ deg (K)

Avogadro number $N_A = 6 \times 10^{23}$ mol^{-1}

Gas constant $N_A K_B = 8.3 \times 10^7$ erg deg^{-1} mol^{-1}

9. Astrophysical Quantities

9.1. Magnitudes

The apparent magnitudes m and m_0 of two objects are related to their apparent intensities (erg cm^{-2} sec^{-1}) l and l_0 as:

$$\frac{l_0}{l} = (2.512)^{m-m_0}.$$

Since the power received from an object is inversely proportional to its distance r, we have

$$\frac{l_0}{l} = (\frac{r}{r_0})^2 \ or \ m - m_0 = 5 \ log \ r/r_0.$$

The apparent magnitude of an object when placed at a distance of 10 pc is known as the absolute magnitude M. Thus if $r_0 = 10pc$ then $m_0 = M$ and we get

$$m - M = 5 \ log \ r/10.$$

If m and M refer to the same object, then m-M is called the modulus of distance. One parsec (pc) unit of distance $= 3.09 \times 10^{18}$ cm.

9.2. Planets

Planets	Mass M_E	Mean density gm cm^{-3}	Rotational period days	Distance from Sun AU	Orbital period years
Mercury	0.06	5.4	58.66	0.39	0.24
Venus	0.82	5.2	242.98	0.72	0.61
Earth	1.00	5.5	1.00	1.00	1.00
Mars	0.11	3.9	1.03	1.52	1.87
Jupiter	317.84	1.3	0.40	5.20	11.86
Saturn	95.15	0.7	0.43	9.54	29.47
Uranus	14.60	1.3	0.89	19.19	84.06
Neptune	17.21	1.7	0.53	30.06	164.81
Pluto	0.18		6.39	39.53	248.54

Mass of Earth $M_E = 6 \times 10^{27}$ gm
1AU $= 1.5 \times 10^{13}$ cm.
Radius of Earth $R_E = 6371$ km
Magnetic field of Earth $\simeq 0.3$ Gauss.

339

9.3. The Sun

Mass $M_\odot = 1.99 \times 10^{33} gm$
Radius $R_\odot = 6.96 \times 10^{10}$ cm
Luminosity $L_\odot = 3.83 \times 10^{33} erg sec^{-1}$
Age $= 4.6 \times 10^9$ year
Surface gravity $g_\odot = 2.74 \times 10^4$ cm sec^{-2}
Radiation flux density $= 6.28 \times 10^{10}$ erg cm^{-2} sec^{-1}
Temperature of the solar photosphere $T_\odot = 5780$ K
Escape speed from sun $= 6.18 \times 10^7$ cm sec^{-1}
Radiation flux density at 1AU = solar constant $= 1.36 \times 10^6$ erg cm^{-2} sec^{-1}.

9.4. The Milky Way

Mass $= 5 \times 10^{10} M_\odot$
Rotation period $= 3 \times 10^8$ year.

9.5. The Hubble Constant

$H = 50h$ km sec^{-1} Mpc^{-1}, $1 < h < 2$.

9.6. Planck's Radiation Law

The brightness B (watt m^{-2} cps^{-1} rad^{-2}) of radiation at a frequency ν from a blackbody at a temperature T is given by the Planck radiation law:

$$B = \frac{2h\nu^3}{c^2} \frac{1}{\exp(h\nu/K_B T) - 1}$$

Integrating the Planck radiation law over all frequencies gives the total brightness B' as

$$B' = \frac{2h}{c^2} \int \frac{\nu^3 d\nu}{\exp(h\nu/K_B T) - 1}$$
$$= \sigma T^4$$

This is the **Stefan-Boltzmann Law** with the constant $\sigma = 1.80 \times 10^{-8}$ watt m^{-2} K^{-4}. The variation of B with ν for different temperatures is shown in Figures (3.13 & 3.14), reproduced from Radio Astronomy by J.D.Kraus, 1966, McGraw Hill.

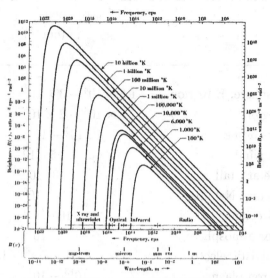

Fig. 3-13. Planck-law radiation curves to logarithmic scales with brightness expressed as a function of frequency $B(\nu)$ (left and bottom scales) and as a function of wavelength B_λ (right and top scales). Wavelength increases to the right.

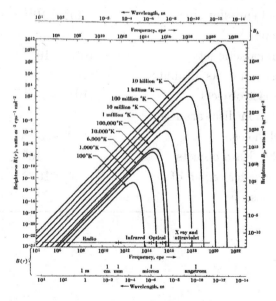

Fig. 3-14. Planck-radiation-law curves with frequency increasing to the right.

* The usual statement gives the integrated value of B' over one hemisphere as obtained by multiplying (3-57) by π. See (3-10).

9.7. Electron Density and Temperature of some of the Astrophysical Plasmas

	Electron density (cm^{-3})	Temperature (Kelvin)
Ionosphere, E Region	$1.5 \times 10^4 - 3 \times 10^4$	$10^4 - 10^5$
F Region	$5 \times 10^4 - 2 \times 10^5$	
Solar corona	$10^9 - 10^{11}$	$10^6 - 10^7$
Solar wind	5 - 10	$10^5 - 10^6$
Cometary tail	10 -100	$10^4 - 10^5$
Planetary Nebulae	10^6	10^5
HII Regions	$10 - 10^4$	10^5
Pulsar atmosphere	10^{10}	10^5
Relativistic electron-		Lorentz factor
positron plasma	$10^{13} - 10^{15}$	$\gamma \simeq 10^2 - 10^3$
in pulsar magnetosphere		
Interstellar medium	10^{-1}	$10^4 - 10^6$
Active galactic Nuclei	$10^{14} - 10^{10}$	$10^5 - 10^8$
Broad emission line	$10^8 - 10^{11}$	$10^4 - 5 \times 10^5$
regions		
Extragalactic jets	$10^{-2} - 10^{-4}$	$10^5 - 10^7$
Accretion disks around		
compact objects	$10^{18} - 10^{21}$	$10^6 - 10^7$
Accretion discs around		
black holes	$10^{10} - 10^{11}$	$10^5 - 10^6$
Herbig-Haro objects	10^8	$10^5 - 10^7$
Intracluster region of	10^{-3}	$1 - 3 \times 10^8$
galaxies		

10. Differential Operators

Cylindrical coordinates (r, θ, z)

Gradient

$$\vec{\nabla} S = \frac{\partial S}{\partial r}\hat{r} + \frac{1}{r}\frac{\partial S}{\partial \theta}\hat{\theta} + \frac{\partial S}{\partial z}\hat{z}$$

Divergence

$$\vec{\nabla}.\vec{V} = \frac{1}{r}\frac{\partial}{\partial r}(rV_r) + \frac{1}{r}\frac{\partial V_\theta}{\partial \theta} + \frac{\partial V_z}{\partial z}$$

Curl

$$\vec{\nabla} \times \vec{V} = \left(\frac{1}{r}\frac{\partial V_z}{\partial \theta} - \frac{\partial V_\theta}{\partial z}\right)\hat{r} + \left(\frac{\partial V_r}{\partial z} - \frac{\partial V_z}{\partial r}\right)\hat{\theta} + \left[\frac{1}{r}\frac{\partial}{\partial r}(rV_\theta) - \frac{1}{r}\frac{\partial V_r}{\partial \theta}\right]\hat{z}$$

Laplacian of a scalar

$$\nabla^2 S = \frac{1}{r}\frac{\partial}{\partial r}\left(r\frac{\partial S}{\partial r}\right) + \frac{1}{r^2}\frac{\partial^2 S}{\partial \theta^2} + \frac{\partial^2 S}{\partial z^2}$$

Laplacian of a vector

$$\nabla^2\vec{V} = \left(\nabla^2 V_r - \frac{2}{r^2}\frac{\partial V_\theta}{\partial \theta} - \frac{V_r}{r^2}\right)\hat{r} + \left(\nabla^2 V_\theta + \frac{2}{r^2}\frac{\partial V_r}{\partial \theta} - \frac{V_\theta}{r^2}\right)\hat{\theta} + \left(\nabla^2 V_z\right)\hat{z}$$

Components of $(\vec{V}.\vec{\nabla})\vec{U}$

$$(\vec{V} \cdot \vec{\nabla})\vec{U}\Big|_r = V_r\frac{\partial U_r}{\partial r} + \frac{V_\theta}{r}\frac{\partial U_r}{\partial \theta} + V_z\frac{\partial U_r}{\partial z} - \frac{V_\theta U_\theta}{r}$$

$$(\vec{V} \cdot \vec{\nabla})\vec{U}\Big|_\theta = V_r\frac{\partial U_\theta}{\partial r} + \frac{V_\theta}{r}\frac{\partial U_\theta}{\partial \theta} + V_z\frac{\partial U_\theta}{\partial z} + \frac{V_\theta U_r}{r}$$

$$(\vec{V} \cdot \vec{\nabla})\vec{U}\Big|_z = V_r\frac{\partial U_z}{\partial r} + \frac{V_\theta}{r}\frac{\partial U_z}{\partial \theta} + V_z\frac{\partial U_z}{\partial z}$$

Divergence of a tensor

$$\vec{\nabla} \cdot \Pi\Big|_r = \frac{1}{r}\frac{\partial}{\partial r}(r\Pi_{rr}) + \frac{1}{r}\frac{\partial \Pi_{\theta r}}{\partial \theta} + \frac{\partial \Pi_{zr}}{\partial z} - \frac{\Pi_{\theta\theta}}{r}$$

$$\vec{\nabla} \cdot \Pi\Big|_\theta = \frac{1}{r}\frac{\partial}{\partial r}(r\Pi_{r\theta}) + \frac{1}{r}\frac{\partial \Pi_{\theta\theta}}{\partial \theta} + \frac{\partial \Pi_{z\theta}}{\partial z} + \frac{\Pi_{\theta r}}{r}$$

$$\vec{\nabla} \cdot \Pi\Big|_z = \frac{1}{r}\frac{\partial}{\partial r}(r\Pi_{rz}) + \frac{1}{r}\frac{\partial \Pi_{\theta z}}{\partial \theta} + \frac{\partial \Pi_{zz}}{\partial z}$$

Spherical coordinates (r, θ, φ)

$$\text{Gradient}\,\vec{\nabla} S = \frac{\partial S}{\partial r}\hat{r} + \frac{1}{r}\frac{\partial S}{\partial \theta}\hat{\theta} + \frac{1}{r\sin\theta}\frac{\partial S}{\partial \varphi}\hat{\varphi}$$

Divergence

$$\vec{\nabla} \cdot \vec{V} = \frac{1}{r^2}\frac{\partial}{\partial r}(r^2 V_r) + \frac{1}{r\sin\theta}\frac{\partial}{\partial\theta}(\sin\theta V_\theta) + \frac{1}{r\sin\theta}\frac{\partial V_\varphi}{\partial\varphi}$$

Curl

$$\vec{\nabla}\times\vec{V}\Big|_r = \frac{1}{r\sin\theta}\frac{\partial}{\partial\theta}(\sin\theta V_\varphi) - \frac{1}{r\sin\theta}\frac{\partial V_\theta}{\partial\varphi}$$

$$\vec{\nabla}\times\vec{V}\,|_\theta = \frac{1}{r\sin\theta}\frac{\partial V_r}{\partial\varphi} - \frac{1}{r}\frac{\partial}{\partial r}(rV_\varphi)$$

$$\vec{\nabla}\times\vec{V}\,|_\varphi = \frac{1}{r}\frac{\partial}{\partial r}(rV_\theta) - \frac{1}{r}\frac{\partial V_r}{\partial\theta}$$

Laplacian of a scalar

$$\nabla^2 S = \frac{1}{r^2}\frac{\partial}{\partial r}\left(r^2\frac{\partial S}{\partial r}\right) + \frac{1}{r^2\sin\theta}\frac{\partial}{\partial\theta}\left(\sin\theta\frac{\partial S}{\partial\theta}\right) + \frac{1}{r^2\sin^2\theta}\frac{\partial^2 S}{\partial\varphi^2}$$

Laplacian of a vector

$$\nabla^2\vec{V}\Big|_r = \nabla^2 V_r - \frac{2V_r}{r^2} - \frac{2}{r^2}\frac{\partial V_\theta}{\partial\theta} - \frac{2\cot\theta V_\theta}{r^2} - \frac{2}{r^2\sin\theta}\frac{\partial V_\varphi}{\partial\varphi}$$

$$\nabla^2\vec{V}\Big|_\theta = \nabla^2 V_\theta + \frac{2}{r^2}\frac{\partial V_r}{\partial\theta} - \frac{V_\theta}{r^2\sin^2\theta} - \frac{2\cos\theta}{r^2\sin^2\theta}\frac{\partial V_\varphi}{\partial\varphi}$$

$$\nabla^2\vec{V}\Big|_\varphi = \nabla^2 V_\varphi - \frac{V_\varphi}{r^2\sin^2\theta} + \frac{2}{r^2\sin\theta}\frac{\partial V_r}{\partial\varphi} + \frac{2\cos\theta}{r^2\sin^2\theta}\frac{\partial V_\theta}{\partial\varphi}$$

Components of $(\vec{V}.\vec{\nabla})\vec{U}$

$$(\vec{V}\cdot\vec{\nabla})\vec{U}\Big|_r = V_r\frac{\partial U_r}{\partial r} + \frac{V_\theta}{r}\frac{\partial U_r}{\partial\theta} + \frac{V_\varphi}{r\sin\theta}\frac{\partial U_r}{\partial\varphi} - \frac{V_\theta U_\theta + V_\varphi U_\varphi}{r}$$

$$(\vec{V}\cdot\vec{\nabla})\vec{U}\,|_\theta = V_r\frac{\partial U_\theta}{\partial r} + \frac{V_\theta}{r}\frac{\partial U_\theta}{\partial\theta} + \frac{V_\varphi}{r\sin\theta}\frac{\partial U_\theta}{\partial\varphi} + \frac{V_\theta U_r}{r} - \frac{\cot\theta V_\varphi U_\varphi}{r}$$

$$(\vec{V}\cdot\vec{\nabla})\vec{U}\,|_\varphi = V_r\frac{\partial U_\varphi}{\partial r} + \frac{V_\theta}{r}\frac{\partial U_\varphi}{\partial\theta} + \frac{V_\varphi}{r\sin\theta}\frac{\partial U_\varphi}{\partial\varphi} + \frac{V_\varphi U_r}{r} + \frac{\cot\theta V_\varphi U_\theta}{r}$$

Divergence of a tensor

$$(\vec{\nabla}\cdot\Pi)_\theta = \frac{1}{r^2}\frac{\partial}{\partial r}(r^2\Pi_{rr}) + \frac{1}{r\sin\theta}\frac{\partial}{\partial\theta}(\sin\theta\Pi_{\theta r}) + \frac{1}{r\sin\theta}\frac{\partial\Pi_{\varphi r}}{\partial\varphi} - \frac{\Pi_{\theta\theta} + \Pi_{\varphi\varphi}}{r}$$

$$(\vec{\nabla}\cdot\Pi)_\theta = \frac{1}{r^2}\frac{\partial}{\partial r}(r^2\Pi_{r\theta}) + \frac{1}{r\sin\theta}\frac{\partial}{\partial\theta}(\sin\theta\Pi_{\theta\theta}) + \frac{1}{r\sin\theta}\frac{\partial\Pi_{\varphi\theta}}{\partial\varphi} + \frac{\Pi_{\theta r}}{r} - \frac{\cot\theta\Pi_{\varphi\varphi}}{r}$$

$$(\vec{\nabla}\cdot\Pi)_\varphi = \frac{1}{r^2}\frac{\partial}{\partial r}(r^2\Pi_{r\varphi}) + \frac{1}{r\sin\theta}\frac{\partial}{\partial\theta}(\sin\theta\Pi_{\theta\varphi}) + \frac{1}{r\sin\theta}\frac{\partial\Pi_{\varphi\varphi}}{\partial\varphi} + \frac{\Pi_{\varphi r}}{r} + \frac{\cot\theta\Pi_{\varphi\theta}}{r}$$

11. Characteristic Numbers for Fluids *

Alfven number or Karman number $= \left[\dfrac{\text{Magnetic force}}{\text{Inertial force}}\right]^{1/2}$

Boussinesq number $= \left[\dfrac{\text{Inertial force}}{\text{Gravitational force}}\right]^{1/2}$

Chandrasekhar number $= \dfrac{\text{Magnetic force}}{\text{Dissipative forces}}$

Cowling number $= \dfrac{\text{Magntic force}}{\text{Inertial force}}$

Eckert number $= \dfrac{\text{Kinetic energy}}{\text{Change in thermal energy}}$

Ekman number $= \left[\dfrac{\text{Viscous force}}{\text{Coriolis force}}\right]^{1/2}$

Euler number $= \dfrac{\text{Pressure drop due to friction}}{\text{Dynamic pressure}}$

Hall coefficient $= \dfrac{\text{Gyrofrequency}}{\text{Collision frequency}}$

Knudsen number $= \dfrac{\text{Hydrodynamic time}}{\text{Collision time}}$

Lundquist number $= \dfrac{J \times B \text{ force}}{\text{Resistive magnetic diffusion force}}$

Mach number $= \dfrac{\text{Flow Speed}}{\text{Sound Speed}}$

Magnetic Mach number $= \left[\dfrac{\text{Inertial force}}{\text{Magnetic force}}\right]^{1/2}$

Magnetic Reynolds number $= \dfrac{\text{Flow Speed}}{\text{Magnetic diffusion speed}}$

Peclet number $= \dfrac{\text{Heat convection}}{\text{Heat conduction}}$

Poisseuille number $= \dfrac{\text{Pressure force}}{\text{Viscous force}}$

Prandtl number $= \dfrac{\text{Momentum diffusion}}{\text{Heat diffusion}}$

Rayleigh number $= \dfrac{\text{Buoyancy force}}{\text{Diffusion force}}$

Reynolds number $= \dfrac{\text{Inertial force}}{\text{Viscous force}}$

Rossby number $= \dfrac{\text{Inertial force}}{\text{Coriolis force}}$

$$\text{Taylor number} = \frac{\text{Centrifugal force}}{\text{Viscous force}}$$

*From NRL Plasma Formulary, 1994, by J.D.Huba.

12. Acknowledgment for Figures

Figure 1.1 Photograph by W.J.S.Lockyer, published in Clouds, Rain and Rainmaking by B.J.Mason, Cambridge University Press, 1975.

Figure 1.2 From Galactic Dynamics by J.Binney and Scott Tremaine, Princeton University Press, 1987.

Figure 1.3 From Rev.Mod.Phys.**56**, 255, 1984.

Figure 1.6 From Plasma Physics Edited by R.Dendy, Cambridge University Press, 1995.

Figure 1.7 Reprinted by permission, From Nature **325**, 696, 1987, Copyright (1987) Macmillan Magazines Ltd.

Figures 1.9 & 7.5 From Our Evolving Universe by M.S.Longair, Cambridge University Press, 1996.

Figure 1.10 Credits: Bruce Balick (University of Washington), Vincent Icke (Leiden University, The Netherlands), Garrelt Mellema (Stockholm University), and NASA.

Figure 1.11 From Solar Physics **66**, 347, 1980.

Figure 2.3 From Science, **246**, 897, 1989.

Figures 3.7 & 3.8 From the Astronomical Journal **69**, 73, 1964.

Figure 4.4 From Plasma Loops in the Solar Corona, Cambridge University Press, 1991.

Figures 4.8 & 4.9 From Gong Website, Courtesy H.M.Antia.

Figure 4.16a From Sky And Telescope, XXXI, 21, 1966.

Figure 4.16b From Rev.Mod.Phys.**56**, 255, 1984.

Figure 4.24 From sky And Telescope **84**, 619, 1992.

Figure 7.4 From Unveiling The Universe by J.E.Van Zyl, 1996, Springer - Verlag London Ltd.

Figure 7.6 From The Realon of the Nebulae by Edwin Hubble, 1982, Yale University Press.

Figure 7.11 From Advances in Physics, 1985, **34**, 1.

Figure 7.13 From Solar and Stellar Phys. 1987, p.55, Eds: F.H.Schroter & M.Schussler, Springer-Verlag.